Handbook of Halal Food Production

Handbook of Halal Food Production

Edited by
Mian N. Riaz and Muhammad M. Chaudry

Boca Raton London New York

CRC Press is an imprint of the
Taylor & Francis Group, an **informa** business

CRC Press
Taylor & Francis Group
6000 Broken Sound Parkway NW, Suite 300
Boca Raton, FL 33487-2742

© 2019 by Taylor & Francis Group, LLC
CRC Press is an imprint of Taylor & Francis Group, an Informa business

No claim to original U.S. Government works

Printed on acid-free paper

International Standard Book Number-13: 978-1-4987-0971-2 (Hardback)

This book contains information obtained from authentic and highly regarded sources. Reasonable efforts have been made to publish reliable data and information, but the author and publisher cannot assume responsibility for the validity of all materials or the consequences of their use. The authors and publishers have attempted to trace the copyright holders of all material reproduced in this publication and apologize to copyright holders if permission to publish in this form has not been obtained. If any copyright material has not been acknowledged please write and let us know so we may rectify in any future reprint.

Except as permitted under U.S. Copyright Law, no part of this book may be reprinted, reproduced, transmitted, or utilized in any form by any electronic, mechanical, or other means, now known or hereafter invented, including photocopying, microfilming, and recording, or in any information storage or retrieval system, without written permission from the publishers.

For permission to photocopy or use material electronically from this work, please access www.copyright.com (http://www.copyright.com/) or contact the Copyright Clearance Center, Inc. (CCC), 222 Rosewood Drive, Danvers, MA 01923, 978-750-8400. CCC is a not-for-profit organization that provides licenses and registration for a variety of users. For organizations that have been granted a photocopy license by the CCC, a separate system of payment has been arranged.

Trademark Notice: Product or corporate names may be trademarks or registered trademarks, and are used only for identification and explanation without intent to infringe.

Library of Congress Cataloging-in-Publication Data

Names: Riaz, Mian N., editor. | Chaudry, Muhammad M., editor.
Title: Handbook of halal food production / edited by Mian N. Riaz and Muhammad M. Chaudry.
Description: Boca Raton, Florida : CRC Press, [2019] | Expanded version of: Halal food production / Mian N. Rias, Muhammad M. Chaudry. Boca Raton, FL ; CRC Press, c2004, with twelve new contributions. | Includes bibliographical references and index.
Identifiers: LCCN 2018023406| ISBN 9781498709712 (hardback : alk. paper) | ISBN 9781315119564 (e-book)
Subjects: LCSH: Food industry and trade--Standards. | Halal food. | Halal food industry.
Classification: LCC TP372.6 .H36 2019 | DDC 297.5/76--dc23
LC record available at https://lccn.loc.gov/2018023406

Visit the Taylor & Francis Web site at
http://www.taylorandfrancis.com

and the CRC Press Web site at
http://www.crcpress.com

Dedication

This book is dedicated to the food scientists and industry professionals, who continuously strive to seek knowledge to serve humankind.

Contents

Preface xi
Acknowledgments xiii
About the Editors xv
Contributors xvii

Chapter 1 Introduction 1
Mian N. Riaz and Munir M. Chaudry

Chapter 2 Halal Food Laws and Regulations 7
Mian N. Riaz and Munir M. Chaudry

Chapter 3 General Guidelines for Halal Food Production 17
Mian N. Riaz and Munir M. Chaudry

Chapter 4 Muslim Demography and Global Halal Trade: A Statistical Overview 29
Mohammed A. Khan, Mian N. Riaz, and Munir M. Chaudry

Chapter 5 Global Halal Economy 61
Rafiuddin Shikoh, Mian N. Riaz, and Munir M. Chaudry

Chapter 6 Animal Welfare 73
Kristin Pufpaff, Mian N. Riaz, and Munir M. Chaudry

Chapter 7 The Religious Slaughter of Animals: A US Perspective on Regulations and Animal Welfare Guidelines 85
Joe Regenstine, Mian N. Riaz, and Munir M. Chaudry

Chapter 8 Halal Production Requirements for Meat and Poultry 105
Kristin Pufpaff, Mian N. Riaz, and Munir M. Chaudry

Chapter 9 Halal Processed Meat Requirements 125
Mustafa Farouk, Mian N. Riaz, and Munir M. Chaudry

Chapter 10 Halal Production Requirements: For Fish and Seafood 145
Mian N. Riaz, Rafiuddin Shaik, and Munir M. Chaudry

Chapter 11 Halal Production Requirements for Dairy Products 155
Mian N. Riaz and Munir M. Chaudry

Chapter 12 Halal Production Requirements for
Cereals and Confectionaries 163
Mian N. Riaz and Munir M. Chaudry

Chapter 13 Enzymes in Halal Food Production 167
Mian N. Riaz and Munir M. Chaudry

Chapter 14 Gelatin in Halal Food Production 177
Mian N. Riaz and Munir M. Chaudry

Chapter 15 Flavors, Flavorings, and Essences in Halal Food 185
*Eric Butrym, Laura LaCourse, Mian N. Riaz, and
Munir M. Chaudry*

Chapter 16 Alcohol in Halal Food Production 201
Roger Ottman, Mian N. Riaz, and Munir M. Chaudry

Chapter 17 Food Additives and Processing Aids 207
Mian N. Riaz and Munir M. Chaudry

Chapter 18 Food Ingredients in Halal Food Production.................................... 213
Mian N. Riaz and Munir M. Chaudry

Chapter 19 Halal Production Requirements for Nutritional Food Supplements 221
Mian N. Riaz and Munir M. Chaudry

Chapter 20 Biotechnology and GMO Ingredients in Halal Foods 225
Mian N. Riaz and Munir M. Chaudry

Chapter 21 Animal Feed and Halal Food 229
Mian N. Riaz and Munir M. Chaudry

Contents	ix

Chapter 22 Halal Cosmetics 233
Mian N. Riaz and Munir M. Chaudry

Chapter 23 Labeling, Packaging, and Coatings for Halal Foods 241
Mian N. Riaz and Munir M. Chaudry

Chapter 24 How to Get Halal Certified 247
Mian N. Riaz and Munir M. Chaudry

Chapter 25 Comparison of Kosher, Halal, and Vegetarianism 261
Mian N. Riaz and Munir M. Chaudry

Chapter 26 Globalization of Halal Certification: From an Industrial
Perspective 283
Jes Knudsen, Mian N. Riaz and Munir M. Chaudry

Chapter 27 Testing Non-Halal Materials 291
Winai Dahlan, Mian N. Riaz, and Munir M. Chaudry

Chapter 28 Potential Hazards and Sanitation of Halal Facilities 299
Mian N. Riaz and Munir M. Chaudry

Chapter 29 Halal Awareness and Education Schemes 303
Mian N. Riaz and Munir M. Chaudry

Chapter 30 Halal Food Model 309
Mian N. Riaz and Munir M. Chaudry

Chapter 31 HACCP and Halal 319
Mian N. Riaz, Sibte Abbas, and Munir M. Chaudry

Epilogue 325
Appendix A: Key Terminology from Other Languages 327
Appendix B: Permitted Food Additives in the European Union 329
Appendix C: Ingredient List 341
Appendix D: Codex Alimentarius 367
Index 371

Preface

The word *"Halal"* has become quite common in the Western food industry in the past two decades, primarily due to the export of food products to the Middle East and Southeast Asia. The meaning of this Arabic word, "permitted" or "lawful," is very clear. Nevertheless, its practical interpretation varies a great deal among food-importing countries, as does its understanding by companies that produce food. Food Science and Technology students generally are not taught about the dietary requirements of different religions and ethnic groups and are only exposed to concepts such as *Kosher*, *Halal*, and *Vegetarian* in industry, when product development, quality assurance, procurement, and other key personnel are forced to learn about these concepts to meet their customer's requirements.

The books currently on the market for people interested in the subject of Halal have been published to help the Muslim consumers decide what to eat and what to avoid among the foods already present in the marketplace. There are very few books that are written to help food industry personnel understand religious food requirements. This book is the result of feedback and industry comments on our first book written on this topic, *Halal Food Production*. It is a combination of looking back at our work and the new research that has been done in this field. Both authors, Mian N. Riaz (working in a university) and Muhammad M. Chaudry (working in the food industry), have recognized this gap in the vital information about Halal available to food professionals. Both authors are food scientists, with collectively more than 65 years of practical experience in this area. This book is the result of their practical experience and knowledge in Halal food requirements and Halal certification.

This book is written to summarize some of the fundamentals to be considered in Halal food production. It is an excellent starting point for the food scientist and technologist and other professionals who are in the Halal food business. There is a wealth of information about Halal food laws and regulations, and general guidelines for Halal food production; and domestic and international Halal food markets, trade, and import requirements for different countries. It also covers specific Halal production requirements for meat, poultry, dairy products, fish, seafood, cereal, confectionary, and food supplements. The role of gelatin, enzymes, alcohol, and other questionable ingredients for Halal food production is addressed in some detail. Guidelines with examples of labeling, packaging, and coatings for Halal food are also presented. The new topic of biotechnology and GMO in Halal food production is explained. A brief discussion of the growing concern about animal feed is also provided. One complete chapter is dedicated to the differences between Halal, Kosher, and vegetarian food production. For the food companies who would like their products to be certified Halal, a procedure is included for obtaining Halal certification. This book also contains appendices, which cover Halal food-related information that can be used as guidelines by Halal food processors.

We believe this book can serve as a source of information to all who are involved or would like to be involved in any aspect of the Halal food business. For the people who are new to this area, this book will serve as a guide for understanding and properly selecting food ingredients for processing Halal foods. In view of the growing Halal food markets worldwide in food service, branded packaged foods, and direct-marketed products as well as food ingredients, both academia and industry will benefit from this work.

We owe a large debt of gratitude to a number of individuals who provided information, inspiration, and guided us in the right direction to complete this book.

Acknowledgments

Completion of this book provides an opportunity to recognize a number of very important individuals. We particularly want to express our heartfelt gratitude to Dr. Joe Regenstein of Cornell University for his exhaustive suggestions and guidance toward the preparation of this book. His combination of sage advice, his work toward editing the chapters, the ability to challenge us to stretch, and a never-ending willingness to share his time will always be greatly valued. He was extremely helpful in providing the most accurate information and critique throughout this work.

We wish to express our thanks and gratitude to Faria Arshad, a PhD student for her help with the final editing and organization of this book. Another person who provided us with very valuable assistance in collecting the pictures and information was Roger Ottman. Also special thanks to all the friends, family members, and colleagues who provided help in numerous ways. We would not have been able to complete this book without the help of these individuals.

Mian N. Riaz
Muhammad M. Chaudry

About the Editors

Mian Nadeem Riaz is a Professor of Food Diversity in the Nutrition and Food Science Department, Director for the Process Engineering R&D Center (Formerly Food Protein R&D Center) and Graduate Faculty in the Food Science and Technology program at Texas A&M University, College Station, Texas, USA. He received his BSc (Honor) and MSc majoring in Food Technology from the University of Agriculture Faisalabad, Pakistan. While earning his Master degree he received the Silver Medal from the University. He joined Texas A&M University 26 years ago after completing his Ph.D in Food Science from the University of Maine at Orono, Maine, He was the first Ph.D. graduate of their Food Science department. His first academic appointment was in 1992 at Texas A&M University. Dr. Riaz received an award which is in honor of Noble Laureate Dr. Bernard Lown from the University of Maine Alumni association. This award was given to Dr. Riaz for his work in reducing food insecurity across the globe and being one of the leading experts in Halal food processing and production. Dr. Riaz is a professional member as well as Certified Food Scientist (CFS) for the Institute of Food Technology (IFT). He was one of the founders and early leaders of the IFT (Institute of Food Technologists) Religious and Ethnic Foods Division that for many years was able to mount important seminars that helped IFT members better understand the changing demographics in the U.S. and the globalization of the food industry. Dr. Riaz has also been active in IFT's Food Science Communicators program, using his understanding of food diversity to help IFT and the audiences. Dr. Riaz published five books and one of the books in the area of Halal food titled *Halal Food Production*. Currently this book is being used in most of universities for Food Science and Nutrition major students. This book has been translated in to Chinese, Persian, and Korean languages. He also published 22 chapters in different books including chapters in the area of Halal food related issues. Currently, he teaches a course on "Religious and Ethnic Food" at Texas A&M University to Food Science and Nutrition majors. Dr. Riaz is a frequent speaker for Halal conferences, seminar and training held all over the world. He has given more than 300 talks in 60 countries.

Mian Nadeem Riaz, Ph.D, CFS
Professor, Food Diversity—Nutrition and Food Science
Director: Process Engineering R&D Center

Muhammad Munir Chaudry is founding member, President and CEO of IFANCA®, The Islamic Food and Nutrition Council of America, headquartered in Chicago/Park Ridge, Illinois USA. Dr. Chaudry, a career food technologist, received his academic education with BSc and MSc in Food Technology from Pakistan, MS in Food Science from the American University of Beirut, Lebanon, and Ph.D. in Food Science from the University of Illinois at Champaign-Urbana, in 1980. Over 15 years in the US food industry, Dr. Chaudry held various technical and management positions where he was responsible for total quality, quality assurance, human resources, and employee training. For the past 28 years, Dr. Chaudry has been managing halal certification and awareness programs for the Islamic Food and Nutrition Council of America. Under his administration, halal certification by IFANCA has expanded to over 4,000 production sites in 70 countries. Halal certified products include meat and poultry, seafood and food products, food ingredients and chemicals, nutritional products and supplements, cosmetics and personal care products, vaccines as well as packaging materials. Dr. Chaudry is a professional member of the Institute of Food Technologists and expert consultant to OIE, the World Organization for Animal Health. He has written several articles and papers in technical journals as well as co-authored a book about Halal Food Production published by CRC Press. The second addition of the book is under publication. He served as an international delegate for food safety workshops conducted in India under the sponsorship of the United Nations International Trade Center and the American Spice Trade Association. He is a graduate of the FBI Citizen Academy and an active member of its alumni association. Dr. Chaudry has been involved in several civil and educational organizations including the Abrahamic Center for Cultural Education, Sabeel Center, Kazi Publications, and American Pan-Islamic Community Council. He is also a member of Chicago Council on Global Affairs and American Civil Liberties Union. Dr. Chaudry is an expert in the management of not-for-profit organizations and industrial corporations in a cross-functional environment.

Muhammad Munir Chaudry, Ph.D
*President: The Islamic Food and
Nutrition Council of America (IFANCA)*

Contributors

Mohammed A. Khan
Researcher,
Mississauga, Canada

Rafiuddin Shikoh
DinarStandard,
Great Neck, NY, USA

Kristin Pufpaff
Independent Contractor,
Nashville, MI, USA

Joe Regenstein
Cornell University,
Ithaca, NY, USA

Mustafa Farouk
Ag Research Limited,
Hamilton, New Zealand

Rafiuddin Shaik
IFANCA,
Park Ridge, IL, USA

Eric Butrym
Independent Contractor,
Pennington, NJ, USA

Laura Lacourse
Independent Contractor,
Hamilton, NJ, USA

Roger Ottman
IFANCA,
Park Ridge, IL, USA

Jes Knudsen
Novozymes
A/S-Denmark

Winai Dahlan
The Halal Science Center
Chulalongkorn University,
Bangkok, Thailand

Sibt-e-Abbas
GC. University,
Sahiwal, Pakistan

1 Introduction

Mian N. Riaz and Munir M. Chaudry

The food industry, like any other industry, responds to the needs and desires of the consumer. People all over the world are now more conscious about foods, health, and nutrition. They are interested in eating healthy foods that are low in calories, cholesterol, fat, and sodium. Many people are interested in foods that are organically produced without the use of synthetic pesticides and other non-natural chemicals. The ethnic and religious diversity in the U.S. and Europe has encouraged the food industry to prepare products, which are suitable for different groups such as the Chinese, Japanese, Italian, Indian, Mexican, Seventh Day Adventist, Vegetarian, Jewish, and Muslim communities.

Islam is the world's second largest religion and also the fastest growing, both globally and in the U.S. (Nakyinsige et al., 2012). The global Muslim population is 1.8 billion and the halal food market is estimated to be $547 billion per year in the U.S. (Dierks, 2011). The expected increase is to $2.1 trillion in tandem with a fivefold increase in the global halal food market (Dagang Asia Net, 2011). The Muslim population in the U.S. is estimated to reach 12.2 million by 2018 (USA Today, 1999). Islam is not merely a religion of rituals—it is a way of life. Rules and manners govern the life of the individual Muslim. There are a set of halal dietary rules that Muslims are expected to follow and are meant to advance their well-being (Bonne and Verbeke, 2008).

In Islam, eating is considered a matter of worship of God, just like religious prayers. Muslims follow the Islamic dietary code and foods that meet that code are called halal (lawful or permitted) (Regenstein et al., 2003). Muslims are supposed to make an effort to obtain halal food of good quality. It is their religious obligation to consume only halal food. For non-Muslim consumers, halal foods are often perceived as specially selected and processed to achieve the highest standards of quality along with being healthier.

Between 300 and 400 million Muslims are estimated to live as minorities in different nations of the world, forming a part of many different cultures and societies. The rest live in countries that have a Muslim majority although some of these countries have many other significant minority groups. In spite of their geographic and ethnic diversity, all Muslims follow their beliefs and the religion of Islam. Halal is a very important and integral part of religious observance for all Muslims. Hence, halal constitutes a universal standard for a Muslim to live by. In the UK there are about 2 to 3 million Muslims but the consumption of halal meat is estimated to include 6 million people. This shows that the halal market's growing trend not only among Muslims but also for non-Muslims (Gregory, 2008).

By definition, halal foods are those that are free from any component that Muslims are prohibited from consuming. According to the Quran (the Muslim scripture), all good and clean foods are halal. Consequently, almost all foods of plant and animal origin are considered halal except those that have been specifically prohibited by the Quran and the Sunnah (the life, actions, and teachings of the Prophet Muhammad [Peace Be Upon Him, PBUH]; Shah Alam and Mohamed Sayuti, 2011).

This book, a revision of the first edition with many new additions, combines the religious and production issues that can help food manufacturers understand halal food production. Producing halal food is similar to producing regular foods, except for certain additional requirements, which will be discussed in this book. Halal foods can be processed by using the same equipment and utensils as regular food, with a few exceptions or changes. In the chapters that follow, food manufacturers will learn the requirements of halal food production and gain some knowledge about Muslims and the Muslim markets.

The book is divided into various chapters covering halal laws in general, production guidelines for various product types (including meat and poultry, fish and seafood, dairy products, cereals, and food ingredients), labeling, biotechnology, and several other areas of concern of the halal consumers.

The book presents the laws and regulations in a format that will be understandable by non-Muslims. Terminology and concepts that are generally associated with religious jurisprudence have been avoided wherever possible. The laws have been translated into general guidelines for the food industry and kindred product industries. Several chapters have been devoted to specific industries in which the authors feel that halal activity is currently the greatest:

- *Chapters 2 and 3*: Halal food laws and halal food guidelines for the food industries.
- *Chapters 4 and 5*: Muslim demographics, halal market, and the global halal economy.
- *Chapters 6 and 7*: Animal welfare and religious slaughter of animals: A U.S. perspective on regulations and animal welfare guidelines.
- *Chapter 8*: Meat and poultry products guidelines, as these are the most highly regulated segment of the food industry with respect to halal requirements. Out of five prohibited food categories, four belong to this group.
- *Chapter 9*: Guidelines for the processed meat industry regarding the production of halal meat products.
- *Chapter 10*: Explanations of the status of various fish, shellfish, crustaceans, and other seafood products as fish and seafood products are subject to more controversy than any other food group among Muslim consumers even though they are not very significant in international halal trade.
- *Chapter 11*: A balanced picture of dairy products requirements, with special emphasis on the use of enzymes. The expanding dairy field includes cheese and whey proteins that have received wide acceptance in non-dairy food products. Controversy over the use of porcine enzymes even with the development of chymosin-type products as rennet replacers or extenders continues among Muslim consumers.
- *Chapter 12*: A brief discussion on the guidelines for cereal-based products, candy, and other products, as there are relatively few controversial issues with these products for different Muslim consumers.

Introduction

- *Chapter 13*: Enzymes and how are they are used in food processing as well as their sources.
- *Chapter 14*: The most important ingredient in the food industry, gelatin, has many guidelines, uses, and sources, which are all discussed.
- *Chapter 15*: The complicated ingredients and their sources that make up the many flavors used in the food industry.
- *Chapter 16*: Alcohol, while not allowed in halal food, is used heavily in food production, and thus its limits and applications are important to discuss.
- *Chapter 17*: Food additives and their use in increasing shelf life and consumer appeal as well as their sources.
- *Chapter 18*: Food ingredients and the many diverse items used all across the food industry. These may be produced from plants, animals, microorganisms, or by synthetic processes. The emphasis is on flavors, amino acids, oils and extracts, and blended products.
- *Chapter 19*: Nutritional supplements and the high visibility and demand for halal certified products throughout the world, specifically in the Southeast Asian countries.
- *Chapter 20*: Information on biotechnology and genetically modified organisms (GMO) issues as they are a major concern for halal food production.
- *Chapter 21*: Poultry and the animal feed industry and its job to provide clean feed with or without the use of animal by-product (especially pork by-product) in feed formulations.
- *Chapter 22*: Animal-based ingredients that are major issues in halal cosmetics as the halal cosmetic market is growing rapidly in the Middle East and Asia. Several companies produce halal cosmetics for these markets.
- *Chapter 23*: How to use the halal symbol on food product labels as labeling is for the benefit of consumers.
- *Chapter 24*: How to get a halal certification and the details that are needed to do so. Halal certification is a growing segment of the market and several food companies are getting or expanding their halal certification program. There are several kinds of certification and a company needs to understand the options and properly fill out the application.
- *Chapter 25*: The similarities and differences between halal, kosher, and vegetarian foods. Most of the food companies consider that halal is similar to kosher. There are several important differences between halal and kosher products.
- *Chapter 26*: The need for the globalization of halal certification and developing universal halal food guidelines for the food industry. Several countries have their own halal food laws (e.g., Indonesia, Malaysia, Singapore, Saudi Arabia, and UAE) and their differences are a major concern of the food industry.
- *Chapter 27*: Methods that are being used to detect the adulteration of halal foods with non-halal ingredients in halal foods. These methods are being developed by several universities that have established validated scientific procedures.

- *Chapter 28*: Potential hazards and sanitation requirements when producing halal foods.
- *Chapter 29*: Halal awareness and education needed from farm to fork for halal food production.
- *Chapter 30*: Halal food models and how halal food companies can learn from these halal food models.
- *Chapter 31*: Hazard analysis and critical control points (HACCP) and how the food industry can use it with their halal program is discussed, as HACCP is an integral part of the food management system. There are several similarities between HACCP and halal control points that could benefit the food industry.

In the chapters covering halal requirements for different products categories, the concepts related to HACCP have been used to provide a framework for identifying halal control points (HCP). The objective is not to replace HACCP, which addresses food safety issues, but to complement these requirements by adding key points for halal compliance. The HCP have been presented in an easy-to-understand flowchart format. By using these guidelines, food companies are encouraged to devise their own HCP and include them in their standard operating procedures to serve as part of their self-compliance efforts.

Marketing and trade aspects of halal foods have been included in two chapters: one covers the domestic and international trade of halal food products, and the second covers import requirements for various Muslim countries as well as the Muslim population.

Finally, information has been included about procedures for getting halal certification. Food manufacturers can obtain supervision from different halal food certifying agencies such as the Islamic Food and Nutrition Council of America (IFANCA) as well as reliable information about Islam and Muslims in North America and their critical food issues. These halal-certifying agencies provide consultation services and help food industry professionals develop products that comply with Islamic food laws. These agencies also offer supervision and certification for halal foods, consumer products, and halal-slaughtered meat and poultry. Their registered trademark certification symbol, for example, the Crescent M®, appears on many product packages. The demand for halal products by many Muslim consumers can easily be an inducement for manufacturers to provide halal products. Halal markings are an important part of the general acceptance of halal products by the Muslim consumer worldwide. Certain key information with respect to halal food production is included in several appendices, such as the relevant section of Codex Alimentarius (the Food and Agricultural Organization [FAO]/World Health Organization [WHO] of the United Nation's [UN] food standards) and the halal status of common and e-numbered (the system used by the European Union [EU]) ingredients.

REFERENCES

Bonne, K. and Verbeke, W. (2008). Religious values informing halal meat production and the control and delivery of halal credence quality. *Agriculture and Human Values*, 25(1), 35–47.

Dagang Asia Net. (2011). The emerging of global halal market. Available from: www.dagangasia.net/articles. Accessed on May 11, 2011.

Dierks. (2011). *Market Watch 2010—The Food Industry.* Malaysia: Malaysian-German Chamber of Commerce and Industry.

Gregory, N. G. (2008). Animal welfare at markets and during transport and slaughter. *Meat Science,* 80(1), 2–11.

Nakyinsige, K., Che Man, Y. B., Sazili, A. Q., Zulkifli, I., and Fatimah, A. B. (2012). Halal meat: A niche product in the food market. Singapore: *2nd International Conference on Economics, Trade and Development IPEDR,* 36, 167–173.

Regenstein, J. M., Chaudry, M. M., and Regenstein, C. E. (2003). The kosher and halal food laws. *Comprehensive Reviews in Food Science and Food Safety,* 2(3), 111–127.

Shah Alam, S. and Mohamed Sayuti, N. (2011). Applying the Theory of Planned Behavior (TPB) in halal food purchasing. *International Journal of Commerce and Management,* 21(1), 8–20.

USA Today. (1999). Facts, figures about Islam: Muslim future: Projected Muslim population in the USA. *USA Today,* June 25, 12B.

2 Halal Food Laws and Regulations

Mian N. Riaz and Munir M. Chaudry

CONTENTS

Principles Regarding Permissibility of Foods 8
Halal and Haram 9
Basis for the Prohibitions 12
How Does One Translate Major Prohibitions into Practice in Today's
Industrial Environment? 13
References 15

The basic guidance about halal food laws is revealed in the Quran (the divine book) from God (the Creator) to Muhammad (the Prophet, PBUH) for all people. The food laws are explained and put into practice through the Sunnah (the life, actions, and teachings of the Prophet Muhammad) as recorded in the Hadith (the compilation of the traditions of the Prophet).

In general, everything is permitted for human use and benefit. Nothing is forbidden except what is prohibited either by a verse of the Quran or an authentic and explicit Sunnah of the Prophet. These rules of Islamic law bring freedom for people to eat and drink anything they like as long as it is not haram (prohibited) (Battour et al., 2010).

There are five fundamental pillars of belief in Islam: (1) to believe that there is no god but Allah, and Muhammad is his last prophet; (2) to pray five times in a day; (3) to give zakat (charity) to the poor; (4) to fast in the month of Ramadan; and (5) to perform the pilgrimage to Mecca once in a lifetime (if one can afford it). In addition, guidelines direct the daily life of a Muslim. Included in these guidelines is a set of dietary laws intended to advance wellness. These laws are binding on the faithful and must be observed at all times, even during pregnancy, periods of illness, or traveling (Twaigery and Spillman, 1989). The life of a Muslim revolves around the concept of halal and haram. The laws are quite comprehensive, because they are applicable not only to eating and drinking, but also to earning one's living, dress code, and dealing with others. This discussion will focus primarily on food.

Food is considered one of the most important factors for interaction among various ethnic, social, and religious groups. All people are concerned about the food they eat: Muslims want to ensure that their food is halal; Jews that their food is kosher; Hindus, Buddhists, and certain other groups that their food is vegetarian. Muslims follow clear guidelines in the selection of their food (Qureshi et al., 2012).

The principles behind halal food are described here.

PRINCIPLES REGARDING PERMISSIBILITY OF FOODS

Eleven generally accepted principles pertaining to halal (permitted) and haram (prohibited) in Islam provide guidance to Muslims in their customary practices (Al-Qaradawi, 1984):

1. The basic principle is that all things created by God are permitted, with a few exceptions that are specifically prohibited.
2. To make lawful and unlawful is the right of God alone. No human being, no matter how pious or powerful, may take this right into his own hands.
3. Prohibiting what is permitted and permitting what is prohibited is similar to ascribing partners to God.
4. The basic reasons for the prohibition of things are impurity and harmfulness. A Muslim is not required to know exactly why or how something is unclean or harmful in what God has prohibited. There might be obvious reasons and there might be obscure reasons.
5. What is permitted is sufficient and what is prohibited is then superfluous. God prohibited only things that are unnecessary or dispensable while providing better alternatives.
6. Whatever is conducive to the "prohibited" is in itself prohibited. If something is prohibited, anything leading to it is also prohibited.
7. Falsely representing unlawful as lawful is prohibited. It is unlawful to legalize God's prohibitions by flimsy excuses. To represent lawful as unlawful is also prohibited.
8. Good intentions do not make the unlawful acceptable. Whenever any permissible action of the believer is accompanied by a good intention, his action becomes an act of worship. In the case of haram, it remains haram no matter how good the intention, how honorable the purpose, or how lofty the goal. Islam does not endorse employing a haram means to achieve a praiseworthy end. Indeed, it insists not only that the goal be honorable, but also that the means chosen to attain it be proper. "The end justifies the means" and "secure your right even through wrongdoing" are maxims not acceptable in Islam. Islamic law demands that the right should be secured through just means only.
9. Doubtful things should be avoided. There is a gray area between clearly lawful and clearly unlawful. This is the area of "what is doubtful." Islam considers it an act of piety for Muslims to avoid doubtful things and for them to stay clear of the unlawful. Prophet Muhammad (PBUH) said (Sakr, 1994): "Halal is clear and haram is clear; in between these two are certain things that are suspected. Many people may not know whether these items are halal or haram. Whosoever leaves them, he is innocent toward his religion and his conscience. He is, therefore, safe. Anyone who gets involved in any of these suspected items, he may fall into the unlawful and the prohibited. This case is similar to the one who wishes to raise his animals next to a restricted area, he may step into it. Indeed the restrictions of Allah are the unlawful."

10. Unlawful things are prohibited to everyone alike. Islamic laws are universally applicable to all races, creeds, and sexes. There is no favored treatment of any privileged class. Actually, in Islam, there are no privileged classes; hence, the question of preferential treatment does not arise. This principle applies not only among Muslims but between Muslims and non-Muslims as well.
11. Necessity dictates exceptions. The range of prohibited things in Islam is very narrow, but emphasis on observing the prohibitions is very strong. At the same time, Islam is not oblivious to the exigencies of life, to their magnitude, or to human weakness and capacity to face them. It permits the Muslim, under the compulsion of necessity, to eat a prohibited food in quantities sufficient to remove the necessity and thereby survive.

Five major terms are used to describe the permissibility of food:

Halal: Means permissible and lawful. It applies not only to meat and poultry, but also to other food products, cosmetics, and personal care products. The term also applies to personal behavior and interaction with the community (Al-Hanafi, 2006).
Haram: Means prohibited. It is directly opposite of halal.
Mashbooh: Is something questionable or doubtful, either due to the differences in scholars' opinions or due to the presence of undetermined ingredients in a food product.
Makrooh: Is a term generally associated with someone's dislike for a food product or, while not clearly haram, is considered dislikeable by some Muslims.
Zabiha or dhabiha: Is a term often used by Muslims in the U.S. to differentiate meat that has been slaughtered by Muslims as opposed to being slaughtered by Ahlul Kitab (Jews or Christians) or without religious connotation.

HALAL AND HARAM

General Quranic guidance dictates that all foods are halal except those that are specifically mentioned as haram. All foods are made lawful according to the Muslim scripture The Glorious Quran (Arabic text and English rendering by Regenstein et al. [2003]):

> O ye who believe! Eat of the good things wherewith We have provided you, and render thanks to Allah, if it is (indeed) He whom ye worship.
>
> **Chapter II, Verse 172**

The unlawful foods are specifically mentioned in The Glorious Quran in the following verses:

> He hath forbidden you only carrion, and blood, and swine flesh, and that which hath been immolated to (the name of) any other than Allah.
>
> **Chapter II, Verse 173**

Forbidden unto you (for food) are carrion and blood and swine flesh, and that which hath been dedicated unto any other than Allah, and the strangled, and the dead through beating, and the dead through falling from a height, and that which hath been killed by (the goring of) horns, and the devoured of wild beasts saving that which ye make lawful (by the death-stroke) and that which hath been immolated unto idols. And (forbidden is it) that ye swear by the divining arrows. This is an abomination.

Chapter V, Verse 3

Consumption of alcohol and other intoxicants is prohibited according to the following verse:

O ye who believe! Strong drink and games of chance, and idols and divining arrows are only an infamy of Satan's handiwork. Leave it aside in order that ye may succeed.

Chapter V, Verse 90

Meat is the most strictly regulated of the food groups. Not only are blood, pork, and the meat of dead animals or those immolated to other than God strongly prohibited, it is also required that halal animals be slaughtered while pronouncing the name of God at the time of slaughter.

Eat of that over which the name of Allah hath been mentioned, if ye are believers in His revelations.

Chapter VI, Verse 118

And eat not of that whereon Allah's name hath not been mentioned, for lo! It is abomination. Lo! The devils do inspire their minions to dispute with you. But if ye obey them, ye will be in truth idolaters.

Chapter VI, Verse 121]

Accordingly, all foods pure and clean are permitted for consumption by Muslims except the following categories, including any products derived from them or contaminated with them:

- Carrion or dead animals
- Flowing or congealed blood
- Swine, including all by-products
- Animals slaughtered without pronouncing the name of God on them
- Animals killed in a manner that prevents their blood from being fully drained from their bodies
- Animals slaughtered while pronouncing a name other than God
- Intoxicants of all types, including alcohol and drugs
- Carnivorous animals with fangs, such as lions, dogs, wolves, or tigers
- Birds with sharp claws (birds of prey), such as falcons, eagles, owls, or vultures
- Land or amphibious animals such as frogs or snakes

From the Quranic verses, the hadith, and their explanations and commentary by Muslim scholars, the Islamic food (dietary) laws are deduced. Additional verses in The Glorious Quran related to food and drinks are as follows:

> O' mankind! Eat of that which is lawful and wholesome in the earth, and follow not the footsteps of the devil. Lo! He is an open enemy for you.
>
> **Chapter II, Verse 168**

> Oh ye who believe! Fulfill your indentures. The beast of cattle is made lawful unto you (for food) except.
>
> **Chapter V, Verse 1**

> They ask thee (O Muhammad) what is made lawful for them. Say: (all) good things are made lawful for you. And those beasts and birds of prey which ye have trained as hounds are trained, ye teach them that which Allah taught you; so eat of that which they catch for you and mention Allah's name upon it, and observe your duty to Allah. Allah is swift to take account.
>
> **Chapter V, Verse 4**

> This day are (all) good things made lawful for you. The food of those who have received Scripture is lawful for you and your food is lawful for them.
>
> **Chapter V, Verse 5**

> Oh ye who believe! Forbid not the good things, which Allah had made lawful for you, and transgress not. Lo Allah loveth not transgressors.
>
> **Chapter V, Verse 87**

> Eat of that which Allah hath bestowed on you as food lawful and good, and keep your duty to Allah in whom ye are believers.
>
> **Chapter V, Verse 88**

> How should ye not eat of that over which the name of Allah hath been mentioned, when He hath explained unto you that which is forbidden into you, unless ye are compelled thereto.
>
> **Chapter VI, Verse 119**

> And eat not of that whereon Allah's name hath not been mentioned, for lo! It is abomination. Lo! The devils do inspire their minions to dispute with you. But if ye obey them, ye will be in truth idolaters.
>
> **Chapter VI, Verse 121**

> And of the cattle He produceth some for burden and some for food; Eat of that which Allah hath bestowed upon you, and follow not the footsteps of the devil, for lo! He is an open foe to you.
>
> **Chapter VI, Verse 142**

Say: "I find not in that which is revealed unto me ought prohibited to an eater that he eat thereof except it be carrion, or blood poured forth, or swine flesh—for that verily is foul—or the abomination which was immolated to the name of other than Allah. But who so is compelled (there to), neither craving nor transgressing, (for him) Lo! Your Lord is forgiving, merciful."

Chapter VI, Verse 145

So eat of the lawful and good food, which Allah has provided for you and thank the bounty of your Lord if it is Him ye serve.

Chapter XVI, Verse 114

He hath forbidden for you only carrion, and blood and the swine flesh, and that which hath been immolated in the name of any other than Allah; but he who is driven thereto, neither craving nor transgressing, Lo! Then Allah is forgiving, merciful.

Chapter XVI, Verse 115

The haram foods are mainly pork, alcohol, blood, dead animals, and animals slaughtered while reciting a name other than that of God. This may also include halal items that have been contaminated or mixed with haram items. In general, most Muslims deem meat and poultry items not slaughtered in the name of God to be haram or makrooh at best.

BASIS FOR THE PROHIBITIONS

In the Islamic faith, Allah is the Almighty God. He has no partners. The first requirement of a Muslim is to declare: "There is no god but God (Allah)." So everything has to be dedicated to God only. There is no challenge to this fact, and no explanations are required or necessary. The basis for the prohibition of the above categories is purely and strictly Quranic guidance. However, some scientists have attempted to explain or justify some of these prohibitions based on their scientific understanding as follows:

Carrion and dead animals are unfit for human consumption because the decaying process leads to the formation of chemicals that are harmful to humans (Awan, 1988).

Blood that is drained from the body contains harmful bacteria, products of metabolism, and toxins (Awan, 1988; Erbil, 2001; Hussaini and Sakr, 1984).

Swine serves as a vector for pathogenic worms to enter the human body. Infection by *Trichinella spiralis* and *Taeniasolium* are not uncommon. Fatty acid composition of pork fat has been mentioned as incompatible with the human fat and biochemical systems (Omojola et al., 2009).

Intoxicants are considered harmful for the nervous system, affecting the senses and human judgment. In many cases they lead to social and family problems and even loss of lives (Al-Qaradawi, 1984; Awan, 1988; Wan Hassan and Awang, 2009).

Although these explanations may or may not be sound scientifically, the underlying principle behind the prohibitions remains the divine order, which appears in the

Halal Food Laws and Regulations

Glorious Quran in several places: "Forbidden unto you are" is what guides a Muslim believer. Thus, the acceptance and accuracy of these "scientific" statements is not crucial to the acceptance of these guiding principles.

HOW DOES ONE TRANSLATE MAJOR PROHIBITIONS INTO PRACTICE IN TODAY'S INDUSTRIAL ENVIRONMENT?

Let us look at how the laws are translated into practice:

- *Carrion and dead animals*: It is generally recognized that eating carrion is offensive to human dignity, and probably nobody consumes it in modern civilized society. However, there is a chance of an animal dying from the shock of stunning before it is properly slaughtered. This is more common in Europe than in North America. The meat of such dead animals is not proper for Muslim consumption (Regenstein et al., 2003).
- *Proper slaughtering*: There are strict requirements for the slaughtering of animals: the animal must be of a halal species, that is, cattle, lamb, and so on; the animal must be slaughtered by an adult Muslim; the name of God must be pronounced at the time of slaughter; and the slaughter must be done by cutting the throat of the animal in a manner that induces rapid, complete bleeding and results in the quickest death (Al-Qaradawi, 2013).
- *Certain other conditions should also be observed*: These include considerate treatment of the animal, giving it water to prevent thirst, and using a sharp knife (preferably without any nicks in the knife). These conditions ensure the humane treatment of animals before and during slaughter. Any by-products or derived ingredients must also be from duly slaughtered animals to be good for Muslim consumption (Apple et al., 2005; Hambrecht et al., 2004; Kannan et al., 2003; Ljungberg et al., 2007; Schaefer et al., 2001).
- *Swine*: Pork, lard, and their by-products or derived ingredients are categorically prohibited for Muslim consumption. All chances of cross-contamination from pork into halal products must be thoroughly prevented. In fact, in Islam, the prohibition extends beyond eating. For example, a Muslim must not buy, sell, raise, transport, slaughter, or in any way directly derive benefit from swine or other haram materials (Omojola et al., 2009).
- *Blood*: Blood that pours forth (liquid blood) is generally not offered in the marketplace or consumed, but products made from blood and ingredients derived from it are available. There is general agreement among religious scholars that anything made from blood is unlawful for Muslims (Mickler, 2000).
- *Alcohol and other intoxicants*: Alcoholic beverages such as wine, beer, and hard liquors are strictly prohibited. Foods containing added amounts of alcoholic beverages are also prohibited because such foods, by definition, become impure. Non-medical drugs and other intoxicants that affect a person's mind, health, and overall performance are prohibited too. Consuming these directly

or incorporating them into foods is not permitted (Wahab, 2004). However, there are certain acceptable allowances for naturally present alcohol or alcohol used in processing of food, as discussed in Chapter 16.

Foods are broadly categorized into four groups for the ease of establishing their halal status and formulating guidelines for the industry.

- *Meat and poultry*: This group contains four out of five haram (prohibited) categories. Hence, more restrictions are observed. Animals must be halal. One cannot slaughter a pig the Islamic way and call it halal. Animals must be slaughtered by a sane, adult Muslim while pronouncing the name of God. A sharp knife must be used to sever the jugular veins, carotid arteries, trachea, and esophagus, and the blood must be drained out completely (Qureshi et al., 2012). Islam places great emphasis on humane treatment of animals, so dismemberment must not take place before the animal is completely dead, as described earlier.
- *Fish and seafood*: To determine the acceptability of fish and seafood, one has to understand the rules of the different schools of Islamic jurisprudence, as well as the cultural practices of Muslims living in different regions. All Muslims accept fish with scales; however, some groups do not accept fish without scales such as catfish. There are even greater differences among Muslims about seafood, such as mollusks and crustaceans. One must understand the requirements in various regions of the world, for example, for exporting products containing seafood or seafood flavors (Mickler, 2000).
- *Milk and eggs*: These products from halal animals are also halal. The predominant source of milk in the West is the cow and the predominant source of eggs is the chicken. All other sources are required to be labeled accordingly. There are a variety of products made from milk and eggs. Milk is used for making cheese, butter, and cream. Most of the cheeses are made with various enzymes, which could be halal if made using microorganisms, plants, or halal-slaughtered animals (Guerrero-Legarreta, 2010). The enzymes could be haram if extracted from porcine sources or questionable when obtained from non-halal-slaughtered animals. Similarly, emulsifiers, mold inhibitors, and other functional ingredients from non-specified sources can make milk and egg products doubtful to consume (Ermis, 2017).
- *Plants and vegetables*: These materials are generally halal except alcoholic drinks or other intoxicants. However, in modern-day processing plants, vegetables and meats might be processed in the same plant and on the same equipment, increasing the chance of cross-contamination. Certain functional ingredients from animal sources might also be used in the processing of vegetables, which make the products doubtful. Hence, processing aids and production methods have to be carefully monitored to maintain the halal status of foods of plant origin (Regenstein et al., 2003).

From this discussion on laws and regulations, it is clear that several factors determine the halal or haram status of a particular foodstuff. It depends on its nature, how it is processed, and how it is obtained. As an example, any product from a pig would be considered as haram because the material itself is haram. Similarly, beef from an animal that has not been slaughtered according to Islamic requirements would still be considered unacceptable. And, of course, a stolen foodstuff or any products that are acquired through means that are incompatible with Islamic teaching would also be haram. Food and drink that are poisonous or intoxicating are obviously haram even in small quantities because they are harmful to health.

REFERENCES

Al-Hanafi. (2006). *KashafIstilahatulFanun*. Beirut, Lebanon: Darul Kutubul Alilmiya.
Al-Qaradawi, Y. (1984). *The Lawful and Prohibited in Islam*. Beirut, Lebanon: The Holy Quran Publishing House.
Al-Qaradawi, Y. (2013). *The Lawful and the Prohibited in Islam*. Selangor, Malaysia: The Other Press.
Apple, J. K., Kegley, E. B., Maxwell, C. V., Rakes, L. K., Galloway, D., and Wistuba, T. J. (2005). Effects of dietary magnesium and short-duration transportation on stress response, postmortem muscle metabolism, and meat quality of finishing swine 1. *Journal of Animal Science*, 83(7), 1633–1645.
Awan, J. A. (1988). Islamic food laws-I: Philosophy of the prohibition of unlawful foods. *Science and Technology in the Islamic World*, 6(3), 151–165.
Battour, M. M., Ismail, M. N., and Battor, M. (2010). Toward a halal tourism market. *Tourism Analysis*, 15(4), 461–470.
Ermis, E. (2017). Halal status of enzymes used in food industry. *Trends in Food Science and Technology*, 64, 69–73.
Erbil, C. (2001). Why is slaughtering animals prescribed as it is. Retrieved on February 28, 2018. http://www.fountainmagazine.com/Issue/detail/Why-is-Slaughtering-Animals-Prescribed-as-It-Is.
Guerrero-Legarreta, I. (2010). *Handbook of Poultry Science and Technology, Volume 2: Secondary Processing*. Hoboken, NJ: Wiley.
Hambrecht, E., Eissen, J. J., Nooijen, R. I. J., Ducro, B. J., Smits, C. H. M., Den Hartog, L. A., and Verstegen, M. W. A. (2004). Preslaughter stress and muscle energy largely determine pork quality at two commercial processing plants. *Journal of Animal Science*, 82(5), 1401–1409.
Hussaini, M. M. and Sakr, A. H. (1984). *Islamic Dietary Laws and Practices*. Bedford Park, IL: Islamic Food and Nutrition Council of America.
Kannan, G., Kouakou, B., Terrill, T. H., and Gelaye, S. (2003). Endocrine, blood metabolite, and meat quality changes in goats as influenced by short-term, preslaughter stress 1. *Journal of Animal Science*, 81(6), 1499–1507.
Ljungberg, D., Gebresenbet, G., and Aradom, S. (2007). Logistics chain of animal transport and abattoir operations. *Biosystems Engineering*, 96(2), 267–277.
Mickler, D. L. (2000). Halal foodways. Retrieved on July 22, 2001. http://www.unichef.com/Halalfood.htm.
Omojola, A. B., Fagbuaro, S. S., and Ayeni, A. A. (2009). Cholesterol content, physical and sensory properties of pork from pigs fed varying levels of dietary garlic (Allium sativum). *World Applied Sciences Journal*, 6(7), 971–975.

Qureshi, S. S., Jamal, M., Qureshi, M. S., Rauf, M., Syed, B. H., Zulfiqar, M., and Chand, N. (2012). A review of Halal food with special reference to meat and its trade potential. *Journal of Animal and Plant Sciences*, 22(2 Suppl), 79–83.
Regenstein, J. M., Chaudry, M. M., and Regenstein, C. E. (2003). The kosher and halal food laws. *Comprehensive Reviews in Food Science and Food Safety*, 2(3), 111–127.
Sakr, A. H. (1994). Halal and Haram defined. In: *Understanding Halal Foods-Fallacies and Facts*, Sakr, A. H. (Ed.). Lombard, IL: Foundation for Islamic Knowledge, 23–30.
Schaefer, A. L., Dubeski, P. L., Aalhus, J. L., and Tong, A. K. W. (2001). Role of nutrition in reducing antemortem stress and meat quality aberrations. *Journal of Animal Science*, 79(suppl_E), E91–E101.
Twaigery, S. and Spillman, D. (1989). An introduction to Moslem dietary laws. *Food Technology*, 43(2), 88–90.
Wahab, A. R. (2004). Guidelines for the preparation of halal food and goods for the Muslim consumers. *AmalMerge (M) Sdn. Bhd.*, (M), 1–12. Retrieved on May 8, 2018. http://www.halalrc.org/images/Research%20Material/Literature/halal%20Guidelines.pdf.
Wan Hassan, W. M. and Awang, K. W. (2009). Halal food in New Zealand restaurants: An exploratory study. *International Journal of Economics and Management*, 3(2), 385–402.

3 General Guidelines for Halal Food Production

Mian N. Riaz and Munir M. Chaudry

CONTENTS

Meat and Poultry 17
    Conditions and Method of Slaughtering (Dhabh or Zabh) 18
    The Slaughterer 18
    The Instrument 18
    The Cut 19
    The Invocation 19
    Abominable Acts in Slaying of Animals 19
    Advantages of Halal Slaughtering 20
Fish and Seafood 21
Milk and Eggs 21
Plant and Vegetable Materials 21
Food Ingredients 22
Gelatin 22
    Glycerin 23
    Emulsifiers 23
    Enzymes 23
    Alcohol 23
    Animal Fat and Protein 24
    Flavors and Flavorings 24
Sanitation 25
Specific Halal Guidelines 25
    State Halal Laws 25
References 26

In Chapter 2, halal laws and regulations were discussed, and in this chapter, we will try to explain how these laws and regulations apply to real situations in the production of halal food. The guidelines in this chapter are general in nature and specific guidelines for different product types appear in subsequent chapters. Here, foods are broadly classified into four groups to establish their halal status and to formulate guidelines for halal production and certification.

MEAT AND POULTRY

It is understood that meat of only halal animals is allowed for consumption by Muslims. An animal must be of a halal species to be slaughtered as halal (Qureshi et al., 2012).

The animal must be slaughtered by a sane adult Muslim while pronouncing the name of God. A sharp knife must be used to cut the throat in a manner that induces thorough removal of blood and quick death (AMI, 2007; Grandin, 2010). Islam places great emphasis on the humane treatment of animals. The animals must be raised, transported, handled, and held under humane conditions. However, these are only desirable actions and mishandling of animals does not make their meat haram (Apple et al., 2005; Hambrecht et al., 2004; Kannan et al., 2003; Ljungberg et al., 2007; Schaefer et al., 2001). Stunning of animals before the non-religious slaughter is generally accepted by Muslims in the U.S. and Canada when the methods of intervention are non-lethal, that is, the animal can recover, and be healthy and functioning sometime after the intervention. In many European countries, the type and severity of stunning usually kills the animals before bleeding, which makes it unacceptable for halal (Gibson et al., 2009; Gregory et al., 2010). Moreover, the start of dismemberment (e.g., cutting off the horns, ears, lower legs) of an animal must not take place before the animal is completely dead (Chaudry, 1997; Small et al., 2013).

Conditions and Method of Slaughtering (Dhabh or Zabh)

Dhabh is a clearly defined method of killing an animal for the sole purpose of making its meat fit for human consumption. The word dhabh in Arabic means purification or rendering something good or wholesome. The dhabh method is also called dhakaat in Arabic, which means purification or making something complete (Jamaludin, 2012).

The following conditions must be fulfilled for dhabh to meet the requirements of Muslim Islamic law (jurisprudence).

The Slaughterer

The person doing the dhabh must be of sound mind and an adult Muslim. The person can be of either sex. If a person lacks or loses competence through intoxication or loss of mental abilities, he or she may not do halal slaughter. The meat of an animal killed by an idolater, a nonbeliever, or someone who has apostatized from Islam is not acceptable (Anil et al., 2000; Limon et al., 2010; Önenç and Kaya, 2004).

The killing of an animal as halal by a person of the book will be discussed separately.

The Instrument

The knife used to do the dhabh must be extremely sharp (and ideally free of any nicks) to facilitate quick cutting of the skin and severing of blood vessels to enable the blood to flow immediately and quickly, in other words, to bring about an immediate and massive hemorrhage (Grandin, 2010). The Prophet Muhammad (PBUH) said: "Verily God has prescribed proficiency in all things. Thus if you kill, kill well; and if you perform dhabh, perform it well. Let each of you sharpen his blade and let him spare suffering to the animal he slays" (Khan, 1991). The Prophet (PBUH) is reported to have forbidden the use of an instrument that killed the animal by cutting its skin but not severing the jugular vein. It is also a tradition not to sharpen the knife in front of the animal about to be slaughtered (AFIC, 2003; AMI, 2007).

THE CUT

The incision should be made in the neck at some point just below the glottis and the base of the neck. Traditionally, camels used to be slayed by making an incision anywhere on the neck followed by a traditional horizontal cut. This process is called nahr, which means spearing the hollow of the neck. With modern restraining methods and reversible stunning techniques, it might be appropriate to consider whether this procedure should be phased out. The trachea and the esophagus must be cut in addition to the jugular veins and the carotid arteries. The spinal cord must not be cut (AMI, 2007; Anil et al., 2002). The head is therefore not to be severed completely. It is interesting to note that the kosher kill is very similar to the traditional method of dhabh described, except that the invocation is not made on each animal although a single prayer is said by the slaughterer before work is begun (Regenstein et al., 2003).

THE INVOCATION

Tasmiyyah or the invocation means pronouncing the name of God by saying Bismillah (in the name of Allah) or Bismillah Allahu Akbar (in the name of God, God is Great) before cutting the neck. Opinions differ somewhat on the issue of the invocation as addressed by three of the earliest jurists. According to Imam Malik, if the name of God is not mentioned over the animal before slaughtering, the meat of the animal is haram or forbidden, whether one neglects to say Bismillah intentionally or unintentionally. According to the jurist Abu Hanifah, if one neglects to say Bismillah intentionally, the meat is haram; if the omission is unintentional, the meat is halal. According to Imam Shaf'ii, whether one neglects to say Bismillah intentionally or unintentionally before slaughtering, the meat is halal so long as the person is competent to perform dhabh (Department of Standards Malaysia, 2009; Khan, 1991). It should also be said that the above traditions do not prove that the pronouncing of God's name is not obligatory when doing dhabh. In fact, the tradition emphasizes that the pronouncing of God's name was a widely known matter and was considered an essential condition of dhabh (Khan, 1991).

ABOMINABLE ACTS IN SLAYING OF ANIMALS

It is abominable to first throw the animal down on its side and sharpen the knife afterward. It is narrated that the Prophet (PBUH) once passed by a person who, having cast a goat to the ground, was pressing its head with his foot and sharpening his knife while the animal was watching. The Prophet (PBUH) said, "Will this goat not die before being slain? Do you wish to kill it twice? Do not kill one animal in the presence of another, or sharpen your knives before them" (Khan, 1991; Micera et al., 2010).

It is abominable to let the knife reach the spinal cord or to cut off the head of the animal. In South Asia, the term used for cutting of the head, usually by hitting the animal from behind the neck, is called jhatka or killing with a blow. There is general abhorrence in the Muslim community to such killing.

It is abominable to break the neck of an animal or begin skinning it or cut any parts while it is convulsing or before its life is completely departed. The Prophet Muhammad (PBUH) said, "Do not deal hastily with the souls (of animals) before their life departs" (Khan, 1991). It is sometimes the practice in fast-paced commercial slaughterhouses to start removing the horns, ears, and front legs while the animal still seems to be alive. This is against the principles and requirements of dhabh and must be avoided (Casoli et al., 2005). Note that the modern scientific signs of animal insensibility are the absence of any head reflex and the tongue going limp. The kicking of the legs is an involuntary action and does not mean the animal is still alive. It is abominable to perform dhabh with a dull instrument. The Prophet Muhammad (PBUH) commanded that knives be sharpened and be concealed from animals to be slain. It is also abominable to slaughter one animal while the next in line is watching the animal being killed. This is against the humaneness of the process of slaughtering (Nakyinsige et al., 2012).

From the foregoing description, it is clear that both intention and a precise method are conditions for the validity of dhabh. The insistence on pronouncing the name of God before slaying an animal is meant to emphasize the sanctity of life and the fact that all life belongs to God. Pronouncing the tasmiyyah induces feelings of tenderness and compassion, and serves to prevent cruelty. It also reinforces the notion that an animal is being slaughtered in the name of God for food and not for recreational purposes. Islam does not allow killing an animal for the sole purpose of receiving pleasure out of killing it (Department of Standards Malaysia, 2009).

Advantages of Halal Slaughtering

The actual method of dhabh has many advantages. To begin with, the speed of the incision made with the recommended sharp knife shortens the total time to slaughter. In a modern slaughterhouse, where animals are stunned before slaying, some of the animals do not become unconscious with one blow and have to be hit more than once, which is very painful.

The method of dhabh allows rapid and efficient bleeding of the animal. It is also obvious that blood being enclosed in a closed circuit can be removed faster by cutting the blood vessels. The force of the beating of the heart keeps the blood circulating. Therefore, the stronger the heartbeat, the greater the quantity of blood that will pour out quickly at the beginning. The body of the dhabh animal convulses involuntarily more than the stunned animal. Convulsions produce the squeezing or wringing action of the muscles of the body on the blood vessels, which helps to get rid of the maximum amount of blood from the meat tissue (Department of Standards Malaysia, 2009; Khan, 1991). The physiological conditions described help with blood removal, but they only operate fully if the animal is bled while alive by cutting across its throat and sparing the vertebral column without stunning the brain of the animal in any way (Khan, 1991). With the type of stunning and the force of the blow or shock used in North America, animals are usually alive for several minutes after stunning. The throat is generally cut within the first two minutes after stunning.

General Guidelines for Halal Food Production

For these reasons, stunning of cattle with a penetrating captive bolt and poultry with electrified water is done in some dhabh slaughter. In some other countries, the blow of stunning is severe enough to kill the animal. In Australia, some organizations contend that stunning renders the animals dead; hence, these organizations do not permit stunning for halal slaughter (AFIC, 2003). Some countries such as New Zealand allow a specifically developed mild reversible stun for halal slaughter of animals. The issue of interventions prior to the traditional halal slaughter of animals remains controversial.

FISH AND SEAFOOD

To determine the acceptability of fish and seafood, one must understand the rules of the different schools of Islamic jurisprudence as well as the cultural practices of Muslims living in different regions. Fish with scales are accepted by all denominations and groups of Muslims. Some groups do not consume fish without scales (e.g., catfish). There are additional differences among Muslims about seafood, especially mollusks (e.g., clams, oysters, and squid) and crustaceans (e.g., shrimp, lobster, and crab) (Regenstein et al., 2003). The requirements and restriction apply not only to fish and seafood but also to flavors as well as other ingredients (e.g., chitosan) derived from such products. Even within a single school, acceptable seafood may vary, often as a reflection of the traditions and customs of the country. A more detailed study of Muslim customs with respect to seafood would be of help to the food industry in providing appropriate seafood to different countries (Mickler, 2000).

MILK AND EGGS

Milk and eggs from halal animals are also halal. Predominantly, milk in the West comes from cows and eggs come from chicken hens. All other sources are required to be labeled accordingly. Numerous products are made from milk and eggs. Milk is used to make cheese, butter, and cream. A variety of enzymes are used in the production of cheeses. Which types of enzymes are used in the making of cheeses is very important (Guerrero-Legarreta, 2010; Hui, 2006). Enzymes can be halal or haram, depending on their origin and their production. Enzymes from microbial and plant sources or halal-slaughtered animals are halal. However, an enzyme from a non-halal-slaughtered animal even of a halal acceptable species or from a porcine source is haram. Depending on the enzymes used in production of cheeses or other dairy products, the products are classified as halal, haram, or makrooh (questionable) (Al-Mazeedi et al., 2013; Khattak et al., 2011). On the same basis, other functional additives such as emulsifiers or mold inhibitors should also be screened to take the doubt out of milk or egg products (Riaz, 2000).

PLANT AND VEGETABLE MATERIALS

Foods from plants are halal, with the exception of khamr (intoxicating drinks). In modern processing plants, however, animal or vegetable products might be processed in the same plant on the same equipment, increasing the chances of contamination.

For example, in some factories, pork and beans as well as corn are canned on the same equipment. When proper cleaning procedures are used, and the halal production segregated from the non-halal, contamination can be avoided. Functional ingredients from animal sources, such as antifoams, must also be avoided in the processing of vegetables. This intentional inclusion of haram ingredients into plant and vegetable products may render them as haram. It is evident that processing aids and production methods have to be carefully monitored to maintain the halal status of vegetable products (Regenstein et al., 2003).

Many fruits and vegetables are coated with "waxes" to make them shine, to retain moisture, or to carry compounds that protect the fruit or vegetable from spoilage. These compounds can include animal components. In the U.S., there is a requirement that these products be labeled with the source of the coating ingredients. These need to be on the master package and a general sign in the vicinity of the fruit and vegetables with such treatments. Generally, these lists do not contain "animal" products as the sellers (e.g., supermarkets) have made it clear they do not want such products. However, in other countries, it is important to check that no animal products are used.

FOOD INGREDIENTS

Food ingredients are one of the main subjects of concern. Vegetable products, as mentioned earlier, are halal unless they have been contaminated with haram ingredients or contain intoxicating substances. The requirements for animal slaughter and the types of seafood permitted for consumption have already been discussed. In this section, a number of commonly used ingredients such as gelatin, glycerin, emulsifiers, enzymes, alcohol, animal fat and protein, and flavors and flavorings will be discussed in more detail. Because most of the products fall into questionable or doubtful categories, they require that the majority of manufacturers have their plants inspected and products certified as halal (Al-Mazeedi et al., 2013; Kamaruddin et al., 2012).

GELATIN

The use of gelatin is very common in many food products. Gelatin can be halal if from dhabh-slaughtered animals, doubtful if from animals not slaughtered in a halal manner, or haram if from prohibited animal sources. The source of gelatin is not required to be identified by the U.S. Food and Drug Administration (FDA) on product labels. When the source is not known, it can be from either halal or haram sources, hence questionable. Muslims avoid products containing gelatin unless they are certified halal. Common sources of gelatin are pigskin, cattle hides, cattle bones, and, to a smaller extent, fish skins and scales. Halal products use gelatin from cattle that have been slaughtered in an Islamic manner or from fish. Some products in the U.S. list "Kosher Gelatin" as an ingredient. When certified using only the letter "K," these products are not acceptable to Muslim consumers (Cheng et al., 2012; Hermanto and Fatimah, 2013). However, products from normative mainstream kosher certifiers will often list gelatin as "Fish Gelatin," which is acceptable or gelatin if it is from kosher-slaughtered cattle. The acceptability of the latter must be decided on by each halal consumer (Jaswir et al., 2009).

Glycerin

Glycerin is another ingredient widely used in the food industry. Products containing glycerin are avoided by Muslims because it could be from animal sources. Currently, glycerin from palm oil and other vegetable oils is available for use in halal products and is again used in all kosher products.

Emulsifiers

Emulsifiers such as mono-glycerides, di-glycerides, poly-sorbates, diacetyl tartaric esters of mono- and di-glycerides (DATEM), and other similar chemicals are another commonly used group of ingredients that can come from halal or haram sources. Some of the companies have started to list the source, especially if it is vegetable, on the labels. If an emulsifier from vegetable sources is used, it is advantageous to indicate that on the label. Emulsifiers from vegetable sources and halal-slaughtered animal sources are halal. All such compounds are from vegetable or microbial sources in kosher products.

Enzymes

Enzymes are used in many food processes. The most common are the ones used in the cheese and the starch industries. Until a few years ago, the majority of the enzymes used in the food industry were from animal sources; now there are microbial alternatives, which are widely used because they are generally less expensive and can be produced more consistently (Birch et al., 2012; Oort, 2009).

Products such as cheeses, whey powders, lactose, whey protein concentrates, and isolates made using microbial enzymes are halal as long as all other halal requirements are met. Some products made with mixed or animal-based enzymes are haram if porcine enzymes are used; otherwise, they fall in the doubtful category. Bovine rennet and other enzymes from non-halal-slaughtered animals have been accepted by some countries. As more and more microbial enzymes become available, such acceptance can be expected to decrease. Use of dairy ingredients in all types of food products is very common, because whey and whey derivatives are an economical source of protein. For the products to be certified halal, dairy ingredients as well as other ingredients must be halal. Again, kosher products avoid animal-based enzymes (Al-Mazeedi et al., 2013; Khattak et al., 2011).

Alcohol

Muslims are prohibited from consuming alcoholic beverages, even in small quantities. Alcoholic drinks such as wine and beer should not be added to other products for flavoring or during cooking. Even a small amount of an alcoholic beverage added to a halal product makes it haram (Riaz, 1997). Cooking with wine, beer, and other alcoholic beverages is quite common in the West as well as in China. In Chinese cooking, rice wine is a common ingredient in many recipes. Product formulators and chefs should avoid the use of beverage alcohol in preparing halal products (Halal Consumer Group, 2012).

Alcohol is so ubiquitous in all biological systems that even fresh fruits contain traces of alcohol. During extraction of essences from fruits, alcohol might get concentrated into the essences. Because such alcohol is naturally present and unavoidable, it does not nullify the halal status of food products in which such essences are present. Furthermore, alcohol in its pure form is used for extracting, dissolving, and precipitating in the food industry. As it is the best solvent or chemical available in many cases to carry out certain processes, religious scholars have realized its importance for use in the industry. Ingredients made with alcohol or extracted by using alcohol have become acceptable as long as the alcohol is evaporated from the final ingredient. Food ingredients with 0.5% residual or technical alcohol are generally acceptable. However, for consumer items, acceptable limits vary from country to country and one group to another. The Islamic Food and Nutrition Council of America accepts a level of 0.1%, which it considers an impurity. In halal food laws, if an impurity is not detectable by taste, smell, or sight, it does not nullify the halal status of a food (Al-Mazeedi et al., 2013).

An article titled Change of State—Istihala (Al-Quaderi, 2001) supports this position in the following words. Wine is haram; however, if the same wine turns to vinegar it becomes halal. The use of vinegar derived from wine is halal as long as no wine remains in it. From these examples, it becomes clear that if an unlawful food item changes state, then the original ruling also changes.

Animal Fat and Protein

Meat and poultry products are not only consumed as staple food items but are also converted into further processed ingredients to be used in formulating a myriad of non-meat food products. In the food industry in the U.S. as well as in other industrialized countries, every part of the animal is used in one way or another. Less desirable parts of the carcass and by-products may be turned into powders that can be used as food ingredients, for example, flavoring agents for soups and snacks. Animal fat is purified and converted into animal shortening and emulsifiers, as well as other functional food ingredients. Feathers and hair can be converted into amino acids. Such ingredients would be halal only if the animals are halal and all precautions are taken to eliminate cross-contamination (Al-Mazeedi et al., 2013; Regenstein, 2012).

Flavors and Flavorings

Flavors and flavorings can be as simple as a single spice or extremely complex in many food systems. Some of the more complex flavorings can contain over 100 different ingredients that have various origins. These ingredients can be from microorganisms, plants, minerals, petroleum, or animals as well as synthetic sources. For formulating halal food products, the manufacturer has to make sure that any flavors, proprietary mixes, or secret formulas are halal and free from doubtful materials (Aris et al., 2012; Jahangir et al., 2016). These formulas will need to be reviewed by the halal certifier. They need to know the origin of each ingredient, but do not need to know the amount of each ingredient used. Even then the halal certifier is expected to maintain the confidentiality of such formulas.

SANITATION

During the manufacture of halal products, it is imperative that all possible sources of contamination be eliminated. This can be accomplished through proper scheduling of products as well as by thoroughly cleaning and sanitizing production lines and equipment. For non-meat products, it is adequate to clean equipment and determine cleanliness by visual observation, that is, the normal expectations of cleaning are usually sufficient. A company might treat haram ingredients similar to allergens and make it part of a control program similar to their allergen control program. Chemicals used for cleaning (especially soaps and foams) should be screened to avoid fats of animal origin. Most kosher certified cleaners meet this standard.

SPECIFIC HALAL GUIDELINES

General guidelines for halal vary somewhat from country to country. The following documents will be helpful to manufacturers of food products for specific markets:

Codex Alimentarius Guidelines for Use of the Term Halal (Appendix A)

Malaysian General Guidelines on the Slaughtering of Animals and the Preparation and Handling of Halal Food

Indonesia Halal Food Laws

Singapore's Halal Regulations and Import Requirements

STATE HALAL LAWS

In 2000, the state of New Jersey passed the Halal Food Protection Act. Minnesota and Illinois followed suit in 2001, enacting their own laws to regulate the halal food industry. These laws address the problem of fraudulent use of the term halal or other such terms to produce and market products that may not be halal. For example, Nimer (2002) contends that some retailers have wrongly placed halal food labels on meat items to attract Muslim customers. The Illinois act, which has the most accommodating language, specifically defines the term halal food as:

> being prepared under and maintained in strict compliance with the laws and customs of the Islamic religion, including, but not limited to, those laws and customs of zabiha/zabeeha (slaughtered according to appropriate Islamic code) and as expressed by reliable recognized Islamic entities and scholars. [And is therefore probably unconstitutional as per the second circuit court of appeals ruling that was not reviewed by the U.S. Supreme Court.]

This law makes it a misdemeanor for any person to make any oral or written statement that directly or indirectly tends to deceive or otherwise would lead a reasonable individual to believe that a non-halal food or food product is halal.

In 2002, the states of California and Michigan also enacted their versions of a halal food laws. It is anticipated that other states with significant Muslim populations, where legislatures are sensitive to their ethnic voters, will also introduce similar bills. In early 2003, a similar bill was introduced into both the Senate and the House of the state of Texas and was passed. Unless subsequent regulations are passed to implement the laws, the passage of the bills provides a false sense of protection and very little actual benefit is received by Muslim consumers. However, in a few cases they provide the state with the legal right to address halal fraud, although in most cases the general laws dealing with consumer fraud, mislabeling, and similar acts will cover intentional halal fraud.

At the federal level there are no comprehensive halal laws; however, the Food Safety and Inspection Service (FSIS) of the U.S. Department of Agriculture (USDA) has issued a directive for the labeling of halal products. The FSIS directive states that the Food Labeling Division, Regulatory Programs, recently approved a standard that is included in the labeling policy book involving the use of "Halal," "Halal Style," or "Halal Brand" on meat and poultry products. They are required to establish that they are prepared under the supervision of an Islamic authority.

Products identified with the term halal must not contain pork or pork derivatives. The Federal Meat and Poultry Inspection does not certify the halal preparation of products, but rather accepts "halal" and similar statements, if the products are prepared under the supervision of an authorized Islamic organization. When "halal" and similar statements are used, plant management is responsible for making the identity of the Islamic organization available to inspection personnel.

REFERENCES

Al-Mazeedi, H. M., Regenstein, J. M., & Riaz, M. N. (2013). The issue of undeclared ingredients in halal and kosher food production: A focus on processing aids. *Comprehensive Reviews in Food Science and Food Safety*, *12*(2), 228–233.

Al-Quaderi, J. (2001). Change of state—Istihala. *The Halal Consum.* 2(7).

AMI 2007. American Meat Institute fact sheet, Washington D.C , USA. https://www.meatinstitute.org/index.php?ht=a/GetDocumentAction/i/130170%20Access%20June%2023" https://www.meatinstitute.org/index.php?ht=a/GetDocumentAction/i/130170. Accessed on June 23, 2018.

Anil, M. H., Love, S., Helps, C. R., & Harbour, D. A. (2002). Potential for carcass contamination with brain tissue following stunning and slaughter in cattle and sheep. *Food Control*, *13*(6–7), 431–436.

Anil, M. H., Raj, A. B. M., & McKinstry, J. L. (2000). Evaluation of electrical stunning in commercial rabbits: effect on brain function. *Meat Science*, *54*(3), 217–220.

Apple, J. K., Kegley, E. B., Maxwell, C. V., Rakes, L. K., Galloway, D., & Wistuba, T. J. (2005). Effects of dietary magnesium and short-duration transportation on stress response, postmortem muscle metabolism, and meat quality of finishing swine 1. *Journal of Animal Science*, *83*(7), 1633–1645.

Aris, A. T., Nor, N. M., Febrianto, N. A., Harivaindaran, K. V., & Yang, T. A. (2012). Muslim attitude and awareness towards Istihalah. *Journal of Islamic Marketing*, *3*(3), 244–254.

Australian Federation of Islamic Council (AFIC). 2003. AFIC, Sydney, NSW, Australia. http://muslimsaustralia.com.au/. Accessed on May 10, 2018.

Birch, G. G., Blakebrough, N., & Parker, K. J. (2012). *Enzymes and food processing*. Germany, Springer Science & Business Media.
Casoli, C., Duranti, E., Cambiotti, F., & Avellini, P. (2005). Wild ungulate slaughtering and meat inspection. *Veterinary Research Communications*, 29(2), 89–95.
Chaudry, M. M. (1997). Islamic foods move slowly into marketplace. *Meat Processing*, 36(2), 34–8.
Cheng, X. L., Wei, F., Xiao, X. Y., Zhao, Y. Y., Shi, Y., Liu, W., & Lin, R. C. (2012). Identification of five gelatins by ultra performance liquid chromatography/time-of-flight mass spectrometry (UPLC/Q-TOF-MS) using principal component analysis. *Journal of Pharmaceutical and Biomedical Analysis*, 62, 191–195.
Food Safety and Inspection Service, United States Department of Agriculture. (1995). Inspection Operations, Northeastern Region, Halal Labeling Requirements, NE-95-06.
Gibson, T. J., Johnson, C. B., Murrell, J. C., Hulls, C. M., Mitchinson, S. L., Stafford, K. J., Johnstone, A. C. & Mellor, D. J. (2009). Electroencephalographic responses of halothane-anaesthetised calves to slaughter by ventral-neck incision without prior stunning. *New Zealand Veterinary Journal*, 57(2), 77–83.
Grandin, T. (2010). Auditing animal welfare at slaughter plants. *Meat Science*, 86(1), 56–65.
Gregory, N. G., Fielding, H. R., von Wenzlawowicz, M., & Von Holleben, K. (2010). Time to collapse following slaughter without stunning in cattle. *Meat Science*, 85(1), 66–69.
Guerrero-Legarreta, I. (2010). *Handbook of poultry science and technology, Volume 2: Secondary processing*. Hoboken: John Wiley & Sons, Inc.
Halal Consumer Group. (2012). http://www.muslimconsumergroup.com. Accessed on June 20, 2012.
Hambrecht, E., Eissen, J. J., Nooijen, R. I. J., Ducro, B. J., Smits, C. H. M., Den Hartog, L. A., & Verstegen, M. W. A. (2004). Preslaughter stress and muscle energy largely determine pork quality at two commercial processing plants. *Journal of Animal Science*, 82(5), 1401–1409.
Hermanto, S., & Fatimah, W. (2013). Differentiation of bovine and porcine gelatin based on spectroscopic and electrophoretic analysis. *Journal of Food and Pharmaceutical Sciences*, 1(3), 68–73.
Hui, Y. H. (Ed.). (2006). *Handbook of food science, technology, and engineering* (Vol. 149). Boca Raton, FL: CRC Press.
Jahangir, M., Mehmood, Z., Bashir, Q., Mehboob, F., & Ali, K. (2016). Halal status of ingredients after physicochemical alteration (Istihalah). *Trends in Food Science & Technology*, 47, 78–81.
Jamaludin, M.A. (2012). Fiqh istihalah: Integration of science and Islamic law. *Revelation and Science*, 2(2), 117–123.
Jaswir, I., Mirghani, M. E. S., Hassan, T., & Yaakob, C. M. (2009). Extraction and characterization of gelatin from different marine fish species in Malaysia. *International Food Research Journal*, 16, 381–389.
Kamaruddin, R., Iberahim, H., & Shabudin, A. (2012). Willingness to pay for halal logistics: The lifestyle choice. *Procedia-Social and Behavioral Sciences*, 50, 722–729.
Kannan, G., Kouakou, B., Terrill, T. H., & Gelaye, S. (2003). Endocrine, blood metabolite, and meat quality changes in goats as influenced by short-term, preslaughter stress 1. *Journal of Animal Science*, 81(6), 1499–1507.
Khan, G.M. (1991). *Al-Dhabh: Slaying animals for food the Islamic way*. Jeddah, Saudi Arabia: Abdul-Qasim Bookstore.
Khattak, J. Z. K., Mir, A., Anwar, Z., Abbas, G., Khattak, H. Z. K., & Ismatullah, H. (2011). Concept of halal food and biotechnology. *Advance Journal of Food Science and Technology*, 3(5), 385–389.
Limon, G., Guitian, J., & Gregory, N. G. (2010). An evaluation of the humaneness of puntilla in cattle. *Meat Science*, 84(3), 352–355.

Ljungberg, D., Gebresenbet, G., & Aradom, S. (2007). Logistics chain of animal transport and abattoir operations. *Biosystems Engineering*, 96(2), 267–277.

Micera, E., Albrizio, M., Surdo, N. C., Moramarco, A. M., & Zarrilli, A. (2010). Stress-related hormones in horses before and after stunning by captive bolt gun. *Meat Science*, 84(4), 634–637.

Mickler, D.L. (2000) Halal Foodways http://www.unichef.com/ Halalfood.htm. Accessed on July 22, 2001.

Malaysia, S. (2009). Halal food-production, preparation, handling and storage—general guidelines. *Department of Standards, Selangor Darul Ehsan, Malaysia, MS1500*.

Nakyinsige, K., Che Man, Y. B., Sazili, A. Q., Zulkifli, I., & Fatimah, A. B. (2012). Halal meat: A niche product in the food market. In *2nd International Conference on Economics, Trade and Development IPEDR* (Vol. 36, pp. 167–173).

Nimer, M. (2002). The North American Muslim resource guide. New York: Routledge, pp. 123–124.

Önenç, A., & Kaya, A. (2004). The effects of electrical stunning and percussive captive bolt stunning on meat quality of cattle processed by Turkish slaughter procedures. *Meat Science*, 66(4), 809–815.

Oort, M. V. (2009). Enzymes in food technology – introduction. In *Enzymes in Food Technology*, 2nd Edn. Eds: Whitehurst, R. and Oort, M. V. pp. 1–17, UK: Wiley-Blackwell.

Qureshi, S. S., Jamal, M., Qureshi, M. S., Rauf, M., Syed, B. H., Zulfiqar, M., & Chand, N. (2012). A review of Halal food with special reference to meat and its trade potential. *Journal of Animal and Plant Sciences*, 22(2 Suppl), 79–83.

Regenstein, J. M. (2012, March). The politics of religious slaughter-how science can be misused. In *65th Annual Reciprocal Meat Conference at North Dakota State University in Fargo, ND*.

Regenstein, J. M., Chaudry, M. M., & Regenstein, C. E. (2003). The kosher and halal food laws. *Comprehensive Reviews in Food Science and Food Safety*, 2(3), 111–127.

Riaz, M. N. (1997). Alcohol: The myths and realities, in *Handbook of Halal and Haram Products*. Ed: Uddin, Z. Richmond Hill, NY: Publishing Center of American Muslim Research & Information, 16–30.

Riaz, M. N. (2000). How cheese manufacturers can benefit from producing cheese for halal market. *Cheese Mark. News*, 20(18), 4, 12.

Schaefer, A. L., Dubeski, P. L., Aalhus, J. L., & Tong, A. K. W. (2001). Role of nutrition in reducing antemortem stress and meat quality aberrations. *Journal of Animal Science*, 79(suppl_E), E91–E101.

Small, A., McLean, D., Keates, H., Owen, J. S., & Ralph, J. (2013). Preliminary investigations into the use of microwave energy for reversible stunning of sheep. *Animal Welfare*, 22(2), 291–296.

4 Muslim Demography and Global Halal Trade
A Statistical Overview

Mohammed A. Khan,
Mian N. Riaz, and Munir M. Chaudry

CONTENTS

Muslim Demography 30
Geographic Spread 30
Age, Life Expectancy, Urbanization 33
Muslim Populations in Europe 34
Muslims in North America and the US Halal Market 34
Global Halal Trade 39
U.S. Trade with the Muslim Countries 44
Conclusion 50
References 59

Globalization with the resultant mass movement of people and goods has greatly changed the contours of international food trade. In an interconnected world, one should not be too surprised to find food products produced in one country being marketed and sold in far off lands. In addition, stores of fast food giants can now be found almost everywhere. Such scenarios are abundantly observable in the Muslim countries and especially so in the affluent countries of the Gulf Cooperation Council (Bahrain, Kuwait, Oman, Qatar, Saudi Arabia, and the United Arab Emirates). The increasing diffusion of culinary tastes, tourism, shifts in family structures, lifestyle changes, rapid urbanization, massive displacement of people due to man-made or natural disasters, and so on, adds new dimensions to the global trade in food. To meet the dietary needs of the changing and growing Muslim population, it is important to have a thorough understanding of the markets where they reside (El-Gohary, 2015). This chapter aims to provide a general overview of the Muslim world and the global halal trade. It will first provide descriptive statistics on the Muslim populations around the world. The focus on demography is necessary because it is demography which drives economy and trade. Second, it will discuss the dynamics of trade in halal products. This will include observable trends in the markets as well as statistical analyses. The two sections are not strictly divided and comments related to trade and other dynamics will be frequently made wherever appropriate. Special attention will be paid to the domestic halal market in North America as well and the United States' trade with the MENA (Middle East and North Africa) region.

MUSLIM DEMOGRAPHY

Precisely documenting the global Muslim population is a challenging task. This is primarily due to the fact that Muslims are now found, in varying numbers, in almost every country and territory of the world. Not all countries methodically collect demographic data. Nor do all national censuses ask their respondents about their faith. Census data methodologies and frequencies are also not uniform. Despite these limitations, we can still get a general view of the Muslim population by adopting a nuanced approach.

There is almost unanimous consensus among demographers that the Muslim population is the fastest growing in the world. The expansive studies of global religious demography published by the Pew Research Center's Forum on Religion and Public Life confirms this assertion (Grim and Karim, 2011; Hackett et al., 2015). This chapter mostly uses this data except in cases where we think that a population has been undercounted or in instances in which more nuance and qualifications are required. For comparative purposes, other sources of data will also be cited. Based on Pew projections, the global Muslim population in 2014 was estimated to be 1.8 billion. The report states that the annual growth rate of the global Muslim population is 1.5%, which is double that of non-Muslims (0.7%). In the 2010 to 2020 decade, the annual growth rate is expected to reach 1.7%. From 2020 to 2030, the population will grow but at a slower rate of 1.4%. In addition, it also reports that Muslims will comprise around 26% of the world's total projected population of 8.3 billion in 2030. By 2050, the global Muslim population is expected to rise to 2.8 billion or 30% of the world's population. The reason behind the growth is the relatively younger population and high fertility rate among the Muslims. The Muslim fertility rate of 3.1 was the highest among religious groups (Hackett et al. 2015).

GEOGRAPHIC SPREAD

Muslims can be found in almost all of the recognized 236 countries and territories of the world. In 72 of them, they number more than a million. They are more than half the population in around 50 countries located in Asia and Africa. The Asia-Pacific (62%) region has the highest proportion of the global Muslim population followed by Middle East-North Africa (20%), Sub-Saharan Africa (15%), Europe (2.7%), and the Americas (0.3%) (Johnson and Grim, 2013).

Table 4.1 shows countries with the largest number of Muslim populations in 2010 and their projected numbers in 2030. Indonesia continues to be the country with the highest number of Muslims. Countries comprising the Indian sub-continent (India, Pakistan, and Bangladesh) have the highest density of Muslims anywhere in the world. The table also demonstrates that the most populous Muslim countries are also growth economies with most of them growing at impressive rates (Grim and Karim, 2011).

TABLE 4.1
Countries with the Largest Number of Muslims

Country	Muslim Population 2010	Muslim (%) 2010	Muslim Population 2030 Projection	Muslim (%) 2030	Predicted GDP Growth 2015–2019+
Indonesia	204,847,000	88.1	238,833,000	88	5.9
Pakistan	178,097,000	96.4	256,117,000	96.4	4.7
India	177, 286,000	14.6	236,182,000	15.9	5.4
Bangladesh	148,607,000	90.4	187,506,000	92.3	6.8
Egypt	80,024,000	94.7	105,065,000	94.7	3.9
Nigeria	75,728,000	47.9	116,832,000	51.5	7.0
Iran	74,819,000	99.7	89,626,000	99.7	2.2
Turkey	74,660,000	98.6	89,127,000	98.6	3.5
Algeria	34,7870,000	98.2	43,915,000	98.2	3.9
Morocco	32, 381,000	99.9	39,259,000	99.9	3.9
Iraq	31,108,000	98.9	48,350,000	98.9	6.5
Sudan	30,855,000	71.4	35,497,000	71.4	4.3
Afghanistan	29,047,000	99.8	50,527,000	99.8	4.5
Ethiopia	28,721,000	33.8	44,466,000	33.8	7.5
Uzbekistan	26,833,000	96.5	32,760,000	96.5	6.5
Saudi Arabia	25,493,000	97.1	35,497,000	97.1	4.4
Yemen	24,023,000	99.0	38,973,000	99.0	4.5
China	23,308,000	1.8	29,949,000	2.1	5.4
Syria	20,895,000	92.8	28,374,000	92.8	N/A
Malaysia	17,139,000	61.4	22,752,000	64.5	5.0
Russia	16,379,000	11.7	18,556,000	14.4	1.6
Niger	15,627,000	98.3	32,022,000	98.3	6.0
Tanzania	13,450,000	29.9	19,463,000	25.8	7.0
Senegal	12,333,000	95.9	18,739,000	95.9	4.8
Mali	12,316,000	92.4	18,840,000	92.1	5.5

Source: Latest data from the Indian government indicates that the Muslim population in the country in the 2011 census was over 180 million (Singh, 2014). Respected scholar Dr. Guha puts the figure at close to 200 million (Guha, 2015). +GDP growth projections 2015 to 2019 have been sourced first from the Global Islamic Investment Gateway website (www.giig2015.com). It lists the top ten Organization of Islamic Conference (OIC) countries. All other data has been sourced from the Conference Board of Canada and the International Monetary Fund. The OIC countries together represent 10% of global GDP. They are estimated to be worth $10.1 trillion by 2019.

The fastest annual growth of Muslims in the 2000 to 2010 decade was in the Gulf countries of Qatar, United Arab Emirates, and Bahrain. These are also among the richest Muslim countries (Johnson and Grim, 2013) (Table 4.2).

The Muslim population is spread out geographically in Muslim majority as well as Muslim minority countries. About 73% of Muslims live as a majority whereas

TABLE 4.2
Countries with the Fastest Annual Population Growth of Muslims, 2000–2010

Country	Muslim Population 2010	Annual Growth Rate 2000–2010	GDP Per Capita Purchasing Power Parity 2013 $
Qatar	1,168,000 (77.5%)	11.8	$136,727
United Arab Emirates	3,577,000 (76.0%)	9.5	$59,845[a]
Bahrain	655,000 (81.2%)	7.2	$43,715
Solomon Islands	<1,000 (<0.1)	6.4	$2,069
Norway	144,000 (3.0%)	5.7	$65,461
Western Sahara	528,000 (99.6%)	5.3	N/A
Rwanda	188,000 (1.8%)	4.9	$1,473
Benin	2,259,000 (24.5%)	4.2	$1,790
Finland	42,000 (0.8%)	3.9	$39,812
Paraguay	1,000 (<0.1)	3.7	$8,092

Source: Muslim population figures are from Hacket, Connor, Stonawski, Skirbekk, Potancoková, and Abel (2015) *The Future of the Global Muslim Population. Projections for 2010–2030*, Pew Research Center's Forum on Religion and Public Life, Washington, DC: 2011. GDP per capita data from the World Bank.

Notes: The percentage in brackets in the second column are the percentage of the Muslim population in a country's entire population. The purchasing power parity conversion factor is the number of units of a country's currency required to buy the same amounts of goods and services in the domestic market as the U.S. dollar would buy in the United States.

[a] The World Bank data did not have an entry for UAE for 2013. The numbers for 2012 were used (http://data.worldbank.org/indicator/PA.NUS.PPP).

27% do as a minority. The percentages, however, should not distract from the fact that even when they comprise a small percentage, they are numerous in absolute numbers. For example, Muslims in China, according to conservative estimates, are 1.8% of the population. When translated into absolute numbers, they are more than 24 million. Similarly, Muslims in Italy comprise only over 2% of the population but are more than 1.5 million in absolute numbers. Small numbers should not deter producers from catering to these niche markets as they are more often willing to pay a premium for the products that carry halal certification (Menipaz and Menipaz, 2011).

The United Nations report, 'World Population Prospects: The 2012 Revision', provides a list of 37 countries whose population is projected to decrease between 2013 and 2050. Only two Muslim countries (Bosnia and Herzegovina, and Albania) are on that list. In all other Muslim countries, the population is expected to grow.

Muslims are spread out in high, middle, and lower income countries. Income status, in addition to other factors, directly affects dietary needs and this aspect should be taken into consideration by food producers. A large number of Muslim populations now face food insecurity as they live in areas that are more prone to climate change, natural disasters, regional conflicts, and political turmoil. The MENA region countries are also prone to price shocks where even slight changes in food prices can lead to turmoil.

A report from the UN Food and Agriculture Organization published in 2015 states that the numbers of chronically undernourished people in the 2010 to 2013 period reached 79.4 million (comprising more than 11% of the population) in the MENA region (Fischer, 2011). Conflicts remained the major source for this food insecurity. The conflict in Syria has made more than 6 million people highly vulnerable to food insecurity. In addition, in 2013, there were more than 2.1 million registered Syrian refugees residing in Lebanon, Jordan, Turkey, Iraq, and Egypt (FAO, 2015). The continuing conflict in the region hints that there will be a need for emergency food items like halal certified humanitarian meals such as those provided by the U.S. government.

On the flip side of food insecurity, the report also found that more than a quarter of this region's population is obese (FAO, 2015). In countries like Saudi Arabia and UAE, there is an alarming increase in diabetes and other health-related issues. In these countries, the demand for health foods such as meal replacements and low-fat dairy products is growing (Alpen Capital, 2011).

AGE, LIFE EXPECTANCY, URBANIZATION

Young people are seen as a major resource of any country and society. They bring in energy, enthusiasm, and new ideas. They also set new trends in lifestyles, customs, and culinary habits. For the past several decades, the global Muslim population has consistently been youthful. In 2012, the median age of Muslims was 23 years. This was the lowest among all religious groups. The median age of the world's overall population for the same year was 28 (Hackett et al., 2015). In the MENA region, it was 15. In Africa, it is 19.2. This trend of a youthful population is expected to continue beyond 2030 when fertility rates will stabilize and the median age for all the Organization of Islamic Conference (OIC) countries will only increase slightly to 30. Surveys indicate that this population wants to experiment with new foods while remaining anchored in the halal framework.

It should also be noted that life expectancy is increasing throughout the world. In the MENA region, it was at 71 in 2012. The dietary and nutrition requirement of this segment of the population also needs to be taken into consideration. About 57% of the population in the MENA region lives in urban areas. Urbanization is catching up fast. The urban growth of the region's 25 largest cities between 2000 and 2010 was 2.7%. In 2015, Egypt, Sudan, and Yemen are the only countries in the region that are less than 50% urbanized (UN Habitat). From a food perspective, urbanization can lead to food insecurity. In the rural areas most people grow their own food. When they move to urban areas they will have to buy their food. The food insecurity is further increased for people who lack a fixed income and rely on purchased foods (FAO, 2015).

On the other hand, a move toward urban areas also opens up possibilities for a wider variety of choice for domestic and imported food for the population. It can also lead to new forms of entrepreneurship with the increased demand for freshly prepared foods.

The entering of women in greater numbers into the region's work force is also leading toward increased demand for prepared and processed foods. There are, however, current disparities within the region of female workforce participation. Only 5% of Yemen's workforce is female, while in the UAE, it is 48% (World Bank, 2013). But all trends indicate the female share in the workforce will only increase (Table 4.3).

MUSLIM POPULATIONS IN EUROPE

Muslims in Europe constitute only a small percentage of the global Muslim population but they represent a significant part of the global halal economy due to their higher incomes. Muslims in Europe now number more than 44 million, comprising 6% of the continent's population. By 2030, they are expected to grow to 58 million or 8% of Europe's population. Eastern Europe is home to the largest number of Muslims on the continent with more than 18 million. They have a historical presence in the area. The Muslim population is also increasing in the western part of the continent. In 2010, the largest number of Muslims in Western Europe were in France, followed by Germany and the United Kingdom (see Table 4.4). With their increasing numbers, the value of the halal market has also grown. In 2009, Frits van Dijk, executive vice president of Nestle, had estimated the total European halal market to be worth $66 billion (Al Arabiya, 2010). The European Halal Expo pegged it at over $70 billion in 2014.

MUSLIMS IN NORTH AMERICA AND THE US HALAL MARKET

Immigration and higher fertility rates have contributed to the growth of the Muslim population in North America. The Canadian census asks the respondents about their faith and in the 2011 census, it reported that just over one million people in country were Muslim. This amounted to 3.2% of the nation's population. They are expected to triple by 2030. Mexico's Muslim population was estimated at 110,000 in 2010 (Grim and Karim, 2011).

The American census does not ask for the respondent's faith and therefore there are disagreements as to what exactly is the Muslim population in the country. The Pew Forum's report cautiously estimated it to be 2.6 million in 2010 and projected that it will grow to 6.2 million by 2020 (Grim and Karim, 2011). But there are other estimates that claim the numbers are much higher. In a June 2009 speech delivered at Cairo University, President Barack Obama mentioned that there are "nearly 7 million American Muslims" (White House, 2009). The Muslimpopulation.com website estimated it at 6.67 million in 2014. A 2015 report of American Muslim Market conservatively estimates it to be 5.7 million with 1.7 million households (*Dinar Standard*, 2015). Based on the growth in mosques, community centers, retail halal businesses, and so on, it appears that this last estimate is a safer one. Muslims are present in every state of the U.S. The cities with the largest number of Muslims are: Detroit, Washington DC, Cedar Rapids, Philadelphia, New York, Atlanta, Peoria (Illinois), San Francisco, Houston, Chicago, Cleveland, Columbus (Georgia), Toledo (Ohio), Boston, Lansing (Michigan), Los Angeles, Buffalo (New York), New Orleans, Albany (Georgia), Columbia (Missouri), Nashville, Dallas, South Bend (Indiana), Grand Rapids (Michigan),

TABLE 4.3
Muslim Populations and Halal Activity in Various Countries

Country or Area	Total Estimated 2015 Population of the Country	Estimated Muslim Population 2010 (%)	Number of Muslims 2010 (Est.)	Notes
Afghanistan[a]	32,007,000	99.8	29,047,000	1, 6
Albania[a]	3,197,000	82.1	2,601,000	1
Algeria[a]	40,633,000	98.2	34,780,000	1
Argentina	42,155,000	2.5	1,000,000	2, 4
Austria	8,558,000	5.7	475,000	1, 4
Australia	23,923,000	1.9	3,99,000	2
Azerbaijan[a]	9,613,000	98.4	8,795,000	1
Bahrain[a]	1,360,000	70.2	866,888	1, 3
Bangladesh[a]	160, 411,000	90.4	148,607,000	1, 4
Belgium	11,183,000	6.0	638,000	2
Benin[a]	10,880,000	24.5	2,259,000	1
Bosnia Herzegovina	3,820,000	41.6	1,564,000	1
Brazil	203,657,000	0.1	204,000	2, 4, 5
Brunei Darussalam[a]	429,000	83.0	211,000	1, 3
Burkina Faso	17,915,000	60.5	9,600,000	2, 6
Cameroon[a]	23,393,000	18.0	3,598,000	1
Canada	35,817,000	3.2	1,146,000	4, 5
Chad[a]	13,606,000	55.7	6,404,000	1, 6
China	1,401,587,000	2.0	23,308,000	1, 4
Comoros[a]	770,000	98.3	679,000	1
Cote d'Ivoire[a]	21,295,000	36.9	7,960,000	1, 6
Djibouti[a]	900,000	97.0	853,000	1, 6
Egypt[a]	84,706,000	94.7	80,024,000	1, 3
Eritrea	6,738,000	36.5	1,909,000	1, 6
Ethiopia	98,942,000	33.8	28,721,000	2, 6
Fiji	893,000	6.3	54,000	2
France	64,983,000	7.5	4,704,000	2, 4, 5
Gabon[a]	1,751,000	9.7	145,000	1
Gambia[a]	1,970,000	95.3	1,669,000	1
Ghana	26,984,000	16.1	3,906,000	2
Greece	11,26,000	4.7	527,000	1
Guinea[a]	12,348,000	84.2	8,693,000	1, 6
Guinea-Bissau[a]	1,788,000	42.8	705,000	1, 6
Guyana[a]	8,08,000	7.2	55,000	2
India	1,282,000	14.6	177,286,000	2, 4, 5
Indonesia[a]	255,709,000	88.1	204,847,000	2, 3, 5
Iran[a]	79,746,000	99.7	74,819,000	1, 3, 5
Iraq[a]	35,767,000	98.9	31,108,000	1, 6

(*Continued*)

TABLE 4.3 (CONTINUED)
Muslim Populations and Halal Activity in Various Countries

Country or Area	Total Estimated 2015 Population of the Country	Estimated Muslim Population 2010 (%)	Number of Muslims 2010 (Est.)	Notes
Jordan[a]	7,690,000	98.8	6,397,000	1, 3
Kazakhstan[a]	16,770,000	56.4	8,887,000	1
Kenya	46,749,000	7.0	2,868,000	6
Kuwait[a]	3,583,000	86.4	2,636,000	1, 3, 5
Kyrgyz Republic[a]	5,708,000	88.8	4,927,000	1
Lebanon[a]	5,054,000	59.7	2,542,000	1
Liberia	4,503,000	12.8	523,000	6
Libya[a]	6,317,000	96.6	6,324,000	1
Malawi	17,309,000	12.8	2,011,000	2
Malaysia[a]	30,651,000	61.4	17,139,000	1, 3, 5
Maldives[a]	358,000	98.4	309,000	1
Mali[a]	16,259,000	92.4	12,316,000	6
Mauritania[a]	4,080,000	99.2	3,338,000	6
Mauritius	1,254,000	16.6	216,000	1
Morocco[a]	33,955,000	99.9	32,381,000	1
Mozambique[a]	27,122,000	22.8	5,340,000	6
Nepal	28,441,000	4.2	1,253,000	1
Netherlands	16,844,000	5.5	914,000	2, 5
New Zealand	4,596,000	0.9	41,000	2, 4, 5
Niger[a]	19,268,000	98.3	15,627,000	6
Nigeria[a]	183,523,000	47.9	75,728,000	1
Oman[a]	4,158,000	87.7	2,547,000	1, 3
Pakistan[a]	188,144,000	96.4	178,097,000	1, 3, 4
Palestine[a]	4,549,000	97.5	4,298,000	2
Philippines	101,803,000	5.1	4,737,000	1, 4
Qatar[a]	2,351,000	77.5	1,168,000	1, 3, 5
Russia	142,098,000	14.0	16,379,000	1, 3
Saudi Arabia[a]	29,898,000	97.1	25,493,000	1, 3, 5
Senegal[a]	14,967,000	95.9	12,333,000	6
Sierra Leone	6,319,000	71.5	4,171,000	6
Singapore	5,619,000	15.0	721,000	1, 3, 5
Somalia[a]	11,123,000	98.6	9,231,000	6
South Africa	53,491,000	1.5	737,000	1, 5
Sudan[a]	39,613,000	97.0	30,850,000	6
Suriname[a]	548,000	15.9	84,000	1
Syria[a]	22,265,000	92.8	20,895,000	6
Tajikistan[a]	8,610,000	99.0	7,006,000	1
Tanzania	52,291,000	29.9	13,952,000	2
Thailand	67,401,000	5.8	3,952,000	2, 4

(*Continued*)

TABLE 4.3 (CONTINUED)
Muslim Populations and Halal Activity in Various Countries

Country or Area	Total Estimated 2015 Population of the Country	Estimated Muslim Population 2010 (%)	Number of Muslims 2010 (Est.)	Notes
Togo[a]	7,171,000	12.2	827,000	1
Tunisia[a]	11,235,000	99.8	10,349,000	1
Turkey[a]	76,691,000	98.6	74,660,000	2, 3, 4, 5
Turkmenistan[a]	5,373,000	93.3	4,830,000	1
Uganda[a]	40,141,000	12.0	4,060,000	6
United Arab Emirates[a]	9,577,000	80.0	3,577,000	1, 3, 4, 5
United Kingdom	63,844,000	4.6	2,869,000	1, 4, 5
United States of America	325,128,000	0.6	2,595,000	2, 4, 5
Uzbekistan[a]	325,128,000	96.5	26,833,000	1
Yemen[a]	25 535,000	99.0	24,023,000	6

[a] Indicates membership in the Organization of Islamic Conference (OIC). It should be noted that not all member countries have Muslim majorities.

Numbers in the notes column are explained below:
1. Net importer of food. A country or territory whose value of imported goods is higher than its value of exported goods over a given period of time.
2. Net exporter of food.
3. Countries that explicitly require halal certificates for importing into that country any food products including meats. This information has been collected from the United States Department of Agriculture's website (http://www.fsis.usda.gov). For more information please directly contact the embassies or consulates of the respective countries.
4. Countries that are major exporters of food directly to Muslim countries.
5. Countries that have very organized halal certification either supported by the community or by their respective governments.
6. Countries requiring assistance for food as classified by the Food and Agriculture Organization of the United Nations in a December 2014 circular (http://www.fao.org/giews/English/hotspots/index.htm).

Hattiesburg (Mississippi), St. Louis, Knoxville (Tennessee), Shreveport (Louisiana), Lincoln (Nebraska), and Anchorage (Alaska) (*Daily Beast*, 2010).

Halal foods comprise one of the fastest growing consumer markets in the U.S. The above cited report found that American Muslims spent $12.5 billion on food and food services in 2014 (*Daily Beast*, 2010). Based on 973 responses from across the country, it gave interesting insights into halal consumer preferences. An overwhelming majority (93%) said that they purchase halal products to eat at home. And 86% of them wanted halal products to be available at their local supermarkets. This should give a hint to the major retailers to make halal products available at their stores. This phenomenal growth is also reflected in the number of stores selling halal meat—from 10 outlets in 1970 to over 2300 in 2012 (based on data compiled by Zabihah.com and the telephone "White Pages"). In addition, there are now over 6900 restaurants serving halal consumers.

TABLE 4.4
Muslims in Select European Countries

Countries	Estimated Muslim Population 2010	Estimated Percentage of Population that is Muslim	Projected Muslim Population 2030	Projected Percentage of Population that is Muslim 2030
Austria	475,000	5.7	799,000	9.3
Belgium	638,000	6.0	1,149,000	10.2
Denmark	226,000	4.1	317,000	5.6
Finland	42,000	0.8	105,000	1.9
France	4,704,000	7.5	6,860,000	10.3
Germany	4,119,000	5.0	5,545,000	7.1
Greece	527,000	4.7	772,000	6.9
Ireland	43,000	0.9	125,000	2.2
Italy	1,583,000	2.6	3,199,000	5.4
Luxembourg	11,000	2.3	14,000	2.3
Netherlands	914,000	5.5	1,365,000	7.8
Norway	144,000	3.0	359,000	6.5
Portugal	65,000	0.6	65,000	0.6
Spain	1,021,000	2.3	1,859,000	3.7
Sweden	433,000	4.9	993,000	8.1
Switzerland	433,000	5.7	663,000	8.2
United Kingdom	2,869,000	4.6	5,567,000	8.2

Source: Grim and Karim (2011) *The Future of the Global Muslim Population. Projections for 2010–2030*, Pew Research Center's Forum on Religion and Public Life, Washington, DC: 2011.

The ethno-cultural diversity of the American Muslim population has led to the emergence of fusion foods and halal versions of various cuisines. Food continued to be an important indicator of religiosity among American Muslims. They experiment with all kinds of food while ensuring that they are halal. Edward Curtis, a leading scholar of American Muslims writes:

> By the beginning of the 21st century, many religiously observant Muslim Americans had begun to nurture a common and nationwide Muslim-American food culture. The cuisines that they consumed continued to reflect their ethnic and racial diversity, though Muslim Americans were also influencing one another's food choices across lines of race and ethnicity. Increasingly, non-Muslim Americans enjoyed the various foods introduced by Muslims to the United States. Of course, none of these cuisines were exclusively Muslim. But soul food, Middle Eastern cuisine, and Indian delicacies, among others, reflected the culinary energy and knowledge of Muslim Americans, and these foods have made an indelible mark on the history of American food as a whole.
> **(Curtis, 2010, p. 209)**

Companies like Saffron Road offer frozen halal dinners with cuisines that span the globe. Their current line offers Chipotle Enchilada to Pad Thai and Korean

Chicken. The popularity of halal cuisine is also on the rise in the non-Muslim population. The Halal Guys, a popular New York food cart business specializing in Middle Eastern food, opened in its first brick and mortar store in 2014. It plans to open dozens of stores across North America and abroad after scoring a deal with the franchise company Fransmart.

The increasing presence of Muslims in schools and colleges is creating a need for institutional dining services to cater to their needs. Hospitals, retirement facilities, hospices, food banks, and so on are also recognizing the need for halal meals for their clients and making appropriate changes to their menus. Prisons across the country are making efforts to make halal meals available for their Muslim inmates.

GLOBAL HALAL TRADE

At 1.8 billion, the global Muslim population is impressive in numbers. Several anthropological, sociological, and cultural fields have found a link between religiosity and consumption (see for instance Fischer, 2011). For now, it has been long established that an overwhelming majority of Muslims, including those who do not strictly follow other religious practices, are very particular in adhering to at least some level of Islamic dietary laws (the most basic one being the avoidance of pork). Apart from religiosity, some segments of the population, especially in minority contexts, might also follow halal as they perceive it to be an identity marker. In addition, there is also a significant population of non-Muslims who are halal consumers for reasons of health, safety, ethics, quality, taste, and animal welfare. All this provides for a large market for halal products.

From a producer's perspective it is very important to understand the current trends and future prospects of food trade in these markets. Adoption of higher halal standards along with competitive pricing is a sure road to success in the Muslim market. On the following pages, we will first take a larger view of the Muslim market and then break it down by region and individual countries. Due to constraints of space, not every country or industry segment can be covered but we will try to be as comprehensive as possible.

From a macro view the global halal market was estimated at $1.3 trillion in 2013. It is expected to grow to $2.5 trillion by 2019. A large number of Muslim-majority countries are net importers and are expected to remain so in the near future (FAO, 2015). Imports into OIC countries comprise 14.4% of global imports. Their exports account for 8.8% of global exports, Table 4.5 shows the annual food imports and self-sufficiency ratios for the entire MENA region. It also shows the FAO projections to 2022. For almost all of the essential food commodities, the region's expected 2022 share of total world consumption is expected to grow or remain the same from the 2010 to 2012 levels. The same is true about the region's share of total world imports. The self-sufficiency data indicates that for a majority of food commodities, the region will not be able to meet the demand and will therefore rely on imports.

Among all the OIC countries, the ones in MENA and Southeast Asia have the largest markets for food trade. They score the highest when it comes to value and quantity. They also have the strictest regulations when it comes to halal imports. Malaysia and Indonesia were the early pioneers in standardizing policies related to

TABLE 4.5
Projections for Food Imports to the Middle East and North Africa Region (MENA)[a]

	Imports Volume (10) 2010–2012	Imports DOMT 2022	Annual Growth Rate (94)	Self-Sufficiency (%) Ratio 2010–2012	Self-Sufficiency (%) Ratio 2022	Share in World Imports (%) 2010–2012	Share in World Imports (%) 2022	Share in World Consumption (%) 2010–2012	Share in World Consumption (%) 2022	Share in World Production (%) 2010–2012	Share in World Production (%) 2022
Wheat	40554.5	49480.7	1.8	50.6	49.9	29.9	33.3	11.5	12.4	5.8	6.1
Coarse grains	32696.6	44938.2	2.9	44.0	41.9	26.5	27.8	5.0	5.6	2.2	2.3
Rice	7700.0	11841.7	4.0	43.2	32.9	20.9	27.3	2.7	3.2	1.2	1.1
Butter	274.0	384.7	3.1	61.6	61.6	33.3	37.7	6.9	7.2	4.2	4.4
Cheese	530.3	761.6	3.3	95.4	95.2	22.5	26.7	6.8	7.4	6.5	7.1
Skim milk powder	342.1	420.8	1.9	2.8	4.9	21.1	19.1	8.5	8.1	0.2	0.4
Whole milk powder	621.7	660.4	0.6	1.4	0.3	27.9	25.4	11.3	9.3	0.2	0.0
Beef and veal	1335.4	1612.6	1.7	63.6	64.3	14.8	15.1	5.2	5.6	3.3	3.6
Poultry meat	2529.7	3332.7	2.6	68.3	68.3	21.1	24.5	7.3	8.1	5.0	5.5
Sheep meat	391.3	431.5	0.9	92.1	92.2	37.2	35.8	18.1	18.8	16.6	17.4
Oilseeds	5037.0	6284.6	2.0	28.9	28.7	4.5	4.4	1.8	1.8	0.5	0.5
Vegetable oils	8207.1	10337.4	2.1	21.8	20.2	12.8	13.2	5.5	5.9	1.2	1.2
Sugar	12522.3	16544.2	2.6	30.0	31.0	25.2	28.7	8.7	9.4	2.5	2.8

Source: FAO (2015). State of Food and Agriculture in the Near East and North Africa Region (www.fao.org/docrep/meeting/030/mj390e.pdf; Accessed on February 28, 2015).

[a] The MENA region classification of FAO includes the following countries: UAE, Kuwait, Qatar, Israel, Bahrain, Iraq, Saudi Arabia, Oman, Lebanon, Syria, Iran, Jordan, Egypt, Pakistan, Yemen, Afghanistan, Somalia, Sudan, Gaza Strip and West Bank, Libya, Tunisia, Algeria, Morocco.

import of halal foods. The GCC countries have individual requirements and they are fast converging toward setting regional standards. There are also attempts by bodies like the Standards and Metrology Institute for Islamic Countries (SMIIC) to develop internationally recognized standards. A spate of scandals related to meat from Europe and elsewhere has increased consumer skepticism about halal authenticity of imported products. Food producers are advised to acquire credible and accepted third-party halal certification when entering the halal markets.

In terms of imports, the bulk agricultural products like wheat, maize, rice, and so on, score the highest among commodities imported into Islamic countries. For purposes of brevity, we will largely only focus on consumer-oriented products. For representative purposes, we have reproduced a table of top 50 commodities which were imported into Middle East in 2012 (Table 4.6). In tune with consumption in the region, cereals, vegetables, fruits, and dairy products dominate the list. However, the share of meat in consumption and imports is expected to increase.

A majority of the food imports to the OIC countries are from non-Muslim countries. The arid climate conditions, water shortages, and lack of arable land in a number of OIC member countries makes them unable to grow and produce food products. They will continue to remain reliant on others for their food consumption needs. Even the small island nation of Maldives imports food products from as many as 70 countries. Table 4.7 shows the top 25 partner countries for imports into the Middle East region. It imports food products from more than 171 countries. Brazil, the U.S., UAE, France, Germany, and Argentina top the list in terms of value. As far as import share is concerned, Brazil and Argentina account for close to half of total food products imported. The UAE is third on the list as it imports a large number of products from throughout the world and then re-exports to the rest of the region. As far as individual countries are concerned, Saudi Arabia is the leader among OIC countries with a share of 2.027% of the total world imports trade. It is followed by Malaysia (0.935%), Egypt (0.734%), Indonesia (0.734%), Algeria (0.591%), and Jordan (0.317%).

In vegetable imports, Saudi Arabia is also the leader with 1.96% share of the world's total. It is followed by Indonesia (1.35%), Egypt (1.34%), Turkey (1.26%), and Malaysia (1.23%) (World Bank, 2013).

Meat is an important component of a healthy diet. A World Health Organization report on global consumption patterns says that wider availability of low priced meat will lead to better nutrition:

> As diets become richer and more diverse, the high-value protein that the livestock sector offers improves the nutrition of the vast majority of the world. Livestock products not only provide high-value protein but are also important sources of a wide range of essential micronutrients, in particular minerals such as iron and zinc, and vitamins such as vitamin A. For the large majority of people in the world, particularly in developing countries, livestock products remain a desired food for nutritional value and taste.

The consumption of meat is rising around the world and the OIC countries are no different than others in this regard. The world's meat consumption from 2014 to 2020 is expected to grow by 10%. According to OECD projections, the highest demand for poultry and beef exports is going to be in Asia followed by MENA (OECD, 2014).

TABLE 4.6
Top 50 Imported Food Commodities in the Middle East

Commodity	Quantity (Tonnes)	Value (1000 $)	Unit Value ($/Tonne)
Wheat	19,360,823.00	6,560,175.00	338.84
Barley	8,019,014.00	2,441,972.00	304.52
Maize	6,939,129.00	2,350,932.00	338.79
Rice—total (rice milled equivalent)	4,674,921.00	4,272,019.00	913.82
Sugar refined	4,022,608.00	3,068,410.00	762.79
Sugar Raw Centrifugal	2,903,212.00	1,912,527.00	658.76
Cake, soybeans	2,844,477.00	1,253,423.00	440.65
Soybeans	2,701,151.00	1,462,397.00	541.40
Oil, palm	2,148,756.00	2,260,487.00	1,052.00
Meat, chicken	2,040,355.00	4,177,221.00	2,047.30
Flour, wheat	1,568,441.00	754,163.00	480.84
Oil, sunflower	1,285,781.00	1,723,792.00	1,340.66
Beverages, non-alcoholic	1,195,804.00	957,632.00	800.83
Forage products	1,164,902.00	353,524.00	303.48
Rapeseed	1,024,779.00	680,279.00	663.83
Sunflower seed	1,016,907.00	690,743.00	679.26
Oranges	987,676.00	569,575.00	576.68
Cake, sunflower	969,746.00	241,803.00	249.35
Bananas	888,714.00	460,812.00	518.52
Onions, dry	852,179.00	316,207.00	371.06
Bran, wheat	847,467.00	169,385.00	199.87
Alfalfa meal and pellets	809,842.00	202,563.00	250.13
Food prep	797,672.00	3,297,051.00	4,133.34
Tomatoes	723,546.00	346,338.00	478.67
Potatoes	637,158.00	301,451.00	473.12
Cotton lint	633,701.00	1,916,115.00	3,023.69
Lentils	606,910.00	491,762.00	810.27
Apples	594,068.00	547,767.00	922.06
Pastry	467,686.00	1,327,014.00	2,837.40
Juice, fruit	416,340.00	380,135.00	913.04
Dregs from brewing, distillation	366,047.00	115,940.00	316.74
Oil, soybean	365,921.00	431,433.00	1,179.03
Waters, ice, etc.	365,762.00	145,148.00	396.84
Milk, whole dried	324,502.00	1,393,236.00	4,293.46
Oil, maize	321,211.00	515,610.00	1,605.21
Meat, cattle, boneless (beef and veal)	283,097.00	1,355,071.00	4,786.60
Lemons and limes	278,289.00	176,262.00	633.38
Chocolate products	277,552.00	1,419,830.00	5,115.55
Eggs, hen, in shell	274,704.00	492,961.00	1,794.52

(Continued)

TABLE 4.6 (CONTINUED)
Top 50 Imported Food Commodities in the Middle East

Commodity	Quantity (Tonnes)	Value (1000 $)	Unit Value ($/Tonne)
Beer of barley	267,785.00	230,434.00	860
Sesame seed	263,907.00	361,654.00	1,370
Cheese, whole cow milk	261,292.00	1,068,851.00	4,090
Potatoes, frozen	258,495.00	281,652.00	1,089
Cake, rapeseed	250,208.00	72,414.00	289
Tomatoes, paste	228,412.00	248,913.00	1,089
Food wastes	225,411.00	273,049.00	1,211
Fruit, prepared	221,119.00	307,993.00	1,392
Chick peas	215,461.00	193,602.00	898
Sugar confectionery	209,926.00	526,134.00	2,506
Flour, maize	208,650.00	73,118.00	350

Source: FAO Stat.

Table 4.8 provides OECD estimates of kilograms per capita consumption of beef/veal in selected Muslim countries from 2013 to 2023. The consumption is expected to grow in most of these countries and so is the demand. The notable exceptions are countries like Sudan, Kazakhstan, and Pakistan, which have enough livestock populations to cover the domestic demand. For most of the OIC member countries their meat intake is less than recommended guidelines.

Table 4.9 provides a list of top ten OIC member importers of meat and live animals in terms of value. Four of them are members of the oil rich GCC. The GCC region meets 56% of its meat requirements through imports (Alpen Capital, 2011). Table 4.10 lists the top ten exporting countries of meat and live animals to OIC countries. Only Pakistan and Turkey are Muslim-majority countries in the exporters list. Brazil with its huge livestock population is the natural leader of the pack. But it is facing increasing competition from Australia, the U.S., and others. India is a major exporter of buffalo meat to the Middle East. But increasing restrictions on trade can adversely affect this trade (Kannan, 2014). In 2013, Malaysia and Saudi Arabia were the third and fifth importers of Indian buffalo meat. Any reduction in its exports (currently 20% of the world total) will create turbulence in the market. But its share can be quickly covered by other players in the market. However, the exit of India from the beef market can significantly increase the prices of beef.

Table 4.11 gives the details of beef and veal trade in the Middle East. The imports consistently outpace domestic production and will continue to do so in the near future. The production levels are expected to remain stagnant while imports grow. Table 4.12 provides the data on the beef trade in North Africa. The production/import imbalance is not as high as in the Middle East, but significant amounts of meat will continue to be imported into the region.

Table 4.13 provides details of the poultry trade in the Middle East. According to FAO projections, per capita consumption in the region will rise from 15.7 kg in 2010 to 17.0 kg in 2015. Imports are expected to account for 43% of total consumption.

TABLE 4.7
Middle East and North Africa 2012 Partner Share Food Products

Partner Name	Import Trade Value US $ Thousand	Import Share (%)
Brazil	3,135,341.89	27.42
United States	1,550,363.20	3.19
United Arab Emirates	1,454,021.16	7.72
France	1,305,641.28	4.64
Germany	1,279,369.21	4.06
Argentina	1,202,404.91	22.35
Netherlands	1,112,2444.93	10.93
Ireland	855,974.31	28.28
Switzerland	840,870.36	6.82
Turkey	787,403.19	4.97
United Kingdom	769,835.20	5.74
Thailand	716,191.47	10.46
Saudi Arabia	659,478.88	4.52
India	647,193.67	4.18
Italy	614,278.07	2.14
Spain	557,607.40	3.26
Egypt	492,521.10	9.12
China	457,947.75	0.90
Belgium	416,065.31	4.59
Poland	272,141.25	13.14
Denmark	182,508.70	8.66
Indonesia	156,922.81	3.99
South Africa	152,433.68	10.68
Austria	148,576.65	5.80
Malaysia	141,955.55	5.53

Source: World Integrated Trade Solution, World Bank (http://wits.worldbank.org/ THE MEAT TRADE).

Domestic production has been rising in this sector but it still unable to meet the demand.

U.S. TRADE WITH THE MUSLIM COUNTRIES

The U.S. has been a major exporting partner to a number of OIC countries. In this section we will provide historical statistics on consumer-oriented products from the U.S. to the MENA region and also to the countries of Indonesia and Malaysia. The data on the MENA region has been further broken down on exports exclusive to the GCC sub-category. Even until the 1960s, the exports from the U.S. to many Muslim countries were paltry: from less than a million to several billion metric tonnes.

TABLE 4.8
Beef and Veal Consumption (est.) in Select OIC Countries

Kilograms per Capita

Country	2013	2014	2015	2016	2017	2018	2019	2020	2021	2022	2023
Algeria	3.8097	4.0864	4.1304	4.1805	4.235	4.2941	4.34	4.3772	4.4207	4.4864	4.5541
Bangladesh	0.9033	0.9148	0.934	0.9504	0.9632	0.9799	0.992	1.0031	1.0158	1.0285	1.0391
Egypt	10.302	10.7173	10.4603	10.2532	10.0794	9.9777	9.8467	9.69	9.5407	9.4191	9.3009
Indonesia	1.7276	1.7656	1.8172	1.8696	1.9005	1.8999	1.9045	1.9336	1.9848	2.0332	2.0692
Iran	3.4392	3.4211	3.4756	3.4548	3.5293	3.5442	3.6354	3.5942	3.6939	3.6468	3.7803
Kazakhstan	18.5643	18.7231	18.8955	19.467	19.8807	19.6222	19.5904	19.7132	19.6847	19.4594	19.185
Malaysia	4.7926	4.8495	4.834	4.8003	4.8012	4.8285	4.8519	4.8564	4.8677	4.8888	4.9104
Nigeria	1.4699	1.4495	1.4356	1.4203	1.4066	1.3848	1.367	1.3526	1.3423	1.3263	1.3115
Pakistan	5.6683	5.7394	5.7359	5.7913	5.809	5.7777	5.7741	5.8193	5.8507	5.8522	5.8432
Saudi Arabia	4.5163	4.561	4.6005	4.6616	4.7292	4.8437	4.9191	4.9928	5.0628	5.1468	5.2261
Sudan	23.4209	23.3975	23.3361	23.2688	23.1818	23.088	22.9893	22.9125	22.8382	22.7731	22.7158
Turkey	5.9611	6.2739	6.3369	6.5781	6.6886	6.8634	6.9713	7.0894	7.159	7.2284	7.2795
World	6.5218	6.493	6.4937	6.4709	6.5055	6.5483	6.585	6.5906	6.612	6.6312	6.6423

Source: OECD/FAO (2014). *OECD-FAO World Agricultural Outlook: 2014–2023.* Paris: OECD.

TABLE 4.9
Top OIC Member Importers of Meat and Live Animals

Country	USD Billion (2013)
Saudi Arabia	$2.70
UAE	$1.51
Egypt	$1.29
Malaysia	$0.90
Iraq	$0.86
Jordan	$0.63
Qatar	$0.60
Indonesia	$0.59
Kuwait	$0.58
Libya	$0.51

Source: Global Islamic Economy Report 2014.

TABLE 4.10
Top Meat and Live Animals Exporters to OIC Countries

Country	USD Billion (2013)
Brazil	$4.7
India	$2.1
Australia	$1.2
U.S.	$1.2
France	$0.8
Turkey	$0.5
New Zealand	$0.3
Netherlands	$0.2
Pakistan	$0.2
Germany	$0.2

Source: Global Islamic Economy Report 2014.

Table 4.14 provides data for U.S. consumer-oriented products to the MENA region for the years 2010 to 2014. In this period, there has been an average change of 12% for the consumer-oriented category in terms of value. The quantity of consumer products has also increased to the region. The biggest change (25%) is in the snack foods category. The prepared foods category has also increased. This is in tune with the overall trend of increasing demand for processed foods. Beef and beef products worth $124 million were exported in 2014. More than $336 million worth of chicken was imported in 2014. Processed foods are among the

TABLE 4.11
Beef and Veal Production Trade in the Middle East

Commodity	Attribute	2010	2011	2012	2013	2014	2015
Meat, beef, and veal	Production (1000 MT CWE)	421	463	467	467	469	469
	Total imports (1000 MT CWE)	717	635	604	618	665	687
	Total supply (1000 MT CWE)	1138	1098	1071	1085	1134	1156
	Total exports (1000 MT CWE)	25	27	41	23	24	24
	Total domestic consumption (1000 MT CWE)	1113	1071	1030	1062	1110	1132

Source: Foreign Agriculture Service, USDA estimates.

TABLE 4.12
Beef and Veal Production, Imports, Exports into/from North Africa

Commodity	Attribute	2010	2011	2012	2013	2014	2015
Meat, beef, and veal	Production (1000 MT CWE)	479	451	423	428	463	443
	Total imports (1000 MT CWE)	364	293	374	309	390	425
	Total supply (1000 MT CWE)	843	744	797	737	853	868
	Total exports (1000 MT CWE)	0	0	0	0	0	0
	Total domestic consumption (1000 MT CWE)	843	744	797	737	853	868

Source: Foreign Agriculture Service, USDA estimates.

TABLE 4.13
Poultry Trade in the Middle East

Commodity	Attribute	2010	2011	2012	2013	2014	2015
Poultry, meat, broiler	Production (1000 MT)	3208	3506	3667	3776	3821	3099
	Total imports (1000 MT)	1854	2017	2066	2220	2125	2173
	Total supply (1000 MT)	5062	5523	5733	5996	5946	5272
	Total exports (1000 MT)	145	260	334	411	465	517
	Total domestic consumption (1000 MT)	4917	5263	5399	5585	5481	4755
	Total use (1000 MT)	5062	5523	5733	5996	5946	5272

Source: Foreign Agriculture Service, USDA estimates.

TABLE 4.14
U.S. Consumer-Oriented Products Exported to the Middle East

Product	UOM	2010 Value	2010 Qty	2011 Value	2011 Qty	2012 Value	2012 Qty	2013 Value	2013 Qty	2014 Value	2014 Qty
Consumer-oriented total	MT	2,229	1,019,273.1	2,713	1,202,297.0	2,837	1,150,111.0	3,549	1,382,016.7	3,462	1,282,315.0
Tree nuts	MT	673	151,017.6	761	162,390.3	755	141,308.4	959	154,804.0	1,003	145,562.9
Dairy products	MT	357	117,880.3	376	102,669.4	432	126,680.4	783	208,216.7	643	153,383.9
Poultry meat and prods. (ex. eggs)	MT	205	234,701.4	330	293,914.2	374	308,290.7	433	364,475.0	353	338,857.6
Beef and beef products	MT	262	134,862.8	355	174,626.7	331	151,115.1	287	152,604.7	276	135,636.6
Prepared food	MT	166	56,406.1	171	53,497.5	172	46,978.2	206	50,963.2	247	60,085.4
Snack foods NESOI	MT	70	28,913.6	113	37,677.2	169	50,667.6	176	48,281.5	188	49,345.9
Fresh fruit	MT	125	108,928.6	124	106,091.8	115	83,075.5	158	116,421.4	166	118,898.4
Condiments and sauces	MT	93	58,548.5	112	64,811.5	121	69,895.9	131	73,686.2	142	77,509.5
Processed vegetables	MT	75	69,574.5	143	143,070.8	113	102,816.0	137	132,777.6	133	123,399.6
Chocolate and cocoa products	MT	37	8,545.8	37	7,794.3	51	11,253.1	63	12,715.7	78	15,115.3
Processed fruit	MT	40	18,998.6	46	19,293.0	54	21,528.7	50	20,322.8	68	27,995.0
Non-alcoholic bev. (ex. juices)	MT	16	1,731.2	18	1,676.9	28	2,443.8	32	2,632.0	36	3,174.4
Fish products	MT	28	11,527.8	38	13,976.0	32	10,888.4	29	8,936.3	36	16,377.0
Fruit and vegetable juices	KL	32	58,029.2	35	54,556.3	16	21,110.6	26	36,776.8	29	49,077.6
Breakfast cereals	MT	9	3,765.7	13	4,573.5	12	4,837.4	16	13,874.9	14	5,273.1
Eggs and egg products	MT	23	1,653.8	22	1,591.9	30	2,129.4	24	2,028.9	13	1,862.7
Other consumer oriented	MT	11	2,736.6	10	2,111.3	10	1,749.5	10	1,526.7	12	1,730.3

(Continued)

TABLE 4.14 (CONTINUED)
U.S. Consumer-Oriented Products Exported to the Middle East

Product	UOM	2010 Value	2010 Qty	2011 Value	2011 Qty	2012 Value	2012 Qty	2013 Value	2013 Qty	2014 Value	2014 Qty
Fresh vegetables	MT	6	5,435.8	6	7,149.2	6	6,199.0	9	7,599.8	8	6,638.1
Meat products NESOI	MT	2	660.1	4	1,276.1	3	828.8	2	393.9	2	444.9

Values in millions of dollars. Appropriate quantities are mentioned in the second column.

Notes: 1. Data Source: U.S. Census Bureau Trade Data [incomplete and not consistent with the earlier formats]. 2. Users should use cautious in interpreting the QUANTITY reports because of the use of mixed units of measure. The QUANTITY line items will only include statistics on the amounts that are reported in units that can be converted to the assigned unit of measure of the grouped commodities.

top-ranking products that are in demand. This is especially so in the GCC countries. Demand for the following products continues to be high: baby food, spreads, canned/preserved food, pasta, sweet and savory snacks, ice cream, chilled processed food, and confectionery. Table 4.15 provides details of the U.S. exports to the GCC countries. The top four exported products are tree nuts, dairy products, poultry, and prepared foods.

Table 4.16 provides statistics on U.S. exports to Indonesia. Top items for this country are dairy products, fresh fruit, prepared foods, processed vegetable, and beef and beef-related products. High growth categories include: meal replacements, ice cream, chilled and frozen processed foods, canned/preserved foods, snack bars, pasta, soups, and dried processed foods. In 2014, consumer-oriented products worth $578 million were exported to Indonesia from the U.S.

Table 4.17 provides statistics on U.S. exports to Malaysia. In 2014, consumer-oriented products worth $528 million were exported to Malaysia from the U.S. The following American made products will have good sales potential in the country: breakfast cereals, snack foods, frozen vegetables fresh (temperate) fruits, dairy, chocolates, non-alcoholic beverages, and pet foods.

CONCLUSION

No man is an island goes a popular saying. Similarly, in today's interconnected world, no country can grow and prosper without trading with others. Domestic production is often constrained for reasons beyond human control like weather and geography. By using their comparative advantages, each nation can trade the products that they have superior resources for to mutual advantage. The Muslim countries stand to benefit from the global trade as they have a variety of choices. To succeed in this diverse market the producers need to produce products which achieve the right balance of price and quality. As far as quality is concerned it should not only be of superior quality but also produced in a manner that meets the halal dietary laws. Companies that tend to adopt this holistic view of production will definitely prosper in the global Muslim market.

TABLE 4.15
U.S. Consumer Oriented Products to the GCC Countries

Product	UOM	2010 Value	2010 Qty	2011 Value	2011 Qty	2012 Value	2012 Qty	2013 Value	2013 Qty	2014 Value	2014 Qty	Period/Period % Change (Value)	Period/Period % Change (Qty)
Consumer-oriented total	MT	1,049	467,196.4	1,317	564,475.3	1,433	556,506.9	1,749	642,582.7	1,870	659,414.4	7	3
Tree nuts	MT	287	63,857.9	320	68,955.2	309	55,872.6	444	68,685.5	470	64,836.5	6	−6
Dairy products	MT	106	33,048.3	155	41,848.9	191	51,563.3	260	67,371.4	271	63,837.4	4	−5
Poultry meat and prods. (ex. eggs)	MT	97	98,470.6	142	121,169.9	176	139,656.5	183	146,947.6	173	157,839.7	−5	7
Prepared food	MT	105	40,225.1	98	35,613.4	97	31,284.4	116	33,750.5	154	42,367.6	32	26
Snack foods NESOI	MT	49	18,504.9	86	28,337.5	132	38,232.1	134	35,482.7	148	36,337.3	10	2
Fresh fruit	MT	91	80,697.4	101	88,618.5	95	68,124.0	131	98,047.6	143	101,820.4	9	4
Beef and beef products	MT	73	17,799.3	109	23,878.3	104	12,328.8	106	10,154.2	114	10,807.8	8	6
Condiments and sauces	MT	67	41,876.5	83	46,258.2	89	50,255.0	97	53,067.4	109	58,493.5	13	10

(*Continued*)

TABLE 4.15 (CONTINUED)
U.S. Consumer Oriented Products to the GCC Countries

Product	UOM	2010 Value	2010 Qty	2011 Value	2011 Qty	2012 Value	2012 Qty	2013 Value	2013 Qty	2014 Value	2014 Qty	Period/Period % Change (Value)	Period/Period % Change (Qty)
Processed vegetables	MT	51	45,987.0	88	79,625.7	84	72,110.7	99	90,146.9	99	81,204.1	—	−10
Chocolate and cocoa products	MT	31	7,068.1	31	6,208.8	42	9,091.6	53	10,650.2	61	12,068.7	17	13
Processed fruit	MT	14	7,456.3	18	7,927.0	27	11,953.1	25	10,136.2	33	12,950.0	29	28
Non-alcoholic bev. (ex. juices)	MT	12	1,295.3	13	1,284.9	21	2,116.7	24	2,335.5	26	2,517.4	5	8
Fruit and vegetable juices	KL	22	51,409.5	19	41,297.5	10	16,788.8	15	25,837.0	20	38,096.8	29	47
Fish products	MT	8	1,445.5	13	2,010.3	14	1,781.6	11	1,370.4	13	2,263.1	12	65
Fresh vegetables	MT	5	4,191.9	5	5,237.3	6	5,791.2	9	7,225.2	8	6,009.1	−12	−17

(Continued)

TABLE 4.15 (CONTINUED)
U.S. Consumer Oriented Products to the GCC Countries

Product	UOM	2010 Value	2010 Qty	2011 Value	2011 Qty	2012 Value	2012 Qty	2013 Value	2013 Qty	2014 Value	2014 Qty	Period/Period % Change (Value)	Period/Period % Change (Qty)
Eggs and egg products	MT	17	864.4	16	881.4	22	860.3	18	886.5	7	1,170.8	−61	32
Breakfast cereals	MT	3	910.2	6	1,677.3	3	1,406.7	3	1,315.6	4	1,540.7	30	17
Meat products NESOI	MT	1	201.0	3	749.8	1	258.6	2	341.2	2	361.6	21	6

Source: U.S. Census Bureau Trade Data.

Values in millions of dollars. Appropriate quantities are mentioned in the second column.

Users should use cautious interpretation on QUANTITY reports using mixed units of measure. QUANTITY line items will only include statistics on the units of measure that are equal to, or are able to be converted to, the assigned unit of measure of the grouped commodities.

TABLE 4.16
U.S. Consumer-Oriented Products to Indonesia

Product	UOM	2010 Value	2010 Qty	2011 Value	2011 Qty	2012 Value	2012 Qty	2013 Value	2013 Qty	2014 Value	2014 Qty	Period/Period % Change (Value)	Period/Period % Change (Qty)
Consumer-oriented total	MT	1,049	467,196.4	1,317	564,475.3	1,433	556,506.9	1,749	642,582.7	1,870	659,414.4	7	3
Tree nuts	MT	287	63,857.9	320	68,955.2	309	55,872.6	444	68,685.5	470	64,836.5	6	−6
Dairy products	MT	106	33,048.3	155	41,848.9	191	51,563.3	260	67,371.4	271	63,837.4	4	−5
Poultry meat and prods. (ex. eggs)	MT	97	98,470.6	142	121,169.9	176	139,656.5	183	146,947.6	173	157,839.7	−5	7
Prepared food	MT	105	40,225.1	98	35,613.4	97	31,284.4	116	33,750.5	154	42,367.6	32	26
Snack foods NESOI	MT	49	18,504.9	86	28,337.5	132	38,232.1	134	35,482.7	148	36,337.3	10	2
Fresh fruit	MT	91	80,697.4	101	88,618.5	95	68,124.0	131	98,047.6	143	101,820.4	9	4
Beef and beef products	MT	73	17,799.3	109	23,878.3	104	12,328.8	106	10,154.2	114	10,807.8	8	6
Condiments and sauces	MT	67	41,876.5	83	46,258.2	89	50,255.0	97	53,067.4	109	58,493.5	13	10

(Continued)

TABLE 4.16 (CONTINUED)
U.S. Consumer-Oriented Products to Indonesia

Product	UOM	2010 Value	2010 Qty	2011 Value	2011 Qty	2012 Value	2012 Qty	2013 Value	2013 Qty	2014 Value	2014 Qty	Period/Period % Change (Value)	Period/Period % Change (Qty)
Processed vegetables	MT	51	45,987.0	88	79,625.7	84	72,110.7	99	90,146.9	99	81,204.1	–	–10
Chocolate and cocoa products	MT	31	7,068.1	31	6,208.8	42	9,091.6	53	10,650.2	61	12,068.7	17	13
Processed fruit	MT	14	7,456.3	18	7,927.0	27	11,953.1	25	10,136.2	33	12,950.0	29	28
Non-alcoholic bev. (ex. juices)	MT	12	1,295.3	13	1,284.9	21	2,116.7	24	2,335.5	26	2,517.4	5	8
Fruit and vegetable juices	KL	22	51,409.5	19	41,297.5	10	16,788.8	15	25,837.0	20	38,096.8	29	47
Fish products	MT	8	1,445.5	13	2,010.3	14	1,781.6	11	1,370.4	13	2,263.1	12	65
Fresh vegetables	MT	5	4,191.9	5	5,237.3	6	5,791.2	9	7,225.2	8	6,009.1	–12	–17

(*Continued*)

TABLE 4.16 (CONTINUED)
U.S. Consumer-Oriented Products to Indonesia

Product	UOM	2010 Value	2010 Qty	2011 Value	2011 Qty	2012 Value	2012 Qty	2013 Value	2013 Qty	2014 Value	2014 Qty	Period/Period % Change (Value)	Period/Period % Change (Qty)
Eggs and egg products	MT	17	864.4	16	881.4	22	860.3	18	886.5	7	1,170.8	−61	32
Breakfast cereals	MT	3	910.2	6	1,677.3	3	1,406.7	3	1,315.6	4	1,540.7	30	17
Meat products NESOI	MT	1	201.0	3	749.8	1	258.6	2	341.2	2	361.6	21	6

Source: U.S. Census Bureau Trade Data.

Values in millions of dollars. Appropriate quantities are in mentioned in second column.

Users should use cautious interpretation on QUANTITY reports using mixed units of measure. QUANTITY line items will only include statistics on the units of measure that are equal to, or are able to be converted to, the assigned unit of measure of the grouped commodities.

TABLE 4.17
U.S. Consumer-Oriented Products to Malaysia

Product	UOM	2010 Value	2010 Qty	2011 Value	2011 Qty	2012 Value	2012 Qty	2013 Value	2013 Qty	2014 Value	2014 Qty	Average % Change (Value)	Average % Change (Qty)
Consumer-oriented total	MT	336	181,919.5	440	221,474.8	481	214,523.0	561	239,598.6	528	215,669.5	11	4
Dairy products	MT	94	49,672.5	137	55,714.5	133	45,438.2	181	63,068.4	182	61,587.1	16	5
Prepared food	MT	55	8,383.9	61	9,122.7	89	8,991.5	110	12,691.5	100	8,538.8	15	—
Fresh fruit	MT	55	55,306.2	63	63,550.0	69	59,437.8	77	63,110.0	70	48,057.9	6	−4
Processed vegetables	MT	25	23,419.7	41	40,609.2	45	40,921.7	51	47,347.0	45	40,415.1	15	14
Tree nuts	MT	16	3,359.4	21	4,070.2	19	3,615.1	29	4,275.5	29	3,760.4	14	3
Processed fruit	MT	23	8,899.2	27	9,891.0	31	10,758.5	27	9,730.9	23	9,043.4	1	—
Fish products	MT	4	3,146.3	7	3,455.7	9	3,900.3	16	12,862.5	19	11,151.1	40	32
Fruit and vegetable juices	KL	14	7,885.6	24	6,977.7	16	4,245.8	10	3,182.1	10	2,429.8	−8	−29
Fresh vegetables	MT	6	11,371.2	8	14,207.4	11	19,433.8	10	18,138.9	9	16,499.5	11	9
Poultry meat and prods. (ex. eggs)	MT	5	5,792.9	9	8,226.5	9	7,507.2	8	6,804.4	9	9,456.3	11	12
Condiments and sauces	MT	6	3,670.4	7	4,410.1	8	4,701.7	4	2,191.6	8	4,779.4	9	7
Snack foods NESOI	MT	6	2,503.8	7	2,684.5	7	2,190.6	8	2,013.6	7	2,839.2	5	3

(*Continued*)

TABLE 4.17 (CONTINUED)
U.S. Consumer-Oriented Products to Malaysia

Product	UOM	2010 Value	2010 Qty	2011 Value	2011 Qty	2012 Value	2012 Qty	2013 Value	2013 Qty	2014 Value	2014 Qty	Average % Change (Value)	Average % Change (Qty)
Chocolate and cocoa products	MT	4	1,009.0	5	1,327.7	6	1,591.4	11	1,450.8	7	1,300.6	14	6
Breakfast cereals	MT	1	236.9	1	326.8	2	1,421.9	2	345.5	1	428.4	17	15
Other consumer oriented	MT	—	102.3	1	112.9	1	140.7	1	211.9	1	249.3	24	22
Meat products NESOI	MT	—	8.6	—	56.6	—	46.2	—	24.4	—	54.8	56	46
Eggs and egg products	MT	—	12.1	—	10.8	—	0.0	—	—	—	0.0	−18	—
Beef and beef products	MT	—	102.4	—	98.3	1	106.7	—	33.1	—	3.0	−90	−89

Source: U.S. Census Bureau Trade Data.

Values in millions of dollars. Appropriate quantities are in mentioned in second column.

Users should use cautious interpretation on QUANTITY. reports using mixed units of measure. QUANTITY line items will only include statistics on the units of measure that are equal to, or are able to be converted to, the assigned unit of measure of the grouped commodities.

REFERENCES

Alpen Capital. (2011). *GCC Food Industry*. Dubai: Alpen Capital.
Curtis, E. E. (2010). Food, In *Encyclopaedia of Muslim American History*, Curtis, E. E. (Ed.). New York, NY: Facts on File, 206–209.
Daily Beast. (2010). America's Muslim Capitals. Accessed on March 1, 2015. Available online: www.thedailybeast.com/galleries/2010/08/10/america-s-muslim-capitals.html.
Dinar Standard. (2015). The Muslim Green: American Muslim Market Study 2014/2015. Accessed on May 17, 2018. Available online: www.dinarstandard.com/american-market-2014.
El-Gohary, H. (2015). *Emerging Research on Islamic Marketing and Tourism in the Global Economy*. Chester, PA: Business Science Reference.
FAO. (2015). State of Food and Agriculture in the Near East and North Africa Region. Accessed on February 28, 2015. Available online: http://www.fao.org/docrep/meeting/030/mj390e.pdf.
Fischer, J. (2011). *The Halal Frontier: Muslim Consumers in a Globalized Market*. London: Palgrave.
Grim, B. J. and Karim, M. S. (2011). *The Future of the Global Muslim Population: Projections for 2010–2030*. Washington DC: Pew Research Center's Forum on Religion and Public Life.
Guha, R. (2015). In absentia: Where are India's conservative intellectuals? *The Caravan Magazine*, 7(3), 43.
Hackett, C., Connor, P., Stonawski, M., Skirbekk, V., Potancoková, M., and Abel, G. (2015). *The Future of World Religions: Population Growth Projections, 2010–20515*. Washington DC: Pew Research Center's Forum on Religion and Public Life.
Hornby, C. (2009). Halal food going mainstream in Europe – Nestlé. November 18, 2009. Accessed on June 23, 2018. Available online: http://blogs.reuters.com/faithworld/2009/11/18/halal-food-going-mainstream-in-europe-nestle/.
Johnson, T. M. and Grim, B. J. (2013). *The World's Religions in Figures: An Introduction to International Religious Demography*. West Sussex: Wiley-Blackwell, John Wiley & Sons.
Kannan, S. (2014). India's elections spark debate on beef exports. *BBC NEWS*. May 14, 2014. Accessed on March 1, 2015. Available online: www.bbc.com/news/business-27251802.
Menipaz, E. and Menipaz, A. (2011). *International Business: Theory and Practice*. New York, NY: Sage Publications Ltd.
OECD-FAO. (2014). OECD-FAO Agricultural Outlook 2014–2023-Commodity Database (OECD, Paris, 2015). https://stats.oecd.org/Index.aspx?DataSetCode=HIGH_AGLINK_2014.
Singh. (2014). Over 180 million Muslims in India. *The Indian Express*. Available online: https://indianexpress.com/article/india/india-others/over-180-million-muslims-in-india-but-they-are-not-part-of-global-terror-groups-govt/.
UN Habitat, N. D. (2015). State of the World's Cities: Trends in Middle East and North Africa. Accessed on February 15, 2015. Available online: ww2.unhabitat.org/mediacentre/documents/sowc/NorthAfrica.pdf.
White House. (2009). Remarks by the President on a New Beginning. Accessed on February 15, 2015. Available online: http://www.whitehouse.gov/the-press-office/remarks-president-cairo-university-6-04-09.
World Bank. (2013). Opening Doors. Accessed on May 17, 2018. Available online: http://data.worldbank.org/indicator/NY.GDP.PCAP.PP.CD.

5 Global Halal Economy[1]

Rafiuddin Shikoh, Mian N. Riaz, and Munir M. Chaudry

CONTENTS

Relevance and Importance of the Global Halal Market Economy 61
Defining the "Halal Market Economy".. 63
Size and Classification of Key Halal Market Economy Sectors Globally 63
Halal Market Economy Key Drivers... 65
Islamic Values and Sector Impact 66
Halal Food Values and Adoption Spectrum 66
Halal Market Economy Industry Participants and Trends 67
Halal Food: From "Farm to Fork" 67
Key Challenges/Trends 68
Pharmaceutical and Cosmetics Sectors.. 69
Key Challenges/Trends 69
The Future of the Global Halal Market Economy 70
Note.. 70
References.. 71

Muslims today are 23% of the global population, estimated at 1.7 billion and growing at a faster rate than any other faith-based group in the world. This large population is increasingly asserting its need for faith-acceptable products and services in the form of religious-compliant products in seven distinct sectors (Sack, 2000). Collectively, the Muslim consumer expenditure in these seven core food and lifestyle sectors being affected by religious-compliance is estimated at $1.8 trillion in 2014. This estimate is in addition to the estimated $1.8 trillion in assets that Islamic finance institutions held in 2014 (Thomson Reuters, 2014). The significance of these markets has already attracted multinationals (Rehman and Shahbaz Shabbir, 2010). In this chapter, a wider market definition and the drivers of the global halal market economy will be presented while sharing an overview of the core seven halal market economy sectors: halal food, Islamic finance, Muslim travel, modest fashion, media/recreation, pharmaceuticals, and cosmetics. This wider perspective provides a context for understanding the drivers and opportunities in the halal food space.

RELEVANCE AND IMPORTANCE OF THE GLOBAL HALAL MARKET ECONOMY

The global Muslim population will grow at about twice the rate of the non-Muslim population over the next two decades—with an average annual growth rate of 1.5%

for Muslims, compared with 0.7% for non-Muslims (Grim and Karim, 2011). Their economic influence is also growing in the global economy as many of the Muslim-majority economies are part of emerging global markets, such as Indonesia, Saudi Arabia, UAE, and Turkey. In addition, Muslims are part of many developed markets (such as the U.S., UK, Germany, and China) as economically significant minority population. While geographically, ethnically, and culturally this is a very diverse market, the common religious values/traditions are influencing all seven sectors' attempt to meet these faith-based consumption/business needs (Al-Harran and Low, 2010).

At the highest level, the values-based Muslim consumer's needs include Islamic/ethical financing, halal (lawful) and tayyab (pure) foods, modest clothing, family friendly travel, gender interaction considerations, and facilitation of religious practices such as prayers, fasting, charity, Hajj/Umrah, and religious education. The needs also extend to business needs for Islamic business financing, investment, and insurance services (Muhamad, 2008).

The Muslim consumers' faith-influenced potential market is estimated at $1.81 trillion for the food and lifestyle sector (travel, clothing, media/recreation, pharmaceuticals/cosmetics) (Dinar Standard Estimates, 2015) (see Table 5.1).

In the food space, Nestle, in 2016, had 150 of its 468 factories world-wide halal certified offering over 300 halal food and beverage items in over 50 countries. Carrefour, Tesco, and other major global retailers are also growing halal food offerings in many markets. In the travel space, the Ritz-Carlton, as part of its strategy to cater to global multicultural travelers, offers Qurans, prayer-carpets, halal food, and bidets in select locations. In other lifestyle segments, Sun Silk, a hair-care products brand, has introduced a line of special shampoos for women that wear the hijab. DKNY has a special Ramadan clothing line.

A key relevance of this wider halal ecosystem to the food sector is the fact that they all share the same core target audiences. A halal food customer is a potential Islamic finance customer and is also a potential halal conscious travel, media, and recreation customer.

TABLE 5.1
Global Muslim Consumer Spending on Food and Lifestyle Sectors

Sector	Existing Muslim Market (2014, USD billions)
Global Muslim food and lifestyle sector expenditures	$1808
Food	$1128
Fashion	$230
Tourism	$142
Cosmetics	$54
Pharmaceuticals	$75
Media and recreation	$179

DEFINING THE "HALAL MARKET ECONOMY"

The halal market economy is being referred to and branded in many way—Muslim lifestyle market, the Islamic market, the Muslim consumer market, a Muslim friendly market, and other variations. The term "Halal Market Economy" will be used in this chapter. Regardless of the labelling, the underlying common element driving this opportunity is a core target audience of Muslims globally whose faith-based values are influencing products and services, and a secondary universal audience with shared-values. However, to present a more focused understanding between sectors that are more or less impacted by this influence, Dinar Standard has developed the following definitions for this economy:

Definition: "The Halal Market Economy represents core sectors and their ecosystem structurally affected by Islamic values driven needs of consumers' lifestyle and business practices."

- The term "Halal" means lawful or permissible in Islam. By default, per Islamic law, all things are halal (permissible) except those identified as prohibited (haram) by Islamic law. So the "Halal Market Economy" refers to those sectors whose core products/services have removed prohibited (haram) elements.
- "Structurally affected" in the definition implies that the core product attributes of a sector has to be adjusted to address Islamic values-driven needs. As an example, in finance, the role of interest has to be changed, which implies the core product attributes have to be changed. In foods, slaughtered animal products and any ingredients derived from such animals have to be halal. Even more remote industries may have to consider the implication of using haram materials, e.g., products made using pig leather will most likely be rejected by many Muslim consumers.
- The seven core sectors are structurally affected. Other sectors such as education and philanthropy may be affect but will not be covered in this chapter.

SIZE AND CLASSIFICATION OF KEY HALAL MARKET ECONOMY SECTORS GLOBALLY

There are six-core halal market economy sectors beyond Islamic finance (Table 5.1) that can be further characterized. The most important for this book is the food market. However, because there can be some overlap technically, the pharmaceutical and cosmetic markets will also be reviewed.

Food market: This represents 17% of global food and non-alcoholic beverage expenditures and is expected to reach $1.6 trillion by 2019. The top countries with the largest Muslim food consumption are Indonesia ($158 billion), Turkey ($109 billion), Pakistan ($100 billion), Iran ($59 billion), and Egypt ($76 billion) based on 2014 data (Grim and Karim, 2011; Dinar Standard Estimates, 2015) (see Figure 5.1).

Pharmaceutical market: The Muslim market represents 6.7% of global expenditure in 2014 and is expected to reach $106 billion by 2020. Top countries for Muslim consumers' pharmaceutical expenditures are Turkey ($9 billion),

Saudi Arabia ($6.0 billion), the U.S. ($6 billion), Indonesia ($5 billion), Algeria ($4 billion), Russia ($3 billion), and Iran ($3 billion) (Grim and Karim, 2011; DinarStandard Estimate, 2015) (see Figure 5.2).

Cosmetics and personal care market: This sector is expected to reach $80 billion by 2020. Countries with the largest Muslim cosmetics expenditures are United Arab Emirates ($5 billion), Turkey ($4 billion), India ($4 billion), Russia ($3 billion), Indonesia ($3 billion), and Malaysia ($2.6 billion) (Grim and Karim, 2011; DinarStandard Estimate, 2015) (see Figure 5.3).

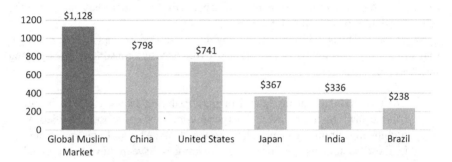

FIGURE 5.1 Top Muslim food and beverage markets (US$ billions, 2014 est.).

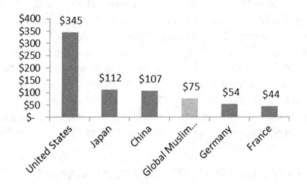

FIGURE 5.2 Muslim markets for pharmaceuticals (US$ billions, 2014 est.).

FIGURE 5.3 Muslim market for personal care/cosmetics globally (US$ billions, 2014 est.).

HALAL MARKET ECONOMY KEY DRIVERS

The halal market economy growth and prominence is being driven by four major Islamic market-based drivers and four major global environment-based drivers. The key Islamic market-based drivers include: large, young, and fast-growing Muslim demographic; their growing economies; Islamic ethos/values increasingly driving lifestyle and business practices; and the growing Intra-OIC trade. Global environment-based drivers include: participation of global multinationals; developed economies seeking growth markets; growing global focus on business ethics and social responsibility; and finally the global revolution in communication technology. Together, these key drivers are shaping the growth of the halal market economy sectors. For the halal market economy sectors, specific Islamic injunctions/values shape how the products and services have to be prepared for Muslim consumers (Figure 5.4).

The four major Islamic market-based drivers:

- *Attractive demographic*: The estimated 1.7 billion Muslim population is also a young population relative to any other large population globally. The median age in Muslim-majority countries is expected to be 30 by 2030. As a comparison, the median age in North America, Europe, and other more-developed regions is projected to be 44 in 2030. By 2030, 29% of the global young population (15–29) is projected to be Muslim.
- *Economic growth*: Muslims live in many global emerging markets such as Indonesia, Saudi Arabia, UAE, and Turkey. The 57 mostly Muslim-majority member countries of the OIC are expected to average a 5.4% annual GDP

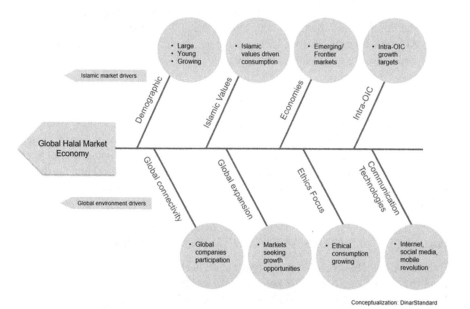

FIGURE 5.4 Key drivers for the growing halal market economy.

growth compared with the global GDP average growth of 3.6% (2015–2019, based on IMF projections).
- *Islamic values-driven consumption*: According to a 2012 study by The Pew Forum, globally 87% of Muslims consider religion "very important" and 93% fast during the month of Ramadan. Comparatively, in Europe, less than 30% rated religion as very important, and in the U.S., 56% rate religion as "very important" in their lives.
- *Intra-OIC trade growth*: The 57 OIC members are working to increase trade among their countries which stood at 18% in 2012. Comparatively, intra-European Union trade reached almost 60% in 2014 (Islamic Banker.com, 2014).

The four major global environment-based drivers:

- *Global multinationals participation*: Global companies from major banks to major global companies are not only participating but leading in the development of the halal market economy given their large scale and stature.
- *Global markets seeking growth opportunities*: Developed economies are seeking growth opportunities. Asia, in particular, has been a focus for growth with many of the OIC countries (which are already mostly in Asia and Africa) as target markets.
- *Growing ethical consumption*: The global driver that connects with the halal market economy's ethical base is the growing global focus on business ethics and social responsibility.
- *Communication technology*: Global communication technologies (social media, mobile, broadband, and others) are revolutionizing lives. The halal market economy lifestyle sectors have been able to gain wide global reach and exposure due to these developments.

ISLAMIC VALUES AND SECTOR IMPACT

Islam is seen by Muslims as a "way of life"; its guidance and values touch upon all aspects of a Muslim's life, including their consumption behavior. Many of these previously discussed values have a universal appeal, and thus many products and services do not have to be exclusively positioned for Muslims. An example is that the majority of Islamic finance customers in Malaysia are non-Muslims attracted by its risk-sharing, ethical, and non-interest-based financing models.

HALAL FOOD VALUES AND ADOPTION SPECTRUM

Halal food and beverage consumption needs derive their foundational principles from the Quran (the holy book believed by Muslims to be the divine words of one God) and the Hadiths (life and sayings of Prophet Mohammad, PBUH). While the focus of the industry today is on what is "Halal" (lawful), much of the guidance is also toward "Tayyab" (pure).

There is a wide diversity in awareness and adoption of the halal concept within the core customer base of Muslims globally. In Muslim-majority countries, most consumers will assume everything is credibly halal while other non-Muslim-majority markets will seek halal verification.

Halal cosmetics and pharmaceutical values and adoption spectrum: Because of the obligation on Muslims to consume what is halal, consumption of pharmaceutical and cosmetic products also falls under the same injunctions as foods.

As narrated by Abu Darda, a companion of the Prophet Muhammad (PBUH), "The Prophet said Allah has sent down both the disease and the cure, and He has appointed a cure for every disease, so treat yourselves medically, but use nothing unlawful."

There is a wide diversity in awareness and adoption of halal cosmetics and pharmaceuticals within the core Muslim customer base globally.

HALAL MARKET ECONOMY INDUSTRY PARTICIPANTS AND TRENDS

The halal market economy sectors operate within the global supply chain of each of these industries.

The extent of the halal process' impact within the value chain varies industry by industry. It is also important to identify key players across different steps of the value chain to better unlock opportunities and understand industry needs.

HALAL FOOD: FROM "FARM TO FORK"

The halal food market is part of the overall global food and agriculture industry, which is one of the largest industries in the world. Broadly, the global value chain covers core suppliers, manufacturing technology, products/services, processed food manufacturers, logistics, and distribution channels. There is also a wider ecosystem of industry players covering its R&D, regulations, compliance, marketing services, and other requirements.

The core of the halal market requirements relates to meat products' requirements for Islamic-law-based animal slaughtering. In addition, halal food requirements extend to all food ingredients as covered in this book. There are now some potentially acceptable technological solutions to test and verify halal ingredients across the food industry value chain. Meanwhile, the logistics of food are also increasingly incorporating halal integrity during transportation.

Standards for requirements, regulations, and monitoring are being managed by various standards setting bodies and halal certifying companies. In most Muslim-majority countries, these requirements are being managed directly by government agencies responsible for import regulations.

A broader halal market dynamic relates to its wider humane ethos relating to fair-treatment of animals, organic/pure (tayyab) foods and environmental friendliness. While this aspect has not taken hold widely, it is a growing trend in select markets as well as a potential growth opportunity.

Most of the food and beverage industry value chain is impacted by the halal food certification process and hence the value-chain-wide potential of the opportunity.

These implications throughout the value chain, along with select players as examples, are described in Table 5.2.

KEY CHALLENGES/TRENDS

Difficulty in obtaining religious-compliant funding is another key challenge. Companies wishing to scale-up or to vertically integrate their supply chain face challenges in obtaining religious-compliant funding.

Leading standardization bodies are taking steps to harmonize halal standards within their region. If these steps are successful within the next few years, although there will not be a unified global standard, at least there will be a limited number of regional standards. Some concerns have been expressed about such activities being carried out under non-Muslim leadership. There are also concerns that these standards may not fully recognize the diversity of Muslim requirements within the geographical area supposedly being covered by the standards.

The absence of any viable international schemes to accredit Halal Certification Bodies (HCBs) has long been a problem for the halal industry. The majority of halal food is being produced in non-Muslim-majority countries and is certified by independent

TABLE 5.2
The Food Industry

Food Industry Segment	Halal Process Impact	Main Companies
Core suppliers	Core feeding, slaughtering, and handling of animals and animal products that adhere to Islamic regulations with halal certification	BRF (Brazil); Allanasons (India); American Foods Group (US)
Technology	Specialized technologies to facilitate slaughtering of animals and related management technologies	
Food processing	Halal certification of food supplies and all ingredients used in producing the food	Nestlé (Switzerland); Al Islami (UAE); Saffron Road (US); Tahira Foods (UK)
Channels	Retailers focus on creating space and positioning of halal food products. This is especially relevant in markets where Muslims are not a majority	TESCO (UK); Carrefour (France); BIM (Turkey); Marrybrown Sdn Bhd (Malaysia)
Logistics	An important aspect of ensuring the purity of the halal concept is ensuring that the global distribution of halal food is kept pure and does not come in contact with non-halal items	Port of Rotterdam (Netherlands)
Ecosystem	Training, R&D, marketing, financial services, regulations, and compliance are all needed to address "halal" needs	SMIIC (OIC/Turkey); International Halal Integrity Alliance (Malaysia); JAKIM (Malaysia); IFANCA (US); ESMA (UAE)

HCBs that operate with little regulatory oversight, although market pressures and cross-acceptability by other HCBs serve as relatively successful gate-keepers. The challenge for manufacturers is to determine which standard will actually provide market access, and in too many cases, multiple certificates are necessary for exporters because of the different standards in different countries. Despite the uncertainties in the regulatory framework, the halal food sector remains vibrant, and is fast becoming an arena of innovation. A new generation of entrepreneurs are driving change in all aspects, from product development to packaging, marketing, and use of social media.

PHARMACEUTICAL AND COSMETICS SECTORS

Broadly, the global value chain of the pharmaceutical industry covers core suppliers, R&D/testing, technology, manufacturing, and distribution channels. There is also a wider ecosystem of industry players covering its regulations, compliance, training services, and others. For established companies in the pharmaceutical industry, a strict adherence to halal principles may mean a total revamping of their value chains. This is why some firms have been opposing regulations for a stricter halal law, such as the case in Indonesia (*The Jakarta Globe*, 2013). Industry impacts for pharmaceuticals are shown in Table 5.3 and for cosmetics in Table 5.4.

KEY CHALLENGES/TRENDS

Pharmaceutical companies have been challenged to develop halal ingredient substitutes as well as ensure the integrity of the full process due to the time required to develop new pharmaceutical products.

TABLE 5.3
The Pharmaceutical Industry

Pharmaceutical Segment	Pharmaceutical Impact	Main Companies
Core suppliers	Use of halal raw material supplies	
Product manufacturing	Ensuring compliance with halal rules during manufacturing	CCM Berhad (Malaysia); Kalbe Farma (Indonesia); Noor Vitamins (U.S.); Julphar (UAE)
Distribution channels	Reaching appropriate markets and supplying demand	Noor Vitamins (U.S.); Julphar (UAE)
Financial services	Supporting Islamic-compliant financing	
Logistics	Transportation and freight	
Marketing	Marketing campaigns that consider unique Muslim values	
Compliance, training	Ensuring compliance and enforcement of various relevant Islamic laws	

TABLE 5.4
The Cosmetic Industry

Cosmetic Segment	"Muslim" Impact	Main Companies
Core suppliers	Use of halal raw material supplies	
Product manufacturing	Ensuring compliance with halal rules during manufacturing	Wardah (Indonesia); Ivy Beauty Corporation (Malaysia); Pure Halal Beauty (UK); OnePure (UAE)
Manufacturing of preservatives	Reaching appropriate markets and supplying demand	
Financial services	Supporting Islamic-compliant financing	
Logistics	Transportation and freight	
Marketing	Marketing campaigns that consider unique Muslim values	
Compliance, training	Ensuring compliance and enforcement of various relevant Islamic laws	

The cosmetics market, or more broadly the personal care products market, is primarily concerned about avoiding non-halal or haram ingredients. Awareness among consumers is growing. Conversely, in many Muslim-majority markets, there is less awareness of this need and consumers mostly rely on the government. Meanwhile, most governments have yet to have strong strategies/regulations in these key areas.

THE FUTURE OF THE GLOBAL HALAL MARKET ECONOMY

The demographic and socio-economic trends among the global Muslim communities, highlighted earlier, point to a continuing strengthening of the global halal market. In addition, given the global dependency of the halal market value chain, the private and governmental agencies from around the world are engaged in building a strong momentum. The Islamic economies are getting strong government support for harmonizing halal food regulations, which should further strengthen its growth. The most exciting opportunity, however, is the convergence of the halal market economy with the global ethical economy. Halal foods naturally market into a wider global ethical consumer market.

NOTE

1. This chapter's research and insights are from DinarStandard's research and advisory work that has focused on global Islamic economies since 2005. An annual report series captures these insights in much detail: State of the Global Islamic Economy Report, Produced by Thomson Reuters, Dubai Islamic Economy Development Center in collaboration with DinarStandard.

REFERENCES

Al-Harran, S. and Low, K. C. (2010). Marketing of halal products: The way forward. *The Halal Journal*, pp. 44–46.
Grim, B. J. and Karim, M. S. (2011). *The Future of the Global Muslim Population: Projections for 2010–2030*. Washington, DC: Pew Research Center's Forum on Religion and Public Life.
Islamic Banker.com. (2014). Available at https://islamicbanker.com/articles/intra-oic-trade-lost-decades. Accessed on February 28, 2018.
The Jakarta Globe. (2013). Halal law "will destroy" Indonesia's pharmaceutical industry: Bio Farma. *The Jakarta Globe*. Available at: www.thejakartaglobe.com/archive/Halal-law-will-destroy-indonesias-pharmaceutical-industry-bio-farma/. Accessed on May 17, 2018.
Muhamad, N. (2008). Muslim consumers' motivation towards Islam and their cognitive processing of performing taboo behaviors. Doctoral dissertation, University of Western Australia, Perth, Australia.
Rehman, A. U. and Shahbaz Shabbir, M. (2010). The relationship between religiosity and new product adoption. *Journal of Islamic Marketing*, 1(1), 63–69.
Sack, D. (2000). *Whitebread Protestants: Food and Religion in American Culture*. New York, NY: St. Martin's Press.
The American Muslim Market Study 2014–15. New York, NY: Dinar Standard. http://www.dinarstandard.com/american-market-2014/. Accessed on June 23, 2018.
Thomson Reuters. (2014). State of the global Islamic economy 2014–2015 Report. In collaboration with *Dinar Standard*. Dubai: Thomson Reuter in Collaboration with Dinar Standard.

6 Animal Welfare

Kristin Pufpaff, Mian N. Riaz, and Munir M. Chaudry

CONTENTS

The History of Animal Welfare in Islam 73
The Science of Animal Welfare 75
Contemporary Issues in Animal Welfare 76
Housing 77
Feeding 78
Breeding 79
Biosecurity 80
Transport 81
Slaughter Handling 82
Conclusions 83
References 84

THE HISTORY OF ANIMAL WELFARE IN ISLAM

Animal welfare is by no means a new issue in the production of halal foods. The welfare of animals has been a fixture of the Islamic view on the raising and slaughter of animals since the time of Prophet Muhammad (PBUH). In recent years, the slaughter of animals in keeping with religious principle has become a contentious point of political and scientific debate. This debate stems largely from a misunderstanding outside of the halal slaughter industry as to the goals of halal slaughter and a lack of vigilance within the industry to strive for better animal welfare. There are many verses in the Quran and teachings in the Hadiths that establish the importance of animal welfare in a true Islamic system of agricultural production. It is important when looking at animal welfare from an Islamic perspective not to focus just on the end of an animal's life but to consider the whole life of the animal up to and including its demise (Department of Standards Malaysia, 2009).

This first section will use a few religious quotes to establish the importance of animal welfare in Islam. The overarching themes within early Islamic teachings as they apply to animal welfare are twofold: Animals are individuals that experience life in a way parallel to those of humans, and thus deserve parallel consideration. However, humans were given a greater gift of cognitive understanding than animals, which gives us certain rights and responsibilities over animals and most particularly over the ones that contribute to our food supply.

The Quran established in the following verse the moral importance of animals.

> There is not an animal (that lives) on the earth, nor a being that flies on its wings, but (forms part of) communities like you. Nothing have we omitted from the Book, and they (all) shall be gathered to their Lord in the end.
>
> **Sura 6 Ayat 38**

By stating that all animals are part of communities and that all will be equally gathered to God at the time of judgment, the verse suggests that any act for or against an animal is equivalent to an act for or against a fellow human. If this line of thinking were left to continue unchecked it would lead to a vegetarian lifestyle. As a result, the Quran gave specific instructions as to what would make an animal unacceptable to eat.

Say,

> I do not find within that which was revealed to me [anything] forbidden to one who would eat it unless it be a dead animal or blood spilled out or the flesh of swine – for indeed, it is impure – or it be [that slaughtered in] disobedience, dedicated to other than Allah. But whoever is forced [by necessity], neither desiring [it] nor transgressing [its limit], then indeed, your Lord is Forgiving and Merciful.
>
> **Sura 6 Ayat 145**

To these simple instructions relating to the moral value of animals and the allowance for humans to eat them, one can add the teachings of the Hadiths. Below are three Hadiths that very clearly show that any form of cruelty, be they either from neglect, misapplication of a husbandry procedure, or aversive training techniques are counter to the teachings of Islam.

> Jabir, may God be please with him, reported: The Prophet (PBUH) passed by an animal that had been branded upon its face. The Prophet (PBUH) said, "Allah has cursed whoever does this. Let not one of you mark the face or strike it."
>
> **Sahih Muslim**

> Ibn Umar, may God be pleased with them, reported: The Messenger of Allah (PBUH) said, "A woman was punished because of a cat she had imprisoned until it died; thus, she entered Hellfire because of it. She did not give it food or water while it was imprisoned, neither did she set it free to eat from the vermin of the earth."
>
> **Sahih Muslim**

> Aisha, may God be pleased with her, reported: I was upon a camel which was misbehaving so I began to strike it, then the Messenger of God (PBUH) said to me, "You must be gentle, for verily, gentleness is not in anything except that it beautifies it, and it is not removed from anything except that it disgraces it."
>
> **Sahih Muslim**

To the above teachings about the treatment of animals during their lifetime one can add the following Hadith on slaughter to complete the picture of how animals should be treated by humans. These teachings make it clear that slaughter may only happen for a valid reason and, even then, it should not be taken lightly.

On the authority of Abu Ya'laaShaddaad bin Aws, may God be pleased with him, reported that the Messenger of God (PBUH) said:

> Verily God has prescribed Ihsaan (proficiency, perfection) in all things. So if you kill then kill well; and if you slaughter, then slaughter well. Let each one of you sharpen his blade and let him spare suffering to the animal he slaughters.
>
> **Sahih Muslim**

> Sharid reported: I heard the Messenger of Allah (PBUH) say, "Whoever kills so much as a sparrow for no reason will have it pleading to Allah on the Day of Resurrection, saying: O Lord, so-and-so killed me for no reason, and he did not kill me for any beneficial purpose."
>
> **Sunan An-Nasa'i**

This section has illustrated that animal welfare was an early and prominent concern in Islam. The question remains: How can a moral system from the time of Muhammad (PBUH) be acted upon in a modern agricultural system? Even this question has a simple answer in Islamic teaching.

> From Abi Abdillah(as), he said: "Seeking knowledge is an obligation in all cases (under all circumstances)."
>
> **Bihar Anwar**

In the spirit of seeking knowledge, the next section has a brief summary of recent developments in animal welfare from a modern scientific perspective.

THE SCIENCE OF ANIMAL WELFARE

The foundation of the modern science of animal welfare was not one event or dependent on one person but the result of the confluence of empathy and scientific reasoning. The most defining moments in the modern science of animal welfare was the result of public pressure on the industry to make changes (Kannan et al., 2003). In 1965, a committee was formed to address animal welfare in the United Kingdom. The Bramble Committee Report listed certain freedoms that all animals should be given. These freedoms have been solidified in the modern concept of the Five Freedoms, which are listed below:

Freedom from thirst and hunger: Have ready access to fresh water and a diet to maintain full health and vigor.
Freedom from discomfort: Be provided an appropriate environment including shelter and a comfortable resting area.
Freedom from pain, injury, and disease: By prevention or rapid diagnosis and treatment.

Freedom to express most normal behaviors: By providing sufficient space, proper facilities, and the company of the animal's own kind.
Freedom from fear and distress: By ensuring conditions and treatment which avoid mental suffering.

These freedoms set the stage for how scientific researchers have approached animal welfare. One of the most difficult aspects of ensuring good animal welfare is determining the true needs and preferences of animals without being anthropomorphic. An example of the conflict between science and human perception is the temperature at which dairy cattle prefer to be kept. Based on human preference, there was a time where dairy cattle were kept in poorly ventilated warm barns because from a human perspective these were the most comfortable, but science has shown that cattle feel the effects of heat stress at much lower temperatures than humans leading to their being most comfortable between 55°F and 65°F. Because it is difficult to measure an animal's mental state, medical and production metrics have served as the primary indicators of "good" animal welfare. However, research has shown that while bad animal welfare often negatively impacts health and production, these effects can be masked by intensive breeding programs and medical advances. Recent studies have helped to determine an animal's preferences and mental state by using carefully controlled preference test studies where an animal is asked to work to achieve a desired goal and how strongly they are willing to work for this goal is taken as a measure of their need, that is, the harder they are willing to work for something, the more important it is to the ANIMAL. Careful behavioral observations of abnormal behaviors and of animal interactions are also being used to evaluate the animal welfare status of a particular set of conditions (Nakyinsige et al., 2012).

Recent work has begun to conceptualize the idea of "coping" as a measure of animal welfare. At very low stress, an animal may actually be bored and show unacceptable behaviors, often referred to as stereotypies. So work is focused on "environmental enrichment." In the general "coping" range, which includes some stress, the animal thrives. Real life is not stress-free! When the stress becomes excessive, the animal becomes subject to disease and even death (Micera et al., 2010).

The above scientific findings support the Islamic philosophy that animals are a part of communities that are parallel to, but not always the same as, human communities. Science has shown that animals are able to feel pain and to remember positive and negative interactions with each other, with humans, and with different environments (Casoli et al., 2005). Modern animal welfare has accepted a set of principles that are easily embraced by an Islamic ethos.

CONTEMPORARY ISSUES IN ANIMAL WELFARE

The modern halal animal production systems cannot rely entirely on historic Islamic teachings to explain what is right and wrong in a modern setting. As the number of people and domesticated animals in the world has increased so have the pressures on agricultural systems. It seems clear from the Islamic teaching discussed previously that any halal production system should take into account the mental and physical state of the animals in that system. Many of the teachings of Islam have been

Animal Welfare

reinforced by science so when there is no clear Islamic ruling on how to resolve an issue, a logical way of dealing with the issue would seem to be to use scientific reasoning and information guided by an Islamic moral compass.

The following sections will deal with a few of the major animal welfare issues that modern agriculture is facing. Not all of the difficulties discussed in the following sections have clear solutions. The mental state of animals is not entirely understood and, thus, its impact on the animal's welfare needs further research. And the need to provide for global food security along with other unintentional consequences of such changes must be carefully weighed against the potential benefit of any change based on welfare considerations. This is not to imply that humans should not make any sacrifices for animal welfare, it is only meant to emphasize that a clear benefit to animals should be demonstrated before changes are made, including addressing other critical issues such as global food security, environmental sustainability, human health and safety, worker health and safety, and the impact on price/affordability. These are all issues that need to be balanced. Because Muslims expect good animal welfare, it is important to discuss these issues so that animal products for the Muslim community are not only halal, but also meet the full range of tayyab and animal welfare standards.

HOUSING

In the time of Prophet Mohammad (PBUH), the primary housing concerns of animal caregivers was how to provide safety from predators for their stock. This was, and often is still, done using dogs and other guard animals to supplement humane watchfulness. Humans were not always successful in their efforts to keep their stock safe but animal welfare is straightforward when it comes to predation. Caregivers either succeed in preventing predation, or fail and animals get injured or killed. The only welfare concern that cannot be immediately addressed by traditional caregivers is if the stock fears predation. Fear is a negative welfare state that can best be prevented by preventing fear inducing circumstances. The animal that is attacked by a predator is in an obvious negative welfare state, but all the animals that can observe the attack may also be in a negative welfare state of fear if they are able to process the information. Thus, outdoor housing systems such as free-range, pasture, and rangeland agricultural systems continue to face this challenge. In the western United States, flocks of sheep historically suffered up to 10% annual loss due to predation, chiefly by coyotes. This is an example of a traditional animal rearing system, range grazing, which can lead to a negative welfare state due to predation and the fear of predation, yet is often the preferred method for raising animals according to many consumers. This percentage has been reduced greatly by the use of guard dogs and other guard animals such as llamas, but it remains an animal welfare concern.

Once animals are brought into confined systems, either fenced pastures or barns, the risk of predation is greatly reduced, as is presumably the associated fear, but other concerns arise. Decreases in space can lead to increased disease, decreased natural movement, and the challenges of adapting to new types of flooring among other issues. The need to have more animals to feed the world's population means that animals began to be kept in closer proximity to their owners through the use of pastures. When animals are confined and can no longer find natural shelter, manmade shelters must be provided for them. These can be temperature controlled for the animals' comfort regardless of what is happening outdoors. But this all has led to an increase in animal interactions that can facilitate the spread of disease. Better methods of disease control such as, but not limited to, modern antibiotics have eliminated many of these problems but others still remain. With the recent emphasis on the reduction of antibiotic use, animal agriculture is looking at greater use of pre-biotics and pro-biotics along with other dietary interventions. Preliminary results suggest that some of these interventions can be successfully used to decrease antibiotic use. On the other hand, the total absence of antibiotic use can lead to situations where a farmer fears the loss of the economic value of his animals and if he does not use antibiotics quickly or completely enough, this leads to undue animal pain and possible deaths. The use of production systems that mandate the absence of antibiotics can therefore lead to poorer animal welfare.

One of the ways that barns have changed to allow for better disease control is to replace dirt floors with other more easily cleaned floors. Often these floors can be slippery, which can cause injury or are unforgiving in ways that can cause lameness. However, with a small investment, the farmer can address this problem. Increased pressures on land use have led to even higher stocking densities in pastures and confinement systems. Some confinement systems, such as those for chickens, have reached stocking densities where there is no longer enough space for animals to perform their full range of movements, but the importance of each of these movements has not always been ascertained. The question that remains for animal agriculture is: How much space and what kind of space do animals need to enjoy a healthy mental state and physical well-being, that is, an ability to cope on an on-going basis?

Making changes in stocking density is not something that can happen overnight without gravely endangering the food supply. Also, given that animal welfare is a relatively young discipline in comparison to the domestication of animals, what steps need to be taken to correct housing problems are not always clear. Many of the housing choices that have led to confinement agriculture were made with the intention of ensuring animal health and safety. Further changes in housing should be made based on scientific research as to what is best for the animals with careful attention to any unintended side effects.

FEEDING

The feeding of animals can be a welfare concern in many different ways. Animals can be given too much food, not enough food, or the wrong kind of food. The welfare issues associated with too much food tend to be more noticeable and show animals where human anthropomorphic behavior negatively impacting the animals is more common, and thus, are rarely a concern in a production setting. Too little food can be a problem for livestock in two ways. First and simply, too little volume of food is a welfare concern. The second way an animal can be starved is less obvious. If a grazing animal is provided with very low quality forage, it is possible that they are prevented from eating enough digestible nutrition even when overfilling their gut. To prevent these problems, animals must be provided with sufficient volume and quality of forage to provide for their nutritional needs or forage should be supplemented during times when the quality is not nutritionally sufficient. It is important that any hay or silage that is produced from forage must be harvested at such a time so that the resulting product provides proper animal nutrition (Gregory et al., 2010).

Another way that feed may become an animal welfare problem is in feeding feeds with the wrong nutritional balance. This can lead to animals with skeletal abnormalities if the wrong balance of minerals is given, growth rates that are not structurally supported if too many carbohydrates are given, or animals with too many hazardous metabolic byproducts in their system when they are fed the wrong foods to maintain high levels of production. This last problem is the most difficult to control in the agriculture industry, because it can be the result of pushing for the best possible production from an animal. Another issue is that feeds may become spoiled. Some of these spoiled feeds are simply devoid of the proper nutrients but in other cases they may be toxic to the animals.

Finding solutions to animal welfare problems associated with feeding is one of the simplest areas of welfare to address but also one of the most complex. It is fairly simple to analyze feed and formulate diets based on an animal's need. However, how is an animal to be fed when the choices are to feed too little for its production requirements or to feed it high calorie feed that could adversely affect its metabolic function? The answer to this question lies with finding a solution to another animal welfare problem, breeding solely for production traits.

An aspect of importance to Muslims is the use of feed that contains najis (filth). This is discussed separately in more detail in Chapter 21.

BREEDING

Farm animal breeding has focused on production traits for many years. The modern dairy cow is an excellent example of both the negatives and positives that can be achieved via breeding programs. In 1910, the average United States dairy cow produced 1320 kg (2910 lb) of milk per year, by 2011 that number had climbed to 10,600 kg (23,300 lb) for the Holstein breed. By any definition, this is a major accomplishment of breeding, but that is only part of the story. The higher production rates have come at the cost of shorter productive life spans, lower reproductive efficiency, and much higher nutrient requirements. As the negative effects of selective breeding tend to become known more slowly and are often much harder to measure than production traits, this can receive too little focus within the agricultural industry. Genetic traits that cause poor animal welfare often get ignored because there is little that the management of a facility can do to improve welfare and the feedback to the breeders is inefficient. An example of this is some broiler chickens that have difficulty moving freely close to the time of slaughter. There is little that a farmer can do to prevent a problem that is caused by extremely rapid and profitable growth rates for these birds that the farmer obtains as one-day-old chicks from the breeders. Production farms and breeding farms must communicate and work better together to breed birds that are physically robust up through the time of slaughter. The use of modern computers and big data sets has allowed the breeding industry, under pressure, to improve animal welfare by beginning to address some of these problems.

The most logical solution to animal welfare problems that have been bred into production animals would be to allow for genetic production to plateau or improve slowly for the time being while addressing animal welfare problems as part of the breeding program without having to reduce production traits. This would mean selecting animals based on a "whole health" basis. This would allow production to be kept at current levels while breeding out negative traits. Some traits that should receive such attention would be skeletal abnormalities in broiler chickens, osteoporosis in laying hens, reproductive failure, lameness, and metabolic illness in dairy cattle, and parasite tolerance in small ruminants. Both good health traits and animals that are temperamentally well suited for the production system being use should be considered in breeding programs.

BIOSECURITY

The concept of biosecurity is a new concern in comparison to the history of animal welfare in Islam. But there is no question that the increased pressure on the global ecosystem and increased ease of travel has made biosecurity a significant concern for animal welfare, and for both human and animal health. The absence of a sound biosecurity system will lead to poorer animal welfare.

The prevention of disease has been one of the primary motivators of confinement agriculture. Confinement allows the persons caring for the animals to inspect them with greater frequency and for one group of animals to be kept physically separate from the diseases of other groups of animals. This is most notably the motivation for keeping chickens in enclosed barns. Chickens that are kept outside are not only at risk for predation but also at risk from diseases spread by wild birds, rodents, and insects. There is currently no practical method for preventing fecal contamination and disease transmission from wild birds to domestic fowl housed in an open air environment. Biosecurity must be considered when any husbandry changes designed to improve animal welfare are considered.

The other areas where biosecurity concerns create animal welfare concerns are in animal transport and human interactions with animals. The transport of animals for long distances allows for widespread disease transmission if proper precautions are not in place. This includes waste management during transport and bio-secure resting places for animals traveling longer distances. Animals being transported long distances should be off-loaded at rest stops to drink and eat as they often will not, or cannot, engage in those activities during transport. In some cases, the rest can be on the transport vehicle, but at other times, offloading may be necessary. Transport welfare will be touched on further in the next section.

Human interaction becomes an issue when people interact with one group of animals and then travel to visit another group. This allows humans to act as vectors for disease. Veterinarians might seem like the greatest risk for farm to farm transmission of disease but they are unlikely to spread disease over a very large area due to the geographical confines of their practice area and their extensive biosecurity education. The global concern is with our rapid and extensive air transportation system that would allow a person to travel from a country with a disease endemic to a disease free country in a matter of hours. The best way to combat this likelihood is to educate the general public about biosecurity steps they should take and encourage full honesty and cooperation with customs and border control agents, and customs forms to ensure that agricultural precautions instigated by governments are followed.

Biosecurity is becoming more complex in the global marketplace but the concern is more widely accepted then many other animal welfare concerns. The ease and high cost of failed biosecurity procedures is of concern to almost all animal agriculture stakeholders. Unfortunately, biosecurity is sometimes not considered by those well-meaning individuals who advocate for more traditional and naturalist approaches to animal husbandry and animal welfare.

TRANSPORT

Moving from one location to another is a highly stressful event in an animal's life. There is no way to make transport a stress-free event for animals but, by taking account of a few behavioral concepts, the stress of relocation can be diminished. Aspects often associated with relocation that are stressful for animals include: changes in the group members (need to re-establish social hierarchies), human interaction, medical procedures, loading into a strange new environment, the environment while in transit, the animal's instability while moving, and unloading into a new environment. Stress can be reduced by better management of relocation stressors (Micera et al., 2010).

The first step in reducing transport stress is to plan ahead. Animals should be sorted into the groups with which they will be transported at least a week in advance if possible. This allows for establishment of a social hierarchy before the other transport stressors occur. Presorting will also allow for correction of another common transport error. Often, as animals are sorted for transport, caregivers will use the opportunity to preform medical procedures such as castrations and vaccinations. This is not only a medically dangerous time to do any surgical procedure but the high stress of transport often causes the animal to have a suboptimal reaction to any vaccines that are given. Giving vaccines sufficiently in advance of transport allows the animals to develop a stronger immune response and to have active antibodies to protect against diseases that they will be at risk for at their destination prior to arrival.

Having appropriate and well maintained loading equipment is imperative to reduce loading stress. The correct equipment will allow animals to be loaded with limited human encouragement. The use of an electric prod is strongly discouraged except in extremely difficult situations. Persons involved with the transport of animals should understand the species specific guidelines for loading and transport equipment, handling, and appropriate space allowances for animals in transit. In many countries, these types of guidelines are available from industry trade groups and agricultural universities that have developed standards for the management of environmental conditions during transit. Some of these guidelines can also be found on the internet.

Once animals reach their destination, it is important that they be given the opportunity to adjust to and explore their new environment at their own pace if possible. This is often not possible in a slaughter setting but should be done in a new farm or feedlot setting. Handling of animals should be limited until they have adjusted to their new environment.

SLAUGHTER HANDLING

The handling of animals in slaughter plants has received at great deal of attention in the United States for the last 50 years or so and has undergone many important improvements. There is still a great deal of work to be done in the area of increasing the industry standard to the level of the current top preforming slaughter plants. However, there are areas of the world that still have very poor animal welfare in slaughter plants and as animal agriculture becomes more global so should the implementation of higher animal welfare standards. Working on global initiatives to improve handling equipment at slaughter plants, to increase welfare training of staff, and to implement effective auditing programs with follow-up by management to correct issues raised by the audit should be a priority for the animal agriculture industry.

There are many useful written and media documents available for those interested in improving animal welfare in slaughterhouses. From an Islamic perspective, slaughterhouse animal welfare and worker welfare should be a top priority. The previously mentioned Hadith about sparing the suffering of animals to be slaughtered sets a precedent for good animal welfare. This should be a goal that every Muslim slaughter person is aware of and constantly working toward. It becomes even more important when slaughter without prior intervention is being used.

There are varying scholarly opinions on what type of prior interventions, if any, should be allowed for halal slaughter, but most Muslims have a preference for meat that is not interfered with prior to traditional slaughter. What follows is an explanation of common welfare problems that can happen during traditional slaughter and some practical ways to prevent them.

The first and most important consideration when doing traditional slaughter is how the animal will be restrained. In the case of poultry manual restraint, restraint using a cone, or shackling (if done carefully without causing pain to the legs) are currently acceptable from a welfare perspective. If using manual restraint, care must be taken not to break any bones prior to slaughter. Gentle handling will prevent damage to the meat and improve animal welfare. If a cone is used to restrain the bird, care should be taken to carefully fold the wings prior to insertion into the cone to prevent wing damage. If the animal is going to be shackled prior to slaughter, it is very important that it be done gently and that the animal is hanging from both legs using shackles that are the proper size for the bird.

For small ruminants such as sheep and goats, acceptable means of restraint include: manual restraint without casting, a squeeze box restraint, a V restraint, and a double rail restraint. Casting or hanging an animal while it is conscience is poor animal welfare and more dangerous for the humans slaughtering the animal. Hanging a conscious sheep or goat by one or two hind legs is still a common practice in the halal industry and alternative practices should be adopted. While casting may be a necessary alternative for more mature small ruminants when proper restraint devices are not available, upright restraint boxes for small animals are not very costly to build and would greatly improve animal welfare.

When considering large ruminants such as cows, oxen, and camels, manual restraint is neither safe nor practical. The ideal form of restraint from an animal welfare context is a smoothly functioning upright restraint box with a head holder. When operating such a box, it is important that the animal not be squeezed too hard and the slaughter should happen as quickly after restraint as possible (Kadim, 2012). Inverting boxes that rotate the cow to some degree prior to slaughter are also fairly common in the halal industry. If properly designed and maintained, they can be acceptable. If these are used, the animal should be slaughtered as quickly as possible after it is rotated as ruminants are distressed by changes in orientation. The inverted restraint boxes must be designed to comfortably accommodate the animal in the inverted position and must be of the correct size for the animal (Ljungberg et al., 2007).

Hoisting conscious cattle is not an acceptable format for slaughter from an animal welfare perspective. Hoisting or dragging conscience cattle still happens in parts of the halal industry and it is a moral imperative that alternative means of slaughter are adopted as soon as possible. Some of the best advice on how to engineer welfare friendly restraint devices can be found on Temple Grandin's website (www.grandin.com/) along with some of her publications.

The next most important part of a traditional slaughter is knife selection and care. Knifes should be as sharp as possible and free of nicks. It is also very important that the right size knife be selected for the animal being slaughtered. It is advisable to have a knife that is twice as long as the neck is wide with a straight blade. Using a smaller than desirable knife can lead to the slaughter person having to make multiple cuts and possibly stabbing the animal, which is more painful to the animal and prohibited in Islamic law, and using too large a knife can lead to human fatigue, which may cause more errors (Anil et al., 2000; Limon et al., 2010; Önenç and Kaya, 2004). The halal industry does not currently have any form of mandatory training in place for its slaughter persons, which sometimes leads to poorly trained individuals using poorly sized dull knives. If traditional slaughter is going to continue to occur, this issue must be addressed. The halal slaughter industry should implement appropriate training and auditing systems to ensure that animal welfare is maintained at the highest possible level currently available, namely, best practices (Gibson et al., 2009).

A final consideration for traditional slaughter is ensuring that no further processing is done until the animal is completely dead. Animals that are still conscious should not be dragged or moved but can be hung on the slaughter line, which also may help improve bleed-out. These stipulations are very consistent with both Islamic law and the animal welfare laws that are found in many countries (Hambrecht et al., 2004).

If all of the above welfare standards are met, it is the opinion of the authors that traditional slaughter without a prior intervention can be done in a manner consistent with good animal welfare. However, slaughter without prior intervention requires that constant attention be paid to the welfare of the animals at all times including using the best equipment the right way. Auditing and training should also emphasize the respect for animals by the halal slaughter industry in keeping with the religious requirements (Apple et al., 2005; Schaefer et al., 2001).

CONCLUSIONS

Many of the animal welfare issues that face the halal animal industry are the same issues that face all of animal agriculture. The moral standards of Islam demand that Muslims and the halal meat industry take precautions to protect animal welfare.

This goal can be most effectively achieved through a partnership with science, technology, and the animal agricultural industry. The path forward for all sectors of animal agriculture should be one of good science informed by ethical and moral considerations for the animals that live in service to mankind.

REFERENCES

Anil, M. H., Raj, A. B. M., and McKinstry, J. L. (2000). Evaluation of electrical stunning in commercial rabbits: Effect on brain function. *Meat Science*, 54(3), 217–220.

Apple, J. K., Kegley, E. B., Maxwell, C. V., Rakes, L. K., Galloway, D., and Wistuba, T. J. (2005). Effects of dietary magnesium and short-duration transportation on stress response, postmortem muscle metabolism, and meat quality of finishing swine. *Journal of Animal Science*, 83(7), 1633–1645.

Casoli, C., Duranti, E., Cambiotti, F., and Avellini, P. (2005). Wild ungulate slaughtering and meat inspection. *Veterinary Research Communications*, 29(2), 89–95.

Gibson, T. J., Johnson, C. B., Murrell, J. C., Hulls, C. M., Mitchinson, S. L., Stafford, K. J., Johnstone, A. C., and Mellor, D. J. (2009). Electroencephalographic responses of halothane-anaesthetised calves to slaughter by ventral-neck incision without prior stunning. *New Zealand Veterinary Journal*, 57(2), 77–83.

Gregory, N. G., Fielding, H. R., von Wenzlawowicz, M., and Von Holleben, K. (2010). Time to collapse following slaughter without stunning in cattle. *Meat Science*, 85(1), 66–69.

Hambrecht, E., Eissen, J. J., Nooijen, R. I. J., Ducro, B. J., Smits, C. H. M., Den Hartog, L. A., and Verstegen, M. W. A. (2004). Preslaughter stress and muscle energy largely determine pork quality at two commercial processing plants. *Journal of Animal Science*, 82(5), 1401–1409.

Kadim, I. T. (2012). *Camel Meat and Meat Products*. Oxfordshire, UK: CABI.

Kannan, G., Kouakou, B., Terrill, T. H., and Gelaye, S. (2003). Endocrine, blood metabolite, and meat quality changes in goats as influenced by short-term, preslaughter stress. *Journal of Animal Science*, 81(6), 1499–1507.

Limon, G., Guitian, J., and Gregory, N. G. (2010). An evaluation of the humaneness of puntilla in cattle. *Meat Science*, 84(3), 352–355.

Ljungberg, D., Gebresenbet, G., and Aradom, S. (2007). Logistics chain of animal transport and abattoir operations. *Biosystems Engineering*, 96(2), 267–277.

Micera, E., Albrizio, M., Surdo, N. C., Moramarco, A. M., and Zarrilli, A. (2010). Stress-related hormones in horses before and after stunning by captive bolt gun. *Meat Science*, 84(4), 634–637.

MS1500, M. S. (2009). Halal food—production, preparation, handling and storage—General guideline. Cyberjaya: Department of Standards Malaysia, 1–13.

Nakyinsige, K., Che Man, Y. B., Sazili, A. Q., Zulkifli, I., and Fatimah, A. B. (2012). Halal meat: A niche product in the food market. *2nd International Conference on Economics, Trade and Development IPEDR*, Singapore, 36, 167–173.

Önenç, A. and Kaya, A. (2004). The effects of electrical stunning and percussive captive bolt stunning on meat quality of cattle processed by Turkish slaughter procedures. *Meat Science*, 66(4), 809–815.

Schaefer, A. L., Dubeski, P. L., Aalhus, J. L., and Tong, A. K. W. (2001). Role of nutrition in reducing antemortem stress and meat quality aberrations. *Journal of Animal Science*, 79(suppl_E), E91–E101.

7 The Religious Slaughter of Animals
A US Perspective on Regulations and Animal Welfare Guidelines

Joe Regenstine, Mian N. Riaz, and Munir M. Chaudry

CONTENTS

7 U.S.C.A. § 1902. Humane Methods 86
7 U.S.C.A. § 1906. Exemption of Ritual Slaughter 87
Section 5: Religious Slaughter (Kosher and Halal) 88
Design of Facilities and Slaughter Process for Religious Slaughter 92
    Handling Procedures at Slaughter Plants for Hoofstock 92
        Steps 1 through 5 92
        Step 6—Restraint 93
        Step 7—Performing the Throat Cut 96
Auditing Religious Slaughter to Improve Animal Welfare for both Kosher
    and Halal Slaughter of Cattle, Sheep, or Goats 99
The Importance of Measurement 100
References 101

The religious slaughter of animals is an issue of contention in many parts of the world, although less so in the U.S. The U.S. experience with this topic might be of some interest. However, before beginning to focus on the religious slaughter of animals, it is important to frame the broader issues of slaughter within the U.S. legal system. As an outgrowth of consumer reaction to the book "The Jungle" written by Upton Sinclair (1906), the U.S. Congress passed various laws to create new approaches to food regulation. These have evolved with time, but at this time the responsibility for most traditional red meat (mammals) and birds used for food, and liquid eggs, is under the jurisdiction of the Food Safety and Inspection Service (FSIS) of the U.S. Department of Agriculture. For most common food animals, FSIS inspection is mandated for interstate sales. For other animals (e.g., ostrich and camel), inspection is available on a fee-for-service basis but is not mandatory. Although a cabinet level department is very similar to a ministry in most other countries, these departments are much more stable and it takes an act of Congress, which is difficult to do, to change the organizational structures. The head of the FSIS reports directly to the Undersecretary of Agriculture for Food Safety. All other food products are under the jurisdiction of the Food and Drug Administration, which is buried deep in the Department of Health and Human Services.

Thus, the handling of meat (meaning in this chapter the species that are under FSIS) in the U.S. is subject to very intensive inspection (continuous for slaughter, at least once a day for processing) and a great deal more direct control (e.g., label approval) than all other food products. The cost of these inspections is borne by the government. An animal covered by FSIS regulations cannot be moved into the market place without having been inspected. (Some states have systems in place for state inspection—until very recently such slaughtered products could only be sold within the state. Under some recently developed procedures, such state meat can sometimes be sold outside of the state, i.e., by basically proving that the plant is equivalent to a directly inspected federal plant.) The other exception to federal inspection is custom slaughter—one can slaughter an animal (or have it slaughtered for you) that one owns following more limited state rules (with each state having its own rules), but then one cannot sell any of the meat or by-products. You can give them away, so that the Muslim tradition of keeping one-third for one's immediate family, one-third for the extended family/community, and one-third for zakat (charity) is possible.

Because of the continuous involvement of FSIS, the practical issues that arise with the religious slaughter of animals and how that process interacts with the government inspectors had to be negotiated. Because all labels are approved by USDA, this also means that at some level they are approving the religious slaughter of animals by putting the mandatory USDA "wholesomeness" seal on products. So FSIS does at least try as part of its label review to determine that the claim of kosher or halal has some broad validity, namely, that there is a reasonable basis for such a claim.

In the late 1950s, the U.S. government, including Congress, began to debate some of the issues that need to be addressed to assure the humane slaughter of animals. This included the religious slaughter of animals.

In 1958, a Humane Slaughter Act was debated and eventually passed and became law. It has two portions that are directly of interest to the religious slaughter of animals.

The first section is a "finding" of the U.S. government that the following methods of slaughter are humane:

7 U.S.C.A. § 1902. HUMANE METHODS

No method of slaughtering or handling in connection with slaughtering shall be deemed to comply with the public policy of the United States unless it is humane. Either of the following two methods of slaughtering and handling are hereby *found to be humane:*

    a. in the case of cattle, calves, horses, mules, sheep, swine, and other livestock, all animals [not poultry] are rendered insensible to pain [unconscious] by a single blow or gunshot or an electrical, chemical or other means that is *rapid and effective,* before being shackled, hoisted, thrown, cast, or cut; or

b. by slaughtering *in accordance with* the ritual requirements of the Jewish faith or any other religious faith that prescribes a method of slaughter whereby the animal suffers loss of consciousness by anemia of the brain caused by the simultaneous and instantaneous severance of the carotid arteries with a sharp instrument and handling in connection with such slaughtering.

Pub. L. 85-765, § 2, Aug. 27, 1958, 72 Stat. 862;
Pub. L. 95-445, § 5(a), Oct. 10, 1978, 92 Stat. 1069.

The second section of the bill provides for an "exemption" for the religious slaughter of animals. The exemption covers both the pre-slaughter handling of the animal and the actual slaughter.

7 U.S.C.A. § 1906. EXEMPTION OF RITUAL SLAUGHTER

Nothing in this chapter shall be construed to prohibit, abridge, or in any way hinder the religious freedom of any person or group. Not with standing any other provision of this chapter, in order to protect freedom of religion, *ritual slaughter and the handling or other preparation of livestock for ritual slaughter are exempted from the terms of this chapter.* For the purposes of this section the term "ritual slaughter" means slaughter in accordance with section 1902(b) of this title.

Pub. L. 85-765, § 6, Aug. 27, 1958, 72 Stat. 864.

There are two important issues to discuss. The first is that the two sections seem to contradict each other. Is the religious slaughter of animals humane or is it exempt from such considerations? Apparently, the history of the legislation, which was supervised by Senator Hubert Humphrey from Minnesota, who later was a vice-president of the U.S. under President Jimmy Carter, was responsible for the bill. A historian, Roger Horowitz, in a book that was recently published (*Kosher USA*) has reviewed the history of this bill and shows that this ambiguity was intentional, namely, that to get acceptance by various stakeholders, these conflicting statements were necessary. Most people when they discuss this bill in terms of the religious slaughter of animals tend to focus on the one of these two approaches that suits their rhetorical need!

The other issue is that of the exemption for pre-slaughter handling of the animal for religious slaughter. This has over time become better defined and essentially creates a "bubble" around the immediate pre-slaughter handling of animals. So the bulk of the handling of the animal is covered by any requirements imposed on all slaughterhouses. But this still exempts the immediate handling prior to the religious slaughter of animals. It is the authors' contention that this exemption with 20-20 hindsight is unfortunate. Progress in improving animal handling for religious slaughter would probably be much further along without this exemption. And as long as handling does not interfere with the religious slaughter of the animal, there is no reason to exempt it from meeting good animal handling practices.

This might also be an appropriate point to discuss the issue of the wording used when describing the religious slaughter of animals. Having done polling of students in classes on animal welfare, and on kosher and halal food regulations, it is clear that the "religious slaughter of animals" is much more acceptable to the (American) students than the term "ritual slaughter of animals," which is often used when a negative reaction is desired. Wording does matter. For more on this issue with respect to other aspects of discussing the slaughter of animals, please see the article by Regenstein (2012) entitled "The Politics of Religious Slaughter—How Science Can Be Misused" (Reciprocal Meat Conference Proceedings).

FSIS, as previously mentioned, reviews the labels of all products under its jurisdiction. Thus, for the religious slaughter of animals, the claim of "halal" on a finished product will fall into two categories. If a generic claim of halal is made, FSIS will permit it without label approval. However, if a claim of halal is made with specific reference to a certifying agency, then FSIS will require that the food company provide documentation to substantiate such claims. The FSIS given U.S. tradition does not try to judge the "quality" of the certification provided. But consumers need to recognize that the generic "halal" is not authenticated while a specific agency certification is at least validated by FSIS.

In recent years, animal welfare has become a more important consideration in the animal foods industry. The issues of animal slaughter have been addressed for many years by the American Meat Institute, which recently became the North American Meat Institute (NAMI). Working with the pre-eminent animal welfare scientist in the U.S., Dr. Temple Grandin, professor of Animal Science at Colorado State University and owner of Grandin Livestock Handling, NAMI has developed a set of animal welfare guidelines. Based on Dr. Grandin's work, the key to a good set of guidelines is to have measurable parameters that are routinely monitored.

It should be noted that Dr. Grandin is possibly the highest functioning autistic person in the U.S. She has used her autism to see the world the way animals see the world. Her website (www.grandin.com) includes a section on the religious slaughter of animals and is highly recommended.

The NAMI guidelines include a section on the religious slaughter of animals. These are found in the following section (courtesy of the North American Meat Association with their permission).

SECTION 5: RELIGIOUS SLAUGHTER (KOSHER AND HALAL)

Cattle, calves, sheep or other animals that are ritually slaughtered without prior stunning should be restrained in a comfortable upright position. For both humane and safety reasons, plants should install modern upright restraining equipment whenever possible. Shackling and hoisting, shackling and dragging, trip floor boxes and leg-clamping boxes should never be used. In a very limited number of glatt kosher plants in the United States and more commonly in South America and Europe, restrainers that position animals on their backs are used. For information about these systems and evaluating animal welfare, refer to www.grandin.com (Ritual Slaughter Section). The throat cut should be made immediately after the head is restrained (within 10 seconds). Small animals such as sheep and goats can be held manually by a person during ritual slaughter. Plants that conduct ritual slaughter should use the same scoring procedures except for stunning scoring, which should be omitted in plants that conduct ritual slaughter without stunning.

Cattle vocalization percentages should be five percent or less of the cattle in the crowd pen, lead up chute and restraint device. A slightly higher vocalization percentage is acceptable because the animal must be held longer in the restraint device compared to conventional slaughter. A five percent or less total vocalization score can be reasonably achieved. Scoring criteria for electric prod use and slipping on the floor should be the same as for conventional slaughter.

Animals must be completely insensible before any other slaughter procedure is performed (shackling, hoisting, cutting, etc.). If the animal does not become insensible within 60 seconds, it should be stunned with a captive bolt gun or other apparatus and designated as non-Kosher or non-Halal.

Upright Pen—This device consists of a narrow stall with an opening in the front for the animal's head. After the animal enters the box, it is nudged forward with a pusher gate and a belly lift comes up under the brisket if needed.

The head is restrained by a chin lift that holds it still for the throat cut.

Vertical travel of the belly lift should be restricted to 28 inches (71.1 cm) so that it does not lift the animal off the floor. The rear pusher gate should be equipped with either a separate pressure regulator or special pilot operated check valves to allow the operator to control the amount of pressure exerted on the animal. Pilot operated check valves enable the operator to stop the air cylinders that control the apparatus at mid-stroke positions.

The pen should be operated from the rear toward the front [author insert: i.e., the back pusher should be set before the belly lift and both before the head holder].

Head restraint is the last step. The operator should avoid sudden jerking of the controls. Many cattle will stand still if the box is slowly closed up around them and less pressure will be required to hold them.

Ritual slaughter should be performed immediately after the head is restrained (*within 10 seconds of restraint*).

An ASPCA pen can be easily installed in one weekend with minimum disruption of plant operations. It has a maximum capacity of 100 cattle per hour and it works best at 75 head per hour *or less*. A small version of this pen could be easily built for calf plants.

Conveyor Restrainer Systems—Either V restrainer or center track restrainer systems can be used for holding cattle, sheep or calves in an upright position during Shehita or Halal slaughter. The restrainer is stopped for each animal and a head holder positions the head for the ritual slaughter official. For cattle, a head holder similar to the front of the ASPCA pen can be used on the center track conveyor restrainer. A bi-parting chin lift is attached to two horizontal sliding doors.

Small Restrainer Systems—For small locker plants that ritually slaughter a few calves or sheep per week, an inexpensive rack constructed from pipe can be used to hold the animal in a manner similar to the center track restrainer. Animals must be allowed to bleed out and become completely insensible before any other slaughter procedure is performed (shackling, hoisting, cutting, etc.). (See Figures 7.1 through 7.3 for details.)

Before continuing, it is important to note that the above material is part of a larger document. For the purpose of this chapter, the information as presented is sufficient,

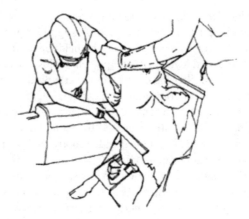

FIGURE 7.1 Restrainer system for religious slaughter of calves and sheep—upright pen.

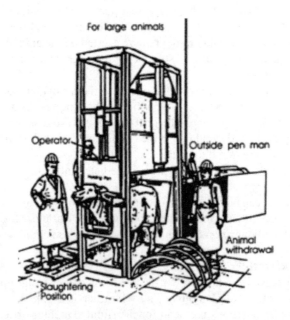

FIGURE 7.2 Upright Pen for religious slaughter of cattle.

but before actually using the guidelines directly in a slaughter plant, it is strongly urged that the entire guidelines be looked at on their website.

With the increased interest in animal welfare, these guidelines are being required by more and more end users, both in the fast food/restaurant industry and by supermarkets and other retail outlets. This has meant that the larger slaughterhouses are meeting these guidelines. The generally smaller religious slaughterhouses have a choice: To only sell to the religious community or to meet these guidelines so meat can be sold beyond into some of those markets requiring such standards. That is a choice each plant has to make, but the author would strongly encourage all slaughterhouses to meet

The Religious Slaughter of Animals

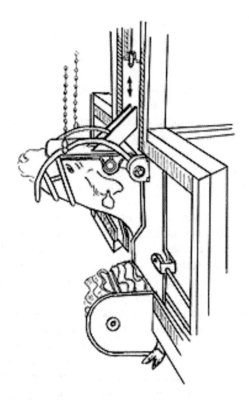

FIGURE 7.3 Center track restrainer being used for ritual slaughter.

these standards, especially with plants doing religious slaughter of animals. Given the controversies, why allow those who would attack religious slaughter to have sympathetic grounds by being able to document breakdowns in animal welfare!

In more recent years, FSIS has also begun to take animal welfare more seriously. So they now have animal welfare inspectors and they do have guidelines that they are enforcing requiring a higher level of animal welfare. So, it is incumbent on slaughterhouses in the U.S. including those doing the religious slaughter of animals to upgrade their activities. This is happening slowly, but is happening.

In practice, this attention to animal welfare has meant that larger animals are being slaughtered religiously in some sort of restraining pen. In the U.S. at this time, it is the authors' understanding that all large animals such as cattle and bison are handled with some sort of restraint system. Small animals can also be handled by high speed upright handling systems such as those developed by Dr. Grandin. Many sheep facilities around the world are now using the "V restrainer" developed by Dr. Grandin. However, the use of shackle and hoisting, that is, the handling of a live sheep and goats by chaining one leg to hoist it off the ground is still being done in the U.S. for both kosher and halal. This is one area where acceptance of proper equipment and making the required investment would improve animal welfare and would demonstrate the commitment of the religious communities to combining the religious requirements with proper animal welfare.

One area that remains controversial in the Muslim community is the issue of pre-stunning of animals prior to the traditional cut across the neck. There is a great deal of literature discussing the various pre-slaughter stunning techniques that might be acceptable for Muslim use. The issue remains of how to be sure that all animals that are stunned are alive at the time of slaughter. In the U.S., both a traditional slaughter and pre-stunned slaughter are used for halal meat. It would be helpful if a system of labeling halal meat were developed to indicate whether meat had been pre-stunned or not so Muslim consumers could make an informed decision. Although there have been surveys in other countries, there are no surveys indicating the percentage of Muslim Americans who accept stunning or of the percentage of Muslim Americans who simply assume that the traditional method was being used for meat and poultry labeled as halal.

The authors are involved in various aspects of improving the religious slaughter of animals focusing currently on the use of a razor sharp knife of the appropriate length (twice the width of the neck and straight) without nicks. However, a detailed discussion of these issues of religious slaughter of animals is beyond the scope of this chapter, but has been covered in the previous chapter.

Another important effort to frame religious slaughter properly is the American Veterinary Medicine Association's (AVMA) work on humane slaughter. The AVMA has been the leader in establishing criteria for euthanasia. Their euthanasia guidelines are generally considered to be the standard that is certainly widely used within the U.S., including most universities that are meeting the federal animal welfare requirements, mainly through the National Institutes of Health (an agency housed within the same organization as the FDA). But they have realized that the euthanasia document does not address two other critical animal welfare issues, namely, depopulation in the case of the need to kill many animals at one time and the issue of the humane slaughter of animals for food. The latter panel has been meeting and their report has been published after extensive reviews within many parts of the veterinary community. Dr. Temple Grandin and the senior author were the main authors of the section on the religious slaughter of animals. A copy of that section of the report is found below (with permission of the AVMA). Again, as with the NAMI guidelines, readers should not use these recommendations directly in a slaughter plant without consulting the entire document, which is now available on the AVMA web site (www.avma.org).

DESIGN OF FACILITIES AND SLAUGHTER PROCESS FOR RELIGIOUS SLAUGHTER

Handling Procedures at Slaughter Plants for Hoofstock

Steps 1 through 5

Refer to the chapter Design of Facilities and Slaughter Process for information on arrival at the plant, unloading, receiving, lairage, and handling. The procedures for these steps are the same regardless of whether the animals will be slaughtered via conventional or religious methods.

Step 6—Restraint

There are various methods used to restrain and position the animal for religious slaughter. In the United States, there is an exemption from the HMSA (1958) for religious slaughter, and methods for restraining the animal for religious slaughter are outside the jurisdiction of USDA FSIS regulations, although Congress has also declared religious slaughter to be humane. The area covered by the handling exemption has been called the area of "intimate" restraint by the FSIS. When an animal is slaughtered in accordance with the ritual requirements of the Jewish faith or any other religious faith that prescribes a method of slaughter whereby the animal suffers loss of consciousness by anemia of the brain caused by the simultaneous and instantaneous severance of the carotid arteries with a sharp instrument, the HMSA specifically declares such slaughter and handling in connection with such slaughter to be humane. However, all procedures outside this area, which many meat inspectors call the "bubble," are beyond the area of intimate restraint and are subject to FSIS oversight the same as conventional slaughter. Unloading animals from transport vehicles, lairage, driving the animals to the restraint point, and insuring that the animal is unconscious with no corneal reflex before invasive dressing procedures begin are under FSIS jurisdiction, the same as conventional slaughter.

Detection of Problems

From an animal welfare standpoint, there are three issues that occur during religious slaughter, which uses a neck cut to create unconsciousness. They are as follows: (1) stress, (2) pain or discomfort caused by how the animal is held and positioned for religious slaughter, and (3) the throat cut itself. Because the HMSA regulations exempt restraint of animals for religious slaughter from the regulations that apply to restraint for conventional slaughter, some small religious slaughter plants use stressful methods of restraint such as shackling and hoisting of live animals even though more welfare-friendly restraint equipment is available. Research has clearly shown that upright restraint is less stressful than shackling and hoisting for sheep and calves (Westervelt et al., 1976). In one study by Dunn (1990) restraining cattle on their backs for over a minute caused more vocalization and a greater increase in cortisol than upright restraint in a standing position for a shorter period of time. Another study by Verlarde (2014) showed that cattle vocalized less in an upright restraint compared to rotating boxes. The OIE also recommends that stressful methods of restraint, such as shackling and hoisting, shackling and dragging, and leg-clamping boxes should not be used, and suspension of live cattle, sheep, goats, or other mammals by their legs is not permitted in the United Kingdom, Canada, Western Europe, and many other countries. Fortunately most mid- to large-size religious slaughter plants in the United States have stopped this practice because of concerns for both animal welfare and worker safety. One study by Grandin (1988) found that conversion of a system that used shackling and hoisting to a conveyor restrainer reduced worker injuries.

Upright restraint is less stressful for both mammals and poultry, compared with being suspended upside down (Westervelt et al., 1976; Bedanova et al., 2007; Kannan et al., 1997). Sheep were less willing to move through a single-file chute after having been subjected to inverted restraint, compared with being put into a restraint device in an upright position (Hutson, 2014). In two different plants where cattle were suspended by one back leg, the percentage of cattle that vocalized varied from 30% to 100% an increased percentages of cattle that vocalize (mooing or bellowing) during restraint are associated with increased cortisol levels (Dunn, 1990). In one study by Grandin (1998) 99% of the cattle vocalizations during handling and restraint were associated with an obvious aversive event such as the use of electric prods or excessive pressure from a restraint device. In cattle, vocalization scoring is routinely used to monitor handling and restraint stress (Grandin, 2013, 2010), and no more than 5% vocalization (3% for non-religious animal slaughter) is acceptable according to the North American Meat Institute standards (Grandin, 2013). The difference in the percentages for acceptability relates to the differences in handling between the two procedures. Vocalization scoring does not work for evaluating the handling and restraint stress in sheep because they usually do not vocalize in response to pain or stress. This may be due to an instinctual inhibition of vocalization in response to the presence of predators (Dwyer, 2004). Research is needed to evaluate vocalization as a method to evaluate stress in goats. The following methods of restraint are highly stressful for conscious mammals and should not be used: hoisting and suspension by one or more limbs, shackling by one or more limbs and dragging, trip floor boxes that are designed to make animals fall, and leg-clamping boxes. Even though suspension is stressful for conscious poultry, such as chickens and turkeys, it is used in a vast majority of all U.S. poultry plants for both conventional and religious slaughter; with attention to handling details and proper equipment, the stress can at least be minimized.

Corrective Action for Problems with Restraint

For the religious slaughter of cattle, restraint devices are available that hold the animal either in an upright position (Figure 7.4) or inverted onto their backs. Smaller

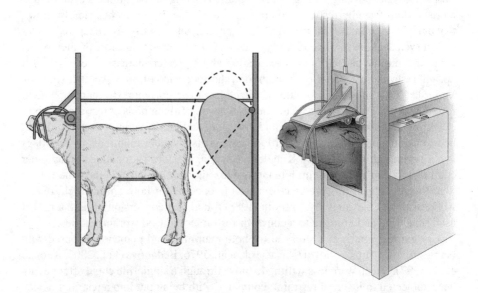

FIGURE 7.4 Recommended restraint of cattle for religious or ritual slaughter.

ruminants, such as sheep or goats, can be held in an upright position by people or placed in a simple restraint device (Westervelt et al., 1976). Large heavy animals, such as cattle or bison, must be held in a mechanical device that holds them in an upright position, holds them in a sideways position, or inverts them onto their backs. Vocalization scoring of cattle can be used both to detect serious welfare problems during restraint of cattle and to document improvements in either design or operation of restraint devices. In cattle, when restraint devices for religious slaughter are operated poorly or have design problems, such as excessive pressure applied to the animal, 25% to 32% of the cattle vocalize (Grandin, 1998; Bourquet et al., 2011). In one study by Grandin (2001) reducing pressure applied by a head-holding device reduced cattle vocalizations from 23% of the cattle to 0%. These problems can occur in both upright and rotating boxes. When the equipment is operated correctly, the percentage of cattle that vocalize will be under 5% (Grandin, 2010, 2012). Inversion for over 90 seconds in a poorly designed rotating box had a higher percentage of cattle vocalizing and higher cortisol levels compared with holding in an upright restraint box (Dunn, 1990).

Information on the correct operation and design of upright restraint devices for religious slaughter can be found in reports by Grandin 1992; 1994, Grandin and Regenstein (1994) and Giger et al (1977). Upright restraint in a comfortable upright position is preferable. When a device that inverts an animal is required by some religious leaders, it should have adjustable sides that support the animal and prevent its body from slipping, twisting, or falling during inversion. Inversion onto the back facilitates the downward cutting stroke. Upright or sideways (lying on the side) restraint may be less aversive than full inversion. Hutson (2014) found that full inversion was more aversive to sheep than being held in an upright position. Sheep can be easily trained to voluntarily enter a tilt table, which tilts them sideways (Grandin, 1989).

It is important to minimize the time that an animal is held firmly by a head restraint. Head restraint using a mechanized device that tightly holds the head is more aversive than body restraint (Grandin, 1992). Before the throat cut, cattle that were held firmly in a head restraint often struggle more than cattle held in a body restraint with no head restraint (Grandin, 1992). Resistance to the head restraint occurs after approximately 30 seconds; therefore, it is important to perform the throat cut before struggling or vocalization begins. When struggling is being evaluated from an animal welfare standpoint, only struggling that occurs before loss of posture should be assessed. When Velarde et al. (2014) evaluated struggling in different types of restraint devices, they did not differentiate between struggling before or after loss of consciousness. Struggling while the animal is conscious is a welfare concern, and struggling from convulsions after an animal loses posture and becomes unconscious has no effect on welfare. Restraint devices should be equipped with pressure-limiting devices to prevent excessive pressure from being applied, which then causes either struggling or vocalization (Grandin, 1992). The percentage of cattle vocalizing (mooing or bellowing) either while in a restraint device or while entering it should be 5% or less (Grandin, 2010, 2012). Restraint devices should not cause animals to struggle or vocalize. For poultry, stress during shackling can be reduced by subdued lighting. Wing flapping can be reduced by installing vertical pieces of conveyor belting with a smooth surface for the breasts of the shackled birds to rub on. A possible future method to reduce bird stress while in shackles is the incorporation of a moving horizontal conveyor that supports the bird's body (Lines et al., 2011). A Dutch poultry plant recently installed a system where each shackled chicken has its body supported in a plastic holder. In both large and small plants, where possible chickens can be held by a person in an upright position for the throat cut and then placed immediately either in a bleeding cone or on the shackle.

Step 7—Performing the Throat Cut

There are three basic ways that religious slaughter is performed: (1) pre-slaughter stunning before the throat cut with either a captive bolt or electric stunning, (2) immediate post-cut stunning with a non-penetrating captive bolt, or (3) slaughter without stunning (traditional hand slaughter). Some religious authorities who supervise either kosher (Jewish) or halal (Muslim) religious slaughter will allow either pre-slaughter or immediate post-slaughter stunning (Nakyinsige et al., 2013). For halal slaughter, electric head-only stunning is used in many large cattle and sheep plants in New Zealand, Australia, and the United Kingdom. Head-only electric stunning is acceptable to many Muslim religious authorities because it is fully reversible and induces temporary unconsciousness (refer to the section Techniques—Physical Methods—Electric). If pre-slaughter stunning is done, there will be no animal welfare concerns about the throat cut in a conscious animal. Since most pre-slaughter stunning methods that are approved for religious slaughter produce a lighter reversible stun, greater attention will be required to the details of procedures to ensure that the animals or birds are and remain unconscious during the throat cut. An effective reversible precut stun in sheep can be easily achieved with 1.25–2 A at a frequency range of 50–400 Hz. According to Grandin (2012, 2015), when the stunner was applied to the head for 1.5 seconds at 300 Hz, it produced a clear tonic rigid phase followed by a clonic kicking phase representative of an epileptic seizure. This pattern is an indicator that it produced unconsciousness. A modified New Zealand head-to-body stunner with the rear body electrode removed worked well because the design of the handle facilitated positioning of the stunner on the sheep's head. In poultry a very light reversible electric water-bath stun is done. The preceding stunning methods are acceptable to a number of halal certifiers. Some halal certifiers will accept non-penetrating captive bolt because the heart will continue to beat after stunning (Vimini et al., 1983). Some religious communities will accept immediate post-cut stunning, and others require slaughter without stunning (traditional hand slaughter). Stunning methods are covered in the Techniques chapter of these guidelines.

Detection of Problems

The greatest welfare concerns may occur during traditional religious hand slaughter. There are two main issues: (1) Does cutting the throat of a conscious animal cause pain? (2) What is the maximum appropriate time that is required for the animal to become unconscious after a properly done throat cut? The throat cut done during both kosher and halal slaughter simultaneously severs both carotid arteries and jugular veins and the trachea. For halal slaughter, a sharp knife is required. Kosher slaughter has more strict specifications for how the cut is performed and the design and sharpening of the knife (Levinger, 2000). A kosher slaughter knife is long enough to span the full width of the neck (i.e., double the width of the neck) and is sharpened on a whetstone. Before and after each animal is cut, the knife is checked for nicks that could cause pain (Levinger, 2000). Any nick in the knife makes the animal non-kosher, so there is a strong incentive to keep the knife razor sharp and nick free.

Painfulness of the Cut

Researchers have reported that cutting the throat of 107- to 109-kg (236- to 240-lb) veal calves with a knife that was 24.5 cm long caused pain comparable to dehorning (Gibson et al., 2009a, 2009b). The knife may have been too short to fully span the throat, and it had been sharpened on a mechanical grinder. A grinder may create nicks on the blade and may not be comparable to a knife sharpened on a whetstone. Slaughter without stunning of cattle with a knife that is too short will result in violent struggling because the tip makes gouging cuts in the wound (Grandin, 1994). One of the rules of kosher slaughter is that the incision must remain open during the cut (Levinger, 2000; Epstein; 1948). When the wound is allowed to close back over the knife, cattle will violently struggle (Grandin, 1994). When an animal is restrained in a comfortable upright position, it becomes possible to observe how the animal reacts to the throat cut. When a kosher knife was used by a skilled slaughter man (shochet), there was little behavioral reaction in cattle during the cut (Grandin, 1994). In calves, there has been a similar observation (Bager et al., 1992). Grandin (1994) reports that people invading the animal's flight zone by getting near to the animal's face caused a bigger reaction. An ear tag punch has also caused a bigger reaction than a good kosher cut (Grandin, 1994). Most chickens slaughtered by shechita exhibited no physical response to the cut, and they lost the ability to stand and eye reflexes at 12–15 seconds (Barnett et al., 2007).

Time to Lose Consciousness

Unconsciousness, as defined in the General Introduction of these Guidelines, is the loss of individual awareness that occurs when the brain's ability to integrate information is blocked or disrupted. Before invasive dressing begins, all signs of brainstem function such as the corneal reflex must be abolished by bleeding. Sheep will lose consciousness as determined by their EEG more quickly than cattle because of differences in the anatomy of the blood vessels that supply the brain (Baldwin and Bell, 1963a). In cattle, when the carotid arteries are severed, the brain can still receive blood from the vertebral arteries (Baldwin and Bell, 1963b). After the cut, sheep will become unconscious and lose posture and no longer be able to stand within 2–14 seconds, while most cattle will lose consciousness and no longer be able to stand within 17–85 seconds (Blackmore et al., 1983, 1984; Nangeroni and Kennet, 1983; Newhook and Blackmore, 1982; Schulz et al., 1978; Daly et al., 1988; Gregory and Wotton, 1984). In these studies, time to onset of unconsciousness was measured by either EEG or loss of the ability to stand (LOP). Allowing the wound to close up after a transverse halal throat cut with a 20-cm-long knife may delay onset of unconsciousness. Electroencephalographic measurements on sheep indicated consciousness could last 60 seconds (Rodriguez et al., 2012). In a study by Lambooij et al., 2012) where a rotating box was used to invert veal calves onto their backs, unconsciousness was measured by EEG. It occurred at an average of 80 seconds. In sheep, unconsciousness as measured by time to eye rotation was 15 seconds (Cranley, 2012).

There is a large amount of biological variability, and a few cattle, calves, or sheep have extended periods of sensibility (>4 minutes) (Gregory and Wotton, 1984, 2012). If the animals can stand and walk they are definitely conscious. In sheep the corneal reflexes, which are a brainstem reflex, may be present for up to 65 seconds after the cut (Cranley, 2012). In veal calves, corneal reflexes were still present at 135 ± 57 seconds after the throat cut (Lambooij et al., 2012). The methods section of Lambooij et al (2012) did not describe the type of knife. However, that study was done in a slaughter plant that performed halal slaughter. Corneal reflexes can also occur in electrically stunned or CO_2-stunned animals where other indicators of return to consciousness, such as the righting reflex, rhythmic breathing, and eye tracking, are absent (Vogel et al., 2011). Corneal reflexes occur during a state of surgical anesthesia (Rumpl et al., 1982) or when visual potentials and SEPs are abolished (Anil and McKinstry, 1991). One of the best indicators for determining onset of unconsciousness is the loss of the ability to stand or walk (LOP). In cattle, a major cause of prolonged periods of consciousness after the throat cut is sealing off of the ends of the severed arteries (false aneurysms) (Gregory et al., 2008). This problem does not occur in sheep.

Aspiration of Blood

Another welfare concern is aspiration of blood into the trachea and lungs after the cut (Agbeniga and Webb, 2012). In one study by Gregory et al (2009) when cattle were held in a well-designed upright restraint, 36% (for kosher) and 69% (for halal) aspirated blood. In 31% of these non-stunned cattle, blood had been aspirated into the bronchi. It is likely that in a rotating box where the animal is held on its back, aspiration of blood will be higher.

Corrective Action for Problems

To reduce painfulness of the act, a knife that is long enough to span the neck where the tip will remain outside the neck during the cut should be used. It is also essential that the knife be extremely sharp, and the use of a whetstone is recommended. A good method for testing a knife for sharpness is the paper test. To perform this test, a single sheet of standard letter-size (8.5 × 11-inch) printer paper is dangled in a vertical position by being held by a thumb and forefinger by one corner. A dry knife held in the other hand should be able to start cutting at the edge of the paper and slice it in half. This method can eliminate the worst dull knives, but it may not detect sharp knives with nicks.

It is also essential to not allow the wound to close back over the knife during the cut. To prevent sealing off of the arteries in cattle, the cut should be angled so it is close to the first cervical vertebra (C1) position (Gregory et al., 2012a) as long as such a cut is accepted by the religious authorities. This will also cut a sensory nerve, which may prevent the cattle from experiencing distressful sensations from aspirating blood (Gregory et al., 2012b). The cut should be located posterior to the larynx and angled toward the C1 position.

Before invasive dressing procedures such as skinning or leg removal are started, the corneal reflexes must be absent. Even though an animal showing only a corneal reflex is unconsciousness, to provide a good margin of safety, it should be absent before dressing procedures start. Absence of the corneal reflex and complete unconsciousness before dressing procedures are started are best practices for all slaughter plants that conduct both conventional slaughter and religious slaughter.

AUDITING RELIGIOUS SLAUGHTER TO IMPROVE ANIMAL WELFARE FOR BOTH KOSHER AND HALAL SLAUGHTER OF CATTLE, SHEEP, OR GOATS

The following audit methods are recommended to maintain an acceptable level of animal welfare when religious slaughter is performed by cutting of the neck.

1. Calm animals will lose sensibility quicker. Follow all procedures for handling that are in other parts of this document (Grandin, 1992, 1994).
2. Conduct collapse-time scoring. When the best methods are employed, 90% of the cattle will collapse and lose the ability to stand within 30 seconds .c Researchers in Europe reported a similar result when they used a well-designed upright restraint device (Gregory et al., 2010). In a rotating box, collapse-time scoring is impossible because the animal is on its back. Alternative measures for determining onset of unconsciousness are time until eye rotation and the amount of time to abolish the presence of natural blinking such as seen with a live animal in the yards (lairage). Natural blinking must not be confused with the corneal reflex. To evaluate natural blinking (menace reflex), a hand is waved within 4 inches (10 cm) of the eye without touching it. A natural blink occurs if the eye does a full cycle of closing and then reopening. Omit scoring of time to unconsciousness if pre- or post-cut stunning is used.
3. The vocalization score should be 5% or less for cattle (Grandin, 2010, 2012). Score on a per-animal basis, as a silent animal or a vocalizer (mooing or bellowing). All cattle that vocalize inside the restraint device are scored. A bovine is also scored as a vocalizer if it vocalizes in direct response to being moved by a person, electric prod, or mechanical device into the restraint device. Do not use vocalization scoring for sheep. Standards for vocalization scoring of goats will need to be developed.
4. In all species, score restraint methods for the percentage of animals that actively struggle before LOP.
5. The percentage of animals (all species) that fall down in the chute (race) leading up to the restraint device or fall before the throat cut in the restraint device should be 1% with a goal of zero. This is the same as conventional slaughter. Restraint devices that are designed to make an animal fall are unacceptable and result in an automatic audit failure. Rotating boxes must fully support the body, and the animal's body should not shift position or fall when the box is rotated.

6. Electric prods should be used judiciously and only in extreme circumstances when all other techniques have failed (Leary et al., 2016). Score prod use using the same criteria as conventional slaughter.
7. Perform the cut quickly, preferably within 10 seconds after the head is fully restrained. Omit this measure if pre-slaughter stunning is used.
8. Reduce the pressure applied by the head holders (but do not remove it), rear pusher gates, and other devices immediately after the cut to promote rapid bleed out.
9. Corneal reflexes, rhythmic breathing, and all other signs of return to sensibility must be absent before invasive dressing procedures such as skinning, leg removal, or dehorning are started. This is a requirement for all methods of slaughter both conventional and religious to be absolutely sure that the animal is completely insensible.
10. Do not use stressful methods of restraint for mammals, such as shackling and hoisting by suspension by one or more limbs, shackling and dragging by one or more limbs, trip floor boxes that are designed to make animals fall, leg-clamping boxes, or other similar devices.
11. If either pre- or post-cut stunning is used, score the same as conventional slaughter.

R3 Auditing Religious Slaughter to Improve Animal Welfare for both Kosher and Halal Slaughter of Chickens, Turkeys, and Other Poultry

1. If stunning is used, audit and monitor the percentage of birds that are effectively stunned using the same criteria as for conventional slaughter.
2. Score the performance of shacklers for faults such as one-legged shackling using the same criteria as for conventional slaughter.
3. There should be 0% uncut red skinned birds that emerge from the de-feathering machine. This is an indicator that a bird entered the scalder alive. This measure is the same as used for conventional slaughter.
4. Score the percentage of birds that wing flap after restraint. In a well-designed shackle line with a breast rub conveyor, the percentage of flapping birds should be very low.

THE IMPORTANCE OF MEASUREMENT

By routinely measuring the performance of religious slaughter procedures, the standards for such slaughter are kept high. Measuring collapse times for unconsciousness or other indicators such as time to eye roll-back or the absence of natural blinking will enable both plant personnel and religious slaughter personnel to improve their procedures.

a. Grandin T, College of Agricultural Sciences, Colorado State University, Ft Collins, Colo: Personal communication, 2012.
b. Grandin T, College of Agricultural Sciences, Colorado State University, Ft Collins, Colo: Personal communication, 2015.
c. Voogd E, Department of Animal Sciences, College of Agricultural, Consumer and Environmental Sciences, Urbana, Ill: Personal communication, 2009.

REFERENCES

Agbeniga, B. and Webb, E. C. (2012). Effect of slaughter technique on bleed-out, blood in the trachea and blood splash in the lungs of cattle. *South African Journal of Animal Science*, 42(5), 524–529.

Anil, M. H. and McKinstry, J. L. (1991). Reflexes and loss of sensibility following head-to-back electrical stunning in sheep. *The Veterinary Record*, 128(5), 106–107.

Bager, F., Braggins, T. J., Devine, C. E., Graafhuis, A. E., Mellor, D. J., Tavener, A., and Upsdell, M. P. (1992). Onset of insensibility at slaughter in calves: Effects of electroplectic seizure and exsanguination on spontaneous electrocortical activity and indices of cerebral metabolism. *Research in Veterinary Science*, 52(2), 162–173.

Baldwin, B. A. and Bell, F. R. (1963a). The anatomy of the cerebral circulation of the sheep and ox. The dynamic distribution of the blood supplied by the carotid and vertebral arteries to cranial regions. *Journal of Anatomy*, 97(Pt 2), 203.

Baldwin, B. A. and Bell, F. R. (1963b). The effect of temporary reduction in cephalic blood flow on the EEG of sheep and calf. *Electroencephalography and Clinical Neurophysiology*, 15(3), 465–473.

Barnett, J. L., Cronin, G. M., and Scott, P. C. (2007). Behavioural responses of poultry during kosher slaughter and their implications for the birds' welfare. *The Veterinary Record*, 160(2), 45–49.

Bedanova, I., Voslarova, E., Chloupek, P., Pistekova, V., Suchy, P., Blahova, J., Dobsikova, R., and Vecerek, V. (2007). Stress in broilers resulting from shackling. *Poultry Science*, 86(6), 1065–1069.

Blackmore, D. K. (1984). Differences in behaviour between sheep and cattle during slaughter. *Research in Veterinary Science*, 37(2), 223–226.

Blackmore, D. K., Newhook, J. C., and Grandin, T. (1983). Time of onset of insensibility in four-to six-week-old calves during slaughter. *Meat Science*, 9(2), 145–149.

Bourguet, C., Deiss, V., Tannugi, C. C., and Terlouw, E. C. (2011). Behavioural and physiological reactions of cattle in a commercial abattoir: Relationships with organisational aspects of the abattoir and animal characteristics. *Meat Science*, 88(1), 158–168.

Cranley, J. (2012). Slaughtering lambs without stunning. *Veterinary Record*, 170(10), 267–268.

Daly, C. C., Kallweit, E., and Ellendorf, F. (1988). Cortical function in cattle during slaughter: Conventional captive bolt stunning followed by exsanguination compared with shechita slaughter. *The Veterinary Record*, 122(14), 325–329.

Dunn, C. S. (1990). Stress reactions of cattle undergoing ritual slaughter using two methods of restraint. *The Veterinary Record*, 126(21), 522–525.

Dwyer, C. M. (2004). How has the risk of predation shaped the behavioural responses of sheep to fear and distress? *Animal Welfare*, 13(3), 269–281.

Epstein, I. Ed. (1948) *The Babylonian Talmud*. London: Soncino Press.

Galvin, J. W., Blokhuis, H., Chimbombi, M. C., Jong, D., and Wotton, S. (2005). Killing of animals for disease control purposes. *Revue Scientifique et Technique-Office International des épizooties*, 24(2), 711.

Gibson, T. J., Johnson, C. B., Murrell, J. C., Chambers, J. P., Stafford, K. J., and Mellor, D. J. (2009a). Components of electroencephalographic responses to slaughter in halothane-anaesthetised calves: Effects of cutting neck tissues compared with major blood vessels. *New Zealand Veterinary Journal*, 57(2), 84–89.

Gibson, T. J., Johnson, C. B., Murrell, J. C., Hulls, C. M., Mitchinson, S. L., Stafford, K. J., Johnstone, A. C., and Mellor, D. J. (2009b). Electroencephalographic responses of halothane-anaesthetised calves to slaughter by ventral-neck incision without prior stunning. *New Zealand Veterinary Journal*, 57(2), 77–83.

Giger, W., Prince, R. P., Westervelt, R. G., and Kinsman, D. M. (1977). Equipment for low-stress, small animal Slaughter. *Transactions of the ASAE*, 20(3), 571–0574.

Grandin, T. (1992). Observations of cattle restraint devices for stunning and slaughtering. *Animal Welfare*, 1(2), 85–90.
Grandin, T. (1994). Euthanasia and slaughter of livestock. *Journal—American Veterinary Medical Association*, 204, 1354–1354.
Grandin, T. (1988). Double rail restrainer conveyor for livestock handling. *Journal of Agricultural Engineering Research*, 41(4), 327–338.
Grandin, T. (1989) Voluntary acceptance of restraint by sheep. *Applied Animal Behaviour Science*, 23, 257–261.
Grandin, T. (1998). The feasibility of using vocalization scoring as an indicator of poor welfare during cattle slaughter. *Applied Animal Behaviour Science*, 56(2–4), 121–128.
Grandin, T. (2001). Cattle vocalizations are associated with handling and equipment problems at beef slaughter plants. *Applied Animal Behaviour Science*, 71(3), 191–201.
Grandin, T. (2010). Auditing animal welfare at slaughter plants. *Meat Science*, 86(1), 56–65.
Grandin, T. (2012). Developing measures to audit welfare of cattle and pigs at slaughter. *Animal Welfare*, 21(3), 351–356.
Grandin, T. (2013). *Recommended Animal Handling Guidelines and Audit Guide: A Systematic Approach to Animal Welfare*. American Meat Institute, Animal Welfare Committee, Washington, DC: American Meat Institute Foundation.
Grandin, T. and Regenstein, J. M. (1994). Religious slaughter and animal welfare: A discussion for meat scientists. *Meat Focus International*, 3(1), 115–123.
Gregory, N. G. and Wotton, S. B. (1984). Time to loss of brain responsiveness following exsanguination in calves. *Research in Veterinary Science*, 37(2), 141–143.
Gregory, N. G., von Wenzlawowicz, M., and von Holleben, K. (2009). Blood in the respiratory tract during slaughter with and without stunning in cattle. *Meat Science*, 82(1), 13–16.
Gregory, N. G., Fielding, H. R., von Wenzlawowicz, M., and Von Holleben, K. (2010). Time to collapse following slaughter without stunning in cattle. *Meat Science*, 85(1), 66–69.
Gregory, N. G., Schuster, P., Mirabito, L., Kolesar, R., and McManus, T. (2012a). Arrested blood flow during false aneurysm formation in the carotid arteries of cattle slaughtered with and without stunning. *Meat Science*, 90(2), 368–372.
Gregory, N. G., von Wenzlawowicz, M., Alam, R. M., Anil, H. M., Yeşildere, T., and Silva-Fletcher, A. (2008). False aneurysms in carotid arteries of cattle and water buffalo during shechita and halal slaughter. *Meat Science*, 79(2), 285–288.
Gregory, N. G., von Wenzlawowicz, M., von Holleben, K., Fielding, H. R., Gibson, T. J., Mirabito, L., and Kolesar, R. (2012b). Complications during shechita and halal slaughter without stunning in cattle. *Animal Welfare*, 21(S2), 81–86.
Humane Methods of Livestock Slaughter Act. (1958). CP.L. 85–765; 7 U.S.C. 1901 et seg.
Hutson, G. D. (2014). 11 behavioural principles of sheep handling. *Livestock Handling and Transport: Theories and Applications*, Ed. Grandin, T. Boston: CABI. 193–217.
Kannan, G., Heath, J. L., Wabeck, C. J., and Mench, J. A. (1997). Shackling of broilers: Effects on stress responses and breast meat quality. *British Poultry Science*, 38(4), 323–332.
Lambooij, E., van der Werf, J. T. N., Reimert, H. G. M., and Hindle, V. A. (2012). Restraining and neck cutting or stunning and neck cutting of veal calves. *Meat Science*, 91(1), 22–28.
Leary, S., Underwood, W., Anthony, R., Corey, D., Grandin, T., and Gwaltney-Brant, S. (2016). *AVMA Guidelines for the Humane Slaughter of Animals: 2016 Edition*. Schaumburg, IL: AVMA.
Levinger, I. M. (1995). *Shechita in the Light of the Year 2000: Critical Review of the Scientific Aspects of Methods of Slaughter and Shechita*. The University of Michigan: Maskil L'David.
Lines, J. A., Jones, T. A., Berry, P. S., Cook, P., Spence, J., and Schofield, C. P. (2011). Evaluation of a breast support conveyor to improve poultry welfare on the shackle line. *The Veterinary Record*, 168(5), 129–129.

Nakyinsige, K., Man, Y. C., Aghwan, Z. A., Zulkifli, I., Goh, Y. M., Bakar, F., Al-Kahtani, H. A., and Sazili, A. Q. (2013). Stunning and animal welfare from Islamic and scientific perspectives. *Meat Science*, 95(2), 352–361.

Nangeroni, L. I. and Kennett, P. D. (1963). *An Electroencephalographic Study of the Effect of Shechita Slaughter on Cortical Function in Ruminants*. Ithaca, NY: Cornell University.

Newhook, J. C. and Blackmore, D. K. (1982). Electroencephalographic studies of stunning and slaughter of sheep and calves: Part 1—the onset of permanent insensibility in sheep during slaughter. *Meat Science*, 6(3), 221–233.

Rodriguez, P., Velarde, A., Dalmau, A., and Llonch, P. (2012). Assessment of unconsciousness during slaughter without stunning in lambs. *Animal Welfare*, 21(2).

Rumpl, E., Gerstenbrand, F., Hackl, J. M., and Prugger, M. (1982). Some observations on the blink reflex in posttraumatic coma. *Electroencephalography and Clinical Neurophysiology*, 54(4), 406–417.

Schulze, W., Schultze-Petzold, H., Hazem, A. S., and Gross, R. (1978). Objectivization of pain and consciousness in the conventional (dart-gun anesthesia) as well as in ritual (kosher incision) slaughter of sheep and calf. *DTW. Deutsche Tierarztliche Wochenschrift*, 85(2), 62–66.

Velarde, A., Rodriguez, P., Dalmau, A., Fuentes, C., Llonch, P., Von Holleben, K. V., and Cenci-Goga, B. T. (2014). Religious slaughter: Evaluation of current practices in selected countries. *Meat Science*, 96(1), 278–287.

Vimini, R. J., Field, R. A., Riley, M. L., and Varnell, T. R. (1983). Effect of delayed bleeding after captive bolt stunning on heart activity and blood removal in beef cattle. *Journal of Animal Science*, 57(3), 628–631.

Vogel, K. D., Badtram, G., Claus, J. R., Grandin, T., Turpin, S., Weyker, R. E., and Voogd, E. (2011). Head-only followed by cardiac arrest electrical stunning is an effective alternative to head-only electrical stunning in pigs. *Journal of Animal Science*, 89(5), 1412–1418.

Westervelt, R. G., Kinsman, D. M., Prince, R. P., and Giger, W. (1976). Physiological stress measurement during slaughter in calves and lambs. *Journal of Animal Science*, 42(4), 831–837.

8 Halal Production Requirements for Meat and Poultry

Kristin Pufpaff, Mian N. Riaz, and Munir M. Chaudry

CONTENTS

Halal Control Points (HCP) in Animal Slaughtering 110
    HCP-1: Allowed Animals 110
    HCP-2: Holding 110
    HCP-3: Stunning 111
        Captive Bolt Stunning 112
        Mushroom-Shaped Hammer Stunner 112
        Electrical Stunning 112
        Carbon Dioxide Stunning or Other Forms of Gas Stunning 113
    HCP-4: Knives 113
    HCP-5: Slaughter Person 113
    HCP-6: Invocation 113
    HCP-7: Slaying/Killing or Bleeding 113
    HCP-8: Post-Slaughter Treatment 114
    HCP-9: Packaging and Labeling 114
Acceptable Birds 114
Halal Control Points (HCP) in Poultry Slaughter 117
    HCP-1: Acceptable Birds 117
    HCP-2: Holding 118
    HCP-3: Stunning 118
    HCP-4: Knives 118
    HCP-5: Slaughter Person 118
    HCP-6: Invocation 118
    HCP-7: Slaughter/Killing or Bleeding 118
    HCP-8: Post-Slaughter Treatment 118
    HCP-9: Packaging and Labeling 119
Halal Requirements for Deboning and Processing Rooms 119
Halal Requirements for Cold Stores 119
Further Processed Meat Items 119
    HCP-1: Meat Source 119
    HCP-2: Equipment 119
    HCP-3: Non-meat Ingredients 120

HCP-4: Casings 121
HCP-5: Packaging and Labeling 121
Industry Perspective on Halal Production 121
References 122

Among all the dietary restrictions or prohibitions placed on Muslims by God, the majority are related to the consumption of meat and animal products, especially land animals (see Chapter 3). Religion is a driving force in determining the consumption of meat and it permits or prohibits the use of certain type of meat by the group of people belong to it. This chapter will address the requirements and procedures surrounding the consumption of meat and poultry in Islam. Restrictions, prohibitions, permissions, and instructions are presented in such a way as to balance two potentially conflicting viewpoints; the requirement that Muslims respect and care for the animals around them, and the permission for Muslims to consume animal flesh for their own subsistence. Chapter 6 has detailed the importance of animal welfare in a holistically Islamic meat production system. This chapter will detail further how those and other requirements relate to the production of meat and poultry in the slaughter plant and during meat processing. This chapter also includes a set of halal control points (HCP) that can be used by the industry to guide in the development of protocols in each area of animal slaughter and meat processing.

The first section of this chapter will address land mammals and go over the flow of the slaughter process from arrival at the plant until the death of the animals. Primary requirements, those required for the meat to be permissible for Muslims to consume, will be shown in bold text while secondary requirements, that is, things that are highly encouraged based of the teachings of Prophet Muhammad (PBUH), and scientific advice will be added to clarify these points. There are some acts that can occur during slaughter that can be considered sins on the soul of the person who engages in them even if they do not make the resulting meat unfit for Muslim consumption. These reprehensible acts will be underlined.

Halal meat is what we get from the slaughter of a halal animal according to the Islamic law, that is, slaughtering of an animal in a prescribed way (Qureshi et al., 2012). Animals slaughtered for Muslim consumption must be of acceptable species and alive at the time of slaughter. Pigs are never acceptable for halal slaughter. Carnivorous animals such as tigers, cats, and dogs are not acceptable for Muslim consumption (Department of Standards Malaysia, 2009). There are different scholarly opinions on the subject of eating horse and other equid meats. Some scholars view the following verse in the Quran as discouraging the eating of horse meat while others say that it could be permissible as it was never strictly forbidden.

And (He has created) horses, mules and donkeys, for you to ride and as an adornment.

al-Nahl 16:8

In terms of practical production opportunities, there is no consumer market for halal horse meat. The most commonly halal-slaughtered red meat animals are cattle (including water buffalo), sheep, goats, and camels. Of these, the first three are all slaughtered in approximately the same way. Only camels have a different prescribed means of Islamic slaughter, which will be defined later in this chapter.

The animal must be safe for human consumption, wholesome, free of gross disease, and should be whole and completely healthy, free of any ailment or visible deformity. It is worth mentioning that a halal animal may become haram during breeding if they are treated with any products that involve the use of ingredient that are obtained from haram sources or derived through genetic engineering involving a component from a haram source (Department of Standards Malaysia, 2009).The first portion of this injunction should, in most countries, be met by governmental bodies responsible for the safety and security of the food supply. The secondary portion of the injunction is the responsibility of the halal slaughter plant in conjunction with farmers, animal welfare auditors, and industry standards overseeing the raising and transport of animals going to the slaughter plant. Animals have a good sense to perceive the signals of danger, such as odor, sights and sounds that can be a cause of pre-slaughter stress, and therefore, require careful pre-slaughter handling (Micera et al., 2010). An animal should be given water and handled humanely before slaughtering so that it is calm and not experiencing unnecessary fear. Poor handling of animals prior to slaughter is not only a bad welfare practice but has negative meat quality consequences (Casoli et al., 2005). This injunction applies to the animal in the holding yard as well as in any shoots leading up to slaughter point. At no point should painful stimuli be applied to sensitive areas of the animal such as the face, genitals, or rectum. At no point should an animal be kept from water or prevented from drinking enough water to sustain itself.

The restraint of animals for slaughter is discussed in Chapters 6 and 7 but there are a few points that are worth repeating. There are some means of restraint that do not directly prohibit meat from being consumed as halal but alternatives should be found in the spirit of halal being designed to treat animals with kindness and respect. Restraint equipment should keep animals upright and quiet, and/or have their bodies completely supported until they have lost consciousness. Hanging or dragging conscience animals is not acceptable from an animal welfare perspective. It is also not acceptable to intentionally trip animals in preparation for slaughter. It is never acceptable to damage the legs of an animal to immobilize them prior to slaughter. Restraint equipment should be designed so that the majority of animals will walk into the equipment for stunning or slaughter without the use of painful driving aids (Apple et al., 2005; Hambrecht et al., 2004; Kannan et al., 2003; Ljungberg et al., 2007; Schaefer et al., 2001). The one time when the best option for slaughter restraint may be to have the animal on the ground is that for camels, which are normally slaughtered while kneeling. The camel should be gently placed in this position by a trained handler. The animals should be accustomed to kneeling on command. There also may be times when cattle may have to be cast prior to slaughter if acceptable handling facilities are not available. With proper use of ropes and trained workmen, this can be done with reasonable respect for animal welfare.

However, if cattle are brought to a facility for the express purpose of slaughtering on a regular basis, then it is important that the plant is willing to make the capital investment needed to have equipment available that negates the need for casting. Any animal that is abused or experiences an excitation phase prior to slaughter is likely to produce poorer quality meat as well as being subjected to poor animal welfare (Nakyinsige et al., 2012). The cost of better handling equipment and better training programs for employees can be recouped in terms of having a safer facility and better meat quality. Once the animal is restrained in a manner so as not to produce pain or fear there are two possible major ways to undertake the slaughter: slaughter with the use of a stunning device prior to the halal slaughter cut or immediate halal slaughter. Slaughtering by hand without prior intervention is still preferred by most Muslims and quite widely followed in Muslim countries and other countries where Muslims control many slaughterhouses (Gibson et al., 2009). No matter whether stunning is used or not in Islam, the animal must die from a cut to the throat that severs the carotid arteries, jugular veins, the trachea, and the esophagus, without severing the spinal cord. If prior stunning is to be done, it must be done in a way that leaves the animal's heart beating throughout the process of exsanguination (i.e., bleeding) (Gregory et al., 2010). The debate surrounding the use of stunning in halal slaughter stems from the following verse:

> Prohibited to you are dead animals, blood, the flesh of swine, and that which has been dedicated to other than God, and [those animals] killed by strangling or by a violent blow or by a head-long fall or by the goring of horns, and those from which a wild animal has eaten, except what you [are able to] slaughter [before its death].
>
> **al-Maa'idah 5:3**

Stunning with gas is religiously equated to killing by strangulation. Captive bolt or mushroom head stunners are violent blows, and, while electric stunners are not directly addressed in the Quran, they may be considered another form of violent blow. By this logic, it is clear why most Muslims prefer meat from animals that have not been stunned but the final line of the above quote allows for a certain degree of theological flexibility. However, it also makes it clear that research into each method of stunning must be done to ensure that no animal experiences heart death before the throat is cut. This is a difficult requirement to meet in a real slaughterhouse where the size, hair covering, and wet/dryness, and so on, of the animals will vary widely from animal to animal.

Research has shown that when non-penetrating captive bolt or mushroom head stunning is applied to a healthy bovine, the animal should live a minimum of two minutes. After that time, there is an increase risk that the heart of animals stunned with a captive bolt may stop or animals stunned with mushroom head stunner may begin to regain sensibility. For these reason, mechanically stunned cattle should be slaughtered within two minutes, preferably sooner, of the stun being applied. The time frames are accelerated in small ruminants so it is recommended that sheep and goats be slaughtered within one minute, preferably sooner, of stunning. There has not been adequate research done on camels to establish a timeframe for post-stun halal slaughtering (Kadim, 2012).

Currently, there is no established form of gas stunning that does not leave a large percentage of the animals dead before bleeding. As a result, gas stunning is not currently recommended for the stunning of red meat animals prior to halal slaughter (Nakyinsige et al., 2012). As further research in this area is conducted, it is possible that a humane and halal compliant form of gas stunning may be achieved (Anil et al., 2002).

Electric stunning at appropriate voltages has been shown to be an effective, fully reversible form of stunning that is particularly effective for sheep and goats. It is important that any plant using these methods be certain that they are using the correct amperages/voltages for the species and size, whether using alternating current (AC) or direct current (DC), maintaining their equipment in the best possible repair, correcting for variation in size and fleece coverage in the case of sheep and wool goats, and using the equipment correctly so as not to "hot wand" the animals. "Hot wanding" an animal occurs when the animal feels the jolt of electricity as pain without the loss of sensibility associated with correct usage. In the worst case, the animal is "frozen" but is still capable of feeling pain (Zivotofsky and Strous, 2012). Sheep often must have their fleeces wetted prior to electric stunning due to the need to overcome the insulation effect of wool. It is very important that head-only electric stunning be used in halal slaughter plants as a head-heart (head-back) stun will stop an animal's heart and cause the meat to be unacceptable for halal consumption. Other forms of stunning such as microwave stunning are being researched, and if those are shown to be able to render animals insensible to pain without causing their death, they should be considered where applicable by religious scholars (Small, 2013). Ideally in plants slaughtering many animals, a Muslim should be responsible for assuring that animals are all alive at the time of halal slaughter and that no animal is further cut until it is insensible (Anil et al., 2000; Limon et al., 2010; Önenç and Kaya, 2004). For the actual slaughtering, the throat cut must be done by a Muslim of sound mind while pronouncing the name of God (Bismillah Allahu Akbar) on each animal. The blessing of each animal must be done at the moment of slaughter and should be undertaken by a person that understands its religious significance. In Islamic teachings, the slaughter person is causing the death of a being with a soul with the permission of God. The blessing is not just said to make the meat permissible to eat, but because it is also an act of worship and should be performed according to the teachings of God and with his name/authorization the slaughter person is killing the animal. But it is also meant to be the last moment of that animal's life so that the animal is also given a chance to submit to its death. By blessing animals before they die, the slaughter person is giving that animal the right to submit their soul as well as providing meat for their community. The right of animals to speak on the Day of Judgment is shown in the following hadith:

> The father of Amru'bnuashareed reported: I heard the Messenger of God (PBUH) say, "Whoever kills so much as a sparrow for no reason will have it pleading to God on the Day of Resurrection, saying: O Lord, so-and-so killed me for no reason, and he did not kill me for any beneficial purpose."
>
> **Sunan An-Nasa'i**

If a person kills an animal without a blessing, they have not used the permission God has given them to take a life. And if an animal is killed but not eaten, the permission has been taken but not for the intended purpose. If a slaughter person intends to bless the animal but forgets it at that moment, God is forgiving.

Slaughter persons should also be trained specifically for the type and size of animal that they are slaughtering to ensure that the process goes as smoothly as possible (Malaysia Standard MS 1500, 2009). Ideally, the slaughter person or the animal should be facing Mecca while slaughtering is being done. This is a secondary requirement and there are multiple opinions as to whether the animal or slaughter person be facing Mecca.

If traditional slaughter is going to be done, care must be taken to follow Islamic teachings closely to prevent any suffering of the animal. The animal should not be hung or dragged prior to slaughter. For more information as to why this is important please refer to Chapter 6 regarding animal welfare. Slaughtering must be carried out using a sharp knife in a swift sweep so that the animal does not feel any unnecessary pain. The most important factor in how much pain an animal will feel during a traditional slaughter is the size, sharpness, and lack of nicks of the knife. If the knife is dull or nicked, it will create drag on the nerves of the tissue of the neck which is painful. If the knife is not long enough, the tip of the knife will enter the wound which will create more pain for the animal. At no point should the animal be fully restrained while the knife is being sharpened nor shall the knife be sharpened within view of the animal. For modern equipment generally this means the head holder should not be applied until the slaughter person is ready to do the slaughter (Grandin, 2010).

The placement of the cut in sheep, goats, and cattle should be as high on the neck as possible without risking contact between the knife and the bones of a larynx (AMI, 2007). In camels, the cut should be placed at the base of the neck between the neck and the thorax. The placement of the cut in camels is lower because higher in the neck the carotid arteries are concealed by the cervical vertebra (Kadim, 2012).

After the neck of the animal is cut it should be allowed to bleed at least to the point when its heart stops pumping before any other processing is done. This is equally important whether the animal has been stunned or not prior to slaughter.

After the animal has died, care must be taken to ensure that a safe product is prepared for consumers. Care should be taken to prevent contamination of the carcass by digesta and debris from the hide. Once an animal has been skinned, halal assurance as well as food safety must be properly taken into account. The following sections lays out the HCP that can be used to assist in the maintenance of food safety as well as ensuring that the final product is halal. A simplified flow chart (Figure 8.1) is given to point out the HCP in red meat slaughterhouses.

HALAL CONTROL POINTS (HCP) IN ANIMAL SLAUGHTERING

HCP-1: Allowed Animals

The animal must be of acceptable halal species such as sheep, goats, cattle, or camels. Swine and carnivorous animals cannot be considered halal even if they are slaughtered in a halal manner.

HCP-2: Holding

Islam advocates merciful treatment of animals. Hence, animals must be treated such that they are not stressed or excited prior to slaughter. Holding areas for animals

Halal Production Requirements for Meat and Poultry

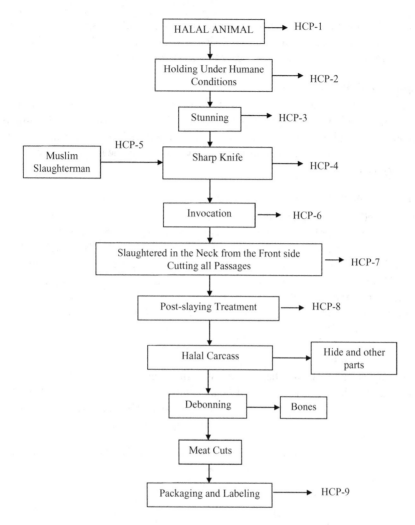

FIGURE 8.1 Halal control points for red meat slaughterhouses.

should be provided with drinking water. Excessive use of electric prods or sticks must be avoided to prevent the animals from becoming stressed (Gregory et al., 2010). Animals should be nourished and well rested. For proper pre-slaughter handling of animals as well as restraining animals for slaughter, the use of the NAMI (North American Meat Association) religious slaughter guidelines in conjunction with the general NAMI guidelines is recommended and is found in Chapter 7 (Regenstein and Grandin, 2002).

HCP-3: STUNNING

It is preferable that animals be slaughtered without stunning but using a proper humane restraining system. However, non-lethal methods of stunning might be used to meet the legal requirement for humane slaughter regulations in some countries. The animal must be alive at the time of slaughter and must die from bleeding rather than from the pre-slaughter stun (Bergeaud-Blackler, 2007).

Captive Bolt Stunning

A mechanical stunner where a bolt enters the head and then retracts, making the animal unconscious, is acceptable provided that the animal is slaughtered within two minutes of stunning. The correct stunning equipment in good repair must be used so as to ensure that all animals lose sensibility without the heart stopping before the neck can be cut.

Mushroom-Shaped Hammer Stunner

This is used in some slaughterhouses for cattle and delivers a more diffuse blow that stuns the animal for a short time. This method is not only acceptable but preferred over captive bold stunning in certain countries. There is a slightly higher risk of incomplete stuns with this method. If it is employed, employees must fully trained and always alert for incompletely stunned animals. The exact placement of the stunner on the animal's head is more critical. Therefore, the use of a head holder to position the head prior to the use of the stunner is recommended (EFSA, 2004; OIE, 2008, Section 7.5).

Electrical Stunning

Head-only stunning using the appropriate amperage/voltage that assures that the stun is reversible is acceptable; head-heart stunning is not acceptable as it stops the heart from pumping. A novel method of fully recoverable stun in development by a cross-disciplinary team in Australia is shown Figure 8.2.

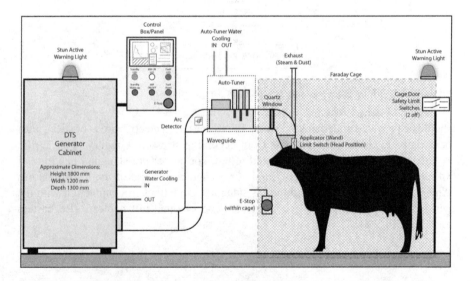

FIGURE 8.2 A novel method of fully recoverable stun in development by a cross-disciplinary team in Australia.

Carbon Dioxide Stunning or Other Forms of Gas Stunning

Gas stunning is not recommended. Further research is required to determine if it is possible to routinely make an animal insensible to pain while not causing its death.

HCP-4: Knives

Knives used for slaughter should be very sharp. The slaughter person should be trained in how to best care for their knives to ensure the best possible edge free of nicks. The size of the knife should be proportional to the size of the neck so that one may minimize the number of back and forth strokes. Dr. Grandin specifically recommends that the knife be twice the width of the animal's neck. The knife must not be sharpened in front of the animal.

HCP-5: Slaughter Person

The slaughter person must be an adult Muslim of sound mind familiar with the process of slaughter. A trained slaughter person will be more efficient, minimize damage to the skin and carcass, and minimize animal suffering (OIE, 2008; Grandin, 2006).The issue of Al-Kitab slaughter, that is, slaughter by people of the book, will be discussed separately.

HCP-6: Invocation

It is mandatory to pronounce the name of God while cutting the throat. It suffices to say Bismillah (in the name of God); however, some slaughter persons prefer to say the full statement Bismillah Allahu Akbar.

HCP-7: Slaying/Killing or Bleeding

The slaughter person must, while pronouncing the name of God, cut the front part of the neck, severing the carotids, jugulars, trachea, and esophagus, without reaching the bone or nerves in the neck. Reaching the bone in the neck both increases the risk that the spinal cord will be severed, which is not allowed for halal slaughter, and increases the risk that the knife will be damaged, which could result in unnecessary pain for future animals slaughtered with the same knife. Ideally, this cut will be done in one stroke without needing any back and forth cutting. In cattle, sheep, and goats the cut should be high up on the neck as it has been suggested that such a position limits aspiration of blood and minimizes the time required to lose consciousness.

Research has shown that up to 69% of cattle slaughtered without stunning may aspirate blood; work should be done to try to reduce blood aspiration (Gregory, 2008). However, the exact circumstances of this research cannot be obtained from the paper as published. Aspiration of the blood into the trachea is not a real animal welfare issue but aspiration into the lung tissue is a major concern (Temple Grandin, personal communication). The anatomy of camels requires that the cut be done where the neck meets the thorax.

HCP-8: Post-Slaughter Treatment

It is abominable to sever parts such as ears, horn, skin, and legs before the animal is completely lifeless. Normally when the bleeding has ceased, the heart stops, and the animal is dead. One may start further acts of processing the carcass after that. The removal of the hide and the internal organs before fabricating the carcass is done in a manner that protects the safety and quality of the meat, and is consistent with the secular laws of the country of origin and/or country of export.

HCP-9: Packaging and Labeling

Packing is then done in clean packages and boxes. Proper labels are affixed to identify the products as halal by using markings, ideally unique and trademarked by the agency certifying the product.

This chapter will now address the slaughter of birds in a step-wise manor. As with the red meat section, primary requirements, those required for the meat to be permissible for Muslims to consume, will be in bold text while secondary requirements, things that are highly encouraged based on the teachings of Prophet Muhammad (PBUH), and scientific advice will be added to clarify these points. There are other acts that can occur during slaughter that can be considered sins of the soul of the person who engages in them even if they do not make the resulting meat unfit for Muslim consumption, these reprehensible acts will be underlined.

ACCEPTABLE BIRDS

Birds for Muslim consumption must be of acceptable species and alive at the time of slaughter. Birds of prey, such as eagles and vultures are not acceptable for halal slaughter. Common birds in the halal market include chickens, ducks, turkeys, quail, and pigeons. For commercially processed poultry, birds are generally acquired from poultry farms that raise chickens specifically for that purpose, or hens may be acquired from poultry farms that raise chickens for eggs when their egg production decreases below a certain level. Chickens of any size, age, and gender may be used for halal production. The preferred feed for halal poultry should be devoid of any animal by-products or other scrap materials, which is a common practice in the West. Some halal slaughterhouses use an integrated approach, for example, where they raise their own chickens on clean feed, but most halal processors do not have any control over the feed. Muslim retailers often prefer free-range farmed chickens that are not fed animal by-products nor treated with growth promotants (illegal in the U.S.) or unnecessary antibiotics (i.e., sub-therapeutic use for growth promotion). From the halal perspective, the use of hormones in chickens for egg or meat production is discouraged; some scholars call it mushbooh (doubtful) or haram whereas others ignore it.

The animal must be safe for human consumption, wholesome, and free of gross disease and should be whole and completely healthy, free of any ailment or deformity. Problems can arise in this area when spent hens (i.e., chickens after finishing their egg laying) are sent to slaughter and may suffer from bone loss. Bone loss makes the birds more delicate at slaughter and increases the risk that legs or wings may be broken during the pre-slaughter and slaughter handling. One of the most common times in a bird's life when they are likely to experience injuries is during the loading process on the farm before going to slaughter. In all types of poultry processing, there should be safeguards in place to prevent injury during this time but it is of paramount importance that birds destined for halal slaughter are protected. Industry recommendations should be consulted so that the most humane and safe method of capture can be used. This is an area in which the halal industry should look for ways to instigate better auditing processes because as the Muslim consumer base becomes more educated about the slaughter process they are increasing likely to demand birds that are undamaged prior to slaughter. Birds that are transported from the farm to a slaughter facility are often taken directly from their transport cages to the slaughter line. It is very important that the time in transit combined with the wait time at the plant not be more than is sustainable by the bird. At no point should an animal be starved for water or prevented from drinking enough water to sustain it. As poultry cannot be given water during transport, temperature control must be maintained to prevent suffering.

Restraint for birds prior to stunning or slaughter is normally done with leg shackles, in cones, or by hand. If leg shackles are used the bird should be placed carefully and gently so as not to injure it and should always be held and hung by two legs. If cones are used the birds should be placed carefully with their wings folded to prevent damage. It is very important that as little time as possible passes between the bird being placed in the cone or on shackles and the slaughter. If restraint is done by hand, it should be with pressure firm enough to stop the bird from injuring itself but gentle enough to prevent human-induced injury or bruising of the meat.

There are some methods of stunning that are acceptable to portions of the Muslim consumer population providing they are non-lethal. Low-voltage stunning for a short time has been shown to leave birds alive. Whether electric stunning is painful to the animal has not been adequately studied and will not be discussed in this section.

Different conditions are used for electrical stunning, depending on the region of the world. Although poultry is not required by law to be stunned before slaughter in the U.S., virtually all commercial poultry is stunned for humane, efficiency, and quality reasons. The birds receive 10 to 20 mA per broiler and 20 to 40 mA per turkey for 10 to 12 seconds. These conditions give an adequate time of unconsciousness for the neck to be cut and sufficient blood to be lost to kill the bird before it regains consciousness. In most European countries, laws require poultry to be stunned with much higher amperages (90+ mA per broiler and 100+ mA per turkeys for four to six seconds). These laws and higher amperages are intended to ensure that the birds are irreversibly stunned so that there is no chance they will be able to recover and sense any discomfort. Essentially, these European electrical stunning conditions kill the bird by electrocution and cardiac arrest, stopping blood flow to the brain. Thus, death is by loss of blood supply to the brain for both stunning conditions, but one is by removal of blood and the other is by stopping blood flow to the brain. The harsher European electrical conditions also result in higher incidences of hemorrhaging and broken bones (Sams, 2001).

Use of low amperage is recommended for halal slaughtering, because low amperage stunning does not kill the birds. Amperages higher than 40 mA for turkeys and higher than 20 mA for broilers must be avoided. Each plant must establish its own working procedures depending on the size, condition, and age of the birds, so that the birds do not die due to electrocution. Carbon dioxide stunning or low atmospheric pressure stunning as it is currently used is normally a stun to kill and thus is equivalent to chemical strangulation. Death by strangulation is prohibited; hence this method is not recommended. If future research should show ways to use gas stunning without killing animals, it may then be possible to consider its use in halal slaughter.

Regardless of whether stunning is used or not, in Islam, the animal must die by the impact of a slit to the throat, cutting the carotid arteries, jugular veins, the trachea, and the esophagus, without severing the spinal cord. Mechanical or machine slaughter of birds, which was initiated in Western countries, is gaining acceptance among Muslims. Almost all countries that import chicken accept machine-killed birds. The method of slaughter by machine approved by the Malaysian government is different in the following aspects from what is usually practiced by the industry in the West.

A Muslim while pronouncing the name of God switches on the machine.

One Muslim slaughter person is positioned after the machine to make a cut on the neck if the machine misses a bird or if the cut is not adequate for proper bleeding. In commercial poultry processing, generally the machine does not properly cut 5% to 10% of the birds. A Muslim then cuts the missed birds. The Muslim backup slaughter person also continuously invokes the name of God on the birds while slaughtering and witnessing the machine kill. The settings of the blade are adjusted to make a cut on the neck, right below the head, and not across the head. A rotary knife should be able to cut at least three of the passages in the neck. Any birds that are not properly cut may be tagged by the Muslim slaughter person/inspector, and used for non-halal purposes. Two slaughter persons might be required to accomplish these requirements, depending on the line speed and efficiency of the operation.

For the actual slaughter, the throat cut must be done or watched (for mechanical slaughter) by a Muslim of sound mind while pronouncing the name of God (Bismillah or Bismillah Allahuu Akbar) on each animal. With higher speed lines, it may be necessary for there to be multiple Muslims blessing a mechanical slaughter line. Slaughtering must be carried out using a sharp, nick free knife in a swift sweep so that the animal does not feel any unnecessary pain. The stipulation applies both to a knife individually used by a slaughter person or any mechanical blade. If the machine is stopped during breaks or for any other reason it must be restarted by the previous procedure involving a Muslim slaughter person/inspector.

Whether slaughtered by hand or machine, the birds must be completely lifeless before they enter the scald tank and/or the defeathering area. The scalder conditions for defeathering, such as water temperature and chlorine level, are the same for halal processing as for regular poultry processing. It is recommended that a Muslim be stationed at the entrance to the scalder to be sure all birds have been properly bled. A bird that has not been properly bled is known as a cadaver and is illegal in many countries to enter the food supply. A halal slaughterhouse should aim to never have a cadaver. However, in poultry processing plants where both halal and non-halal birds are processed, halal birds must be completely segregated during the defeathering,

Halal Production Requirements for Meat and Poultry

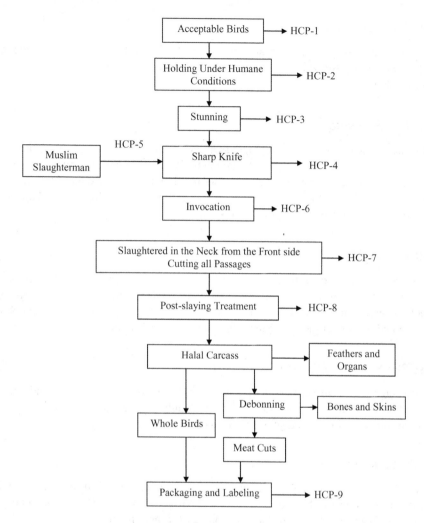

FIGURE 8.3 Halal control points (HCP) in poultry processing.

chilling, eviscerating, processing, and storing processes. It is a common practice after eviscerating to chill the birds in cold water, where they might pick up water in varying percentages. Air chilling rather than chilling with water is used by some companies that prefer it over water chilling. A simplified flow chart (Figure 8.3) is given to point out the HCP in poultry slaughter.

HALAL CONTROL POINTS (HCP) IN POULTRY SLAUGHTER

HCP-1: ACCEPTABLE BIRDS

The animal must be an acceptable halal species such as chickens, ducks, turkeys, quail, or pigeons. Eagles, vultures, and the like cannot be considered halal even if they are slaughtered in a halal manner.

HCP-2: Holding

Islam advocates the merciful treatment of animals. Hence, birds must be treated such that they are not stressed or excited prior to slaughter. Birds should be nourished and well rested.

HCP-3: Stunning

It is preferable that birds be slaughtered without stunning but using a proper humane restraining system. However, non-lethal methods of stunning might be used to meet the legal requirement for humane slaughter regulations in some countries. The animal must be alive at the time of slaughter and must die of bleeding rather than electrocution or other stunning system (EFSA, 2004).

HCP-4: Knives

The knife or mechanical blade must be sharp and free of nicks so that the bird does not feel the pain of the cut. It is even more important for the knife to be sharp and nick free when the birds are slaughtered without any stunning. The knife must not be sharpened in front of the bird. Must a halal bird not see the slaughter of another bird?

HCP-5: Slaughter Person

The slaughter person must be an adult male or female Muslim of sound mind familiar with the process of slaughter. A trained slaughter person will be more efficient and minimize pain or damage to the animal.

HCP-6: Invocation

It is mandatory to pronounce the name of God while cutting the throat or watching the throat cut on a mechanical line. It suffices to say Bismillah (in the name of God); however, Bismillah Allahu Akbar is encouraged if there is time.

HCP-7: Slaughter/Killing or Bleeding

The slaughter person must, while pronouncing the name of God and with a swift cut, cut the front part of the neck severing carotid arteries, jugular veins, the trachea, and the esophagus, without reaching the bone in the neck. This procedure is the same for the person cutting missed birds on a mechanical slaughter line.

HCP-8: Post-Slaughter Treatment

It is abominable to sever any part of the bird or allow entrance to the scald tank before the bird is completely lifeless. Normally when the bleeding has ceased, the heart stops, and the bird is dead. At that point, further acts of processing may start (Zivotofsky and Strous, 2012).

HCP-9: PACKAGING AND LABELING

Packing is then done in clean packages and boxes, and proper labels are affixed to identify the products with preferably a trademarked symbol of the halal certifying agency.

The following section refers to all meat products.

HALAL REQUIREMENTS FOR DEBONING AND PROCESSING ROOMS

The same conditions that apply to abattoirs also apply to boning and cutting establishments, with respect to segregation of halal from non-halal, cleaning, certification, packing, and labeling products with a proper symbol of halal certification. It is imperative that all halal products coming into a processing facility from other establishments be accompanied by a halal batch certificate that is received and reviewed by a Muslim inspector (Ayan, 2001).

HALAL REQUIREMENTS FOR COLD STORES

All incoming halal load must be received and reviewed by a Muslim inspector if the products are not sufficiently sealed either individually or in a bulk container. Halal products must be clearly separated from other products during blast freezing. The freezer cannot be used to freeze halal and non-halal products at the same time.

Frozen halal products must remain isolated from non-halal products in the freezer unless totally sealed so cross-contamination is impossible (Khawanjah, 2001).

Halal products should be loaded separately from non-halal products under the supervision of a Muslim inspector. In a mixed halal and non-halal area or container, Halal products should be placed above non-halal products to avoid potential cross-contamination. All halal products transported from the cold store must be accompanied by a batch certificate for any bulk-packed containers. All halal products loaded for export must be accompanied by a halal certificate.

FURTHER PROCESSED MEAT ITEMS

Meat and poultry products can be marketed fresh or frozen, and can also be used for further processing. A simplified flow chart (Figure 8.4) is given to point out the HCP in further processing.

HCP-1: MEAT SOURCE

Halal inspectors and quality assurance personnel must make sure that meat received from a slaughterhouse or fabricator is acceptable as halal according to the standards of the halal certifying agency of the receiving plant.

HCP-2: EQUIPMENT

The equipment used for halal production must be clean and then inspected by the halal inspector as well as by the plant's quality assurance personnel and the USDA

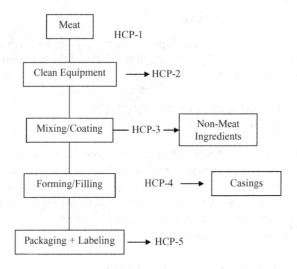

FIGURE 8.4 Further processed meat products.

FSIS or equivalent agency in other countries. The company may use equipment that is used for non-halal meat only after thorough cleaning, but it may not use any equipment that has been used for pork processing unless a DNA PCR test is run on the equipment and product prior to issuing the certificate. If a PCR test is not possible then religious cleaning before production may sometimes be an acceptable alternative. It is almost impossible to clean meat processing equipment under normal operational conditions, which is why pork and non-pork equipment must almost always be segregated. If one has to convert the equipment that has been used for pork to make halal products, the equipment must be religiously cleansed.

An acceptable method for religious cleaning includes the following: (1) Thorough washing of the equipment with hot water and detergent to get rid of visible traces of non-halal meat. (2) Rinsing of the equipment thoroughly with clean water by a Muslim inspector to make it acceptable for halal production. This is sometimes known as religious cleansing. All halal products should be the first to be produced after a sanitation shift if it is a shared facility.

HCP-3: NON-MEAT INGREDIENTS

Thousands of ingredients are approved for use in meat products. One must ensure that prohibited materials do not become part of halal products. Some of the ingredients to be avoided include, but are not limited to, non-halal gelatin, lard, pork extract, natural bacon flavor, other ingredients derived from non-halal animals or halal animals that were not halal slaughtered, and ingredients containing more than 0.5% alcohol. Gelatin is a particularly complex material with respect to its halal status (please see Chapter 14).

Appendix C lists halal, doubtful, and haram ingredients. Meat packers generally receive non-meat ingredients from spice companies or directly from ingredient manufacturers. It is advisable to ask suppliers for halal certificates for all blended products or complex materials such as seasonings and spice blends, batters and breading, and smoke and other flavorings. The halal certificate needs to be acceptable to the plant's halal certifying agency.

HCP-4: CASINGS

Casings can be edible or inedible. Some meat products use casings whereas others do not. Three types of casings are available and used according to the type of product.

> *Natural casings*: These are made from animal guts. These can be from lambs, sheep, goats, cows, or even pigs. Pig casings must not be used for halal products. Among other animals, casing can be from halal-slaughtered animals, generally from Muslim countries, or both halal and from non-halal-slaughtered animals, generally from Western countries. It is required that only casings from halal-slaughtered animals be used (Nakyinsige et al., 2012).
> *Collagen casings*: These are made from finely ground cattle skins or theoretically can be made from pork skins. Because these are edible casings, they should be from halal-slaughtered animals.
> *Cellulose casings*: These are not edible casings. They are peeled off after the product is formed and cooked. Cellulose casings are made with cellulose (a plant material) and other ingredients such as glycerin, which may or may not be halal (see Chapter 22). Halal-certified cellulose casings are available from major manufacturers.

HCP-5: PACKAGING AND LABELING

The final step in the manufacture of processed meat is packing the product in the right containers and labeling them to accurately identify them with respect to the halal certifying agency.

INDUSTRY PERSPECTIVE ON HALAL PRODUCTION

> According to Jackson (2000), until now many Muslims accepted kosher meat products because they believed the slaughter was similar to their requirements and because the animals at least received a blessing at the time of slaughter. However, the blessing is traditionally a group blessing, that is, not for individual animals. However, in some cases, a Jewish slaughter person may say Bismillah Allahu Akbar over each animal while a Muslim observes the slaughter. In other cases, a Muslim may say the Bismillah at the time of slaughter either separate from the Jewish slaughter person or while touching the slaughter person. Some scholars accept this meat as halal based on the permission in the Quran to eat from the table of the Al-Kitab (people of the book).
>
> **surah al-ma'idah 5:5**

This day [all] good foods have been made lawful, and the food of those who were given the Scripture is lawful for you and your food is lawful for them.

Until recently, commercial halal-certified meat products were virtually nonexistent in U.S. supermarkets except for imported products and locally slaughtered meat. Muslim consumers are increasingly looking for convenient halal options available in supermarkets which require that meat be certified as halal and properly packaged because in a supermarket setting word of mouth assurances are not possible. Also, as the Muslim consumer base expands, it is beginning to see halal certification as integral to retaining a chain of custody and accountability for their meat suppliers. Many Muslim-majority countries require prior certification by an accredited halal certifier for imported meats.

Some meat producers think that to be halal, they only have to follow a set of procedures such as those in this chapter. Companies following this policy will encounter significant marketing problems. Proper halal certification by a reputable halal certifying agency is becoming more important in the marketplace.

Companies following a policy of simply labeling their products as halal without backing by a certification agency will encounter marketing problems. The following are some corrections to notions on what can be considered halal meat (adapted from Jackson, 2000):

- Muslim inspectors cannot say a blessing on a truck as it passes their houses on its way from the slaughterhouse to qualify the resultant meat as halal acceptable.
- Inspectors cannot say a blessing only at the start of the slaughtering process. It must be said throughout the process as each animal is slaughtered.
- A Muslim cannot say a blessing after all slaughtering is completed to cover all animals slaughtered that day.
- Inspectors cannot use recordings of blessing to substitute for the devotion of an observant Muslim.
- Meat producers cannot accept the word of the slaughterhouse that halal methods were used and the meat therefore should be considered halal.
- Further processors cannot accept that a product labeled as halal is indeed produced halal. It must be certified by a halal certifying agency and that certification accepted by the plant's certifying agencies.
- Producers must never label a meat product as halal if there is no on-site Muslim participation.
- Producers cannot simultaneously process any pork or pork-derived product while producing halal-labeled meat.
- Producers cannot process any pork or pork-derived product immediately prior to the processing of any halal-labeled meat product without a full, comprehensive, and detailed religious cleaning. They ideally should also perform a pork DNA PCR test. Sampling should be done in accordance with guidelines set up by their halal certification agency.

REFERENCES

American Meat Institute. (2007). *Animal handling: Religious slaughter.* Washington, DC: American Meat Institute.

Anil, M. H., Raj, A. B. M., and McKinstry, J. L. (2000). Evaluation of electrical stunning in commercial rabbits: Effect on brain function. *Meat Science*, 54(3), 217–220.

Anil, M. H., Love, S., Helps, C. R., and Harbour, D. A. (2002). Potential for carcass contamination with brain tissue following stunning and slaughter in cattle and sheep. *Food Control*, 13(6–7), 431–436.

Apple, J. K., Kegley, E. B., Maxwell, C. V., Rakes, L. K., Galloway, D., and Wistuba, T. J. (2005). Effects of dietary magnesium and short-duration transportation on stress response, postmortem muscle metabolism, and meat quality of finishing swine 1. *Journal of Animal Science*, 83(7), 1633–1645.

Ayan, A. H. (2001). Halal food with specific reference to Australian exports. *Food Australia*, 53(11), 498–500.

Bergeaud-Blackler, F. (2007). New challenges for Islamic ritual slaughter: A European perspective. *Journal of Ethnic and Migration Studies*, 33(6), 965–980.

Casoli, C., Duranti, E., Cambiotti, F., and Avellini, P. (2005). Wild ungulate slaughtering and meat inspection. *Veterinary Research Communications*, 29(2), 89–95.

European Food Safety Authority. (2004). A welfare aspects of animal stunning. Available online: www.halalfoodauthority.co.uk. Accessed on May 17, 2018.

Gibson, T. J., Johnson, C. B., Murrell, J. C., Hulls, C. M., Mitchinson, S. L., Stafford, K. J., Johnstone, A. C., and Mellor, D. J. (2009). Electroencephalographic responses of halothane-anaesthetised calves to slaughter by ventral-neck incision without prior stunning. *New Zealand Veterinary Journal*, 57(2), 77–83.

Grandin, T. (2006). Progress and challenges in animal handling and slaughter in the US. *Applied Animal Behaviour Science*, 100(1), 129–139.

Grandin, T. (2010). Auditing animal welfare at slaughter plants. *Meat Science*, 86(1), 56–65.

Gregory, N. G., Von Wenzlawowicz, M., and Von Holleben, K. (2008). Blood in the respiratory tract during slaughter with and without stunning in cattle. *Meat Science*, 82(1), 13–16.

Gregory, N. G., Fielding, H. R., von Wenzlawowicz, M., and Von Holleben, K. (2010). Time to collapse following slaughter without stunning in cattle. *Meat Science*, 85(1), 66–69.

Hambrecht, E., Eissen, J. J., Nooijen, R. I. J., Ducro, B. J., Smits, C. H. M., Den Hartog, L. A., and Verstegen, M. W. A. (2004). Preslaughter stress and muscle energy largely determine pork quality at two commercial processing plants. *Journal of Animal Science*, 82(5), 1401–1409.

Jackson, M. A. (2000). Getting religion: For your products, that is. *Food Technology*, 54(7), 60–66.

Kadim, I. T. (2012). Camel meat and meat products. Oxfordshire, UK: CABI.

Kannan, G., Kouakou, B., Terrill, T. H., and Gelaye, S. (2003). Endocrine, blood metabolite, and meat quality changes in goats as influenced by short-term, preslaughter stress 1. *Journal of Animal Science*, 81(6), 1499–1507.

Khawanjah, M. (2001). Part of transcript of the talk given, Killing methods (Question No. EFSA-Q-2003-093). *Scientific Report of the Scientific Panel for Animal Health and Welfare*. Available online: www.halalfoodauthority.co.uk.

Limon, G., Guitian, J., and Gregory, N. G. (2010). An evaluation of the humaneness of puntilla in cattle. *Meat Science*, 84(3), 352–355.

Ljungberg, D., Gebresenbet, G., and Aradom, S. (2007). Logistics chain of animal transport and abattoir operations. *Biosystems Engineering*, 96(2), 267–277.

Micera, E., Albrizio, M., Surdo, N. C., Moramarco, A. M., and Zarrilli, A. (2010). Stress-related hormones in horses before and after stunning by captive bolt gun. *Meat Science*, 84(4), 634–637.

Malaysia Standard MS1500 (2009). *Halal food—Production, preparation, handling and storage—General guideline*. Cyberjaya: Department of Standards Malaysia, 1–13.

Nakyinsige, K., Che Man, Y. B., Sazili, A. Q., Zulkifli, I., and Fatimah, A. B. (2012). Halal meat: A niche product in the food market. *2nd International Conference on Economics, Trade and Development IPEDR*, Singapore, 36, 167–173.

OIE. (2008). Slaughter of animals. Paris, France: *Terrestrial Animal Health Code World* on a request from the Commission related to welfare aspects of animal stunning.

Önenç, A. and Kaya, A. (2004). The effects of electrical stunning and percussive captive bolt stunning on meat quality of cattle processed by Turkish slaughter procedures. *Meat Science*, 66(4), 809–815.

Qureshi, S. S., Jamal, M., Qureshi, M. S., Rauf, M., Syed, B. H., Zulfiqar, M., and Chand, N. (2012). A review of Halal food with special reference to meat and its trade potential. *Journal of Animal and Plant Sciences*, 22(2 Suppl), 79–83.

Regenstein, J. M. and Grandin, T. (2002). Animal welfare: Kosher and halal, *IFT Religious Ethnic Food Div. News Letter*, 5(1), 2.

Sams, A. R. (2001). *Poultry Meat Processing*, Boca Raton, FL: CRC Press.

Schaefer, A. L., Dubeski, P. L., Aalhus, J. L., and Tong, A. K. W. (2001). Role of nutrition in reducing antemortem stress and meat quality aberrations. *Journal of Animal Science*, 79(Suppl_E), E91–E101.

Small, A., McLean, D., Keates, H., Owen, J. S., and Ralph, J. (2013). Preliminary investigations into the use of microwave energy for reversible stunning of sheep. *Animal Welfare*, 22(2), 291–296.

Zivotofsky, A. Z. and Strous, R. D. (2012). A perspective on the electrical stunning of animals: Are there lessons to be learned from human electro-convulsive therapy (ECT)? *Meat Science*, 90(4), 956–961.

9 Halal Processed Meat Requirements

Mustafa Farouk, Mian N. Riaz, and Munir M. Chaudry

CONTENTS

Halal Processed Meats 125
    Halal and Wholesome 125
    Halal and Wholesome Processed Meats 126
    Meeting the Halal and Wholesome Processed Meats Criteria 126
Halal and Wholesome Processed Meats Ingredients 126
    Inherently Halal and Wholesome 129
    Generally Accepted as Halal and Wholesome 129
    Halal Animal-Derived Halal Ingredients 129
    Halal Animal-Derived Non-Halal Ingredients 129
    Plant-Derived Halal Ingredients 129
    Mineral/Petroleum-Derived Halal Ingredients 129
    Other Additives 130
Categories, Processing, and HCP for Halal Processed Meats 130
    Halal Processed Meats Categories 130
    Halal Processed Meats Manufacture and HCPs 130
Wholesome Aspects of Halal Processed Meats 134
    Processed Meats Wholesome-Related Incidents and Regulations 136
    Halal and Wholesome Processed Meats Certification, Accreditation, and Auditing 139
    Processed Meats Accreditation by International Standards Organizations (ISO) 142
References 142

HALAL PROCESSED MEATS

HALAL AND WHOLESOME

According to Hussaini and Sakr (1982), "halal" is an Arabic word meaning "allowed" or "lawful." The word "wholesome" means to be good, clean, gentle, excellent, fair, and lawful. Halal should be viewed legalistically as any action, product, or food that a Muslim is allowed to consume or partake in. And from the food industry perspectives, "wholesome" can be equated to food quality and safety.

Both the halal and wholesome aspects of processed meats should be taken into consideration in establishing the Halal Control Points (HCPs) for any processed meat product.

Halal and Wholesome Processed Meats

Processed meats are meats whose inherent characteristics have been altered by further processing involving more than the simple act of grinding, cutting, or mixing. This include meats that have undergone processes, such as curing, smoking, dehydration, or where certain additives such as enzymes have been used (Pearson and Gillett, 1999). When such meats are produced to comply with halal and wholesome requirements, they are termed "Halal and Wholesome Processed Meats." Thus, the only difference between conventional processed meats and their halal counterparts is that the latter is produced using halal and wholesome meat and ingredients throughout the production process up to the finished product and the supply chain to the consumer.

Exact data on the global production of processed meats is difficult to obtain due to the wide range of products that fall under this category and the many ways they are produced, used, and merchandized. Therefore, the importance of processed meats can be best gauged by the volume and value of these products in the major markets including halal markets and the caliber of the companies that manufacture these products globally (Farouk, 2010, 2011).

Meeting the Halal and Wholesome Processed Meats Criteria

It is not difficult to produce processed meats that meet the halal and wholesome requirements as long as processors use halal and wholesome meat, plant-based ingredients wherever possible and avoided doubtful ingredients (Farouk, 1997). Ensuring that halal and wholesome criteria are met should be built into each processor's quality control (QA) and auditing procedures. The ingredients listed on processed meats' packages must be clear and unambiguous, consistent with national regulations, and the design and graphics chosen should not be offensive to avoid doubts in the minds of halal consumers and to ensure the success of the products. To avoid costly mistakes and delays, it is advisable to work with reputable and recognized halal certifying bodies accepted in the processed meats target markets along with an internal HCP program.

HALAL AND WHOLESOME PROCESSED MEATS INGREDIENTS

Only halal and wholesome ingredients can be used to produce halal processed meats. The slaughter processes involved in the production of halal meat from halal livestock are discussed in other chapters of this book and elsewhere (Farouk et al., 2013, 2014, 2015). Meat from fresh and salt water animals that live in water at all times is halal unless proven harmful to health (see the more detailed discussion of seafood in Chapter 10. Other ingredients and aids commonly used in processed meats and their HCP are listed in Table 9.1. These ingredients can be categorized into the following for the ease of meeting the halal requirements during the formulation and production of halal processed meats.

TABLE 9.1
Processed Meats Ingredients and their Halal Critical Considerations

Primary Processing Function	Sources/Ingredients/Organisms	Halal Control Points/Considerations
Meat–main ingredient	Goat, sheep, cattle, camel, buffalo, deer, fish, poultry, etc.	Livestock have to be slaughtered or harvested according to the way taught in Islam with due consideration for the welfare of all species concerned
Cure/curing	Salt; sugar/corn syrup solids/honey; nitrites and/or nitrates; phosphates; sodium ascorbate and erythorbate; potassium sorbate; sodium and potassium lactate	Curing generally requires the addition of salt, sugar, and either sodium nitrite ($NaNO_2$), sodium nitrate ($NaNO_3$), potassium nitrite (KNO_2), or potassium nitrate (KNO_3). These chemicals could be toxic and are often colored pink to distinguish them from regular salt. Making sure the curing agent are used in amounts that do not exceed regulatory limits
Spices/flavors/flavor enhancers	Ginger; mustard; pepper; chilies; coriander; fennel; allspice; nutmeg; mace; fenugreek; garlic; onion; turmeric; rosemary; paprika; thyme; clove; oregano; cardamom; monosodium glutamate (MSG); etc.	Spices are a key to defining the character and the uniqueness of processed meat products. They are supplied singly or as premixes, and as oleoresins or extracts. As long as they are from nontoxic plant sources and are not contaminated they are halal. It is important to ensure that in case of premixed or ground spices, extracts, and oleoresins, that they are not mixed with any other functional additives that may not be halal or are extracted/diluted with non-halal solvents (HCP)
Binders/fillers/extenders/emulsifiers/stabilizers	Cereal; starch; vegetable flour; soy protein concentrate; non-fat dry; textured vegetable proteins (TVP); milk (NFDM); soy flour; whey proteins; eggs; dried yeast; yeast extract; potato and tapioca flours; gelatin; carrageenan; alginates; transglutaminase; etc.	These are non-meat products used in processed meats for several reasons including to reduce cost, to improve yield, to increase protein content, to improve emulsion stability and slicing characteristics, and to increase fat and water binding. Their level of additions in formulations are specified in regulations. Ensuring they are from halal sources and safe is an HCP

(Continued)

TABLE 9.1 (CONTINUED)
Processed Meats Ingredients and their Halal Critical Considerations

Primary Processing Function	Sources/Ingredients/Organisms	Halal Control Points/Considerations
Processing aids	Casings; molds; cans; equipment lubricants; etc.	Casings are commonly used in meat processing. Casings can be natural or artificial. Natural casings are from the gastrointestinal tract of animals and artificial casings are mostly from collagen and cellulose. Regardless of the source of the casing, it must be halal and nontoxic. Cans and other processing aids must not be laminated or coated with non-halal materials (HCP). Lubricants used in the plant must also be halal certified (HCP)
Antioxidants/preservatives	Tocopherols (vitamin E, alpha-tocopherol); fat-soluble antioxidants (butylated hydroxyanisole, BHA; butylated hydroxytoluene, BHT); etc.	Some spices such as sage, mustard, rosemary, and ginger provide natural antioxidant activity, however, the artificial antioxidants are generally more effective. The use of harmful chemicals such as sulfur dioxide, nisin, or formalin in processed meats to mask product spoilage or to artificially enhance certain product qualities must be avoided and strict adherence to specified levels must be observed
Starter cultures	*Pediococcus cerevisiae; P. acidilactici; Lactobacillus plantarum;* etc.	Starter cultures may be obtained from commercial sources in frozen or freeze-dried forms and usually include a blend of two or more types or strains of organisms. The organisms most commonly used for fermentation belong to the genera Pediococcus, Micrococcus, and Lactobacillus. Starter cultures of yeast (*Debaryomyces hansenii*) or mold (*Penicillium nalgiovense*) are available for ripening sausages. Large processors may have custom-developed starters whose content is a trade secret, but generic blends are available. These must be verified halal (HCP). The use of transgenic microbes should be from a halal source

Halal Processed Meat Requirements

INHERENTLY HALAL AND WHOLESOME

These are fresh, unprocessed, nontoxic ingredients that are naturally halal to be used in any amounts and combinations. Examples include fresh fruits, vegetables, honey, milk, acceptable fish, eggs, onions, garlic, herbs, and spices. Halal certificates may not be required for ingredients in this category.

GENERALLY ACCEPTED AS HALAL AND WHOLESOME

These are the unadulterated individually further processed inherently halal and wholesome ingredients or a mixture of these ingredients. These include chopped, dried, liquefied fruits and vegetables, condensed, concentrated, skimmed forms of milk, or liquefied honey. Halal certificates may still be required for ingredients in this category to ensure they are not contaminated post-processing and by packaging.

HALAL ANIMAL-DERIVED HALAL INGREDIENTS

Ingredients derived from animals can include gelling agents (e.g., gelatin), casings (collagen and natural casings), fats (mono and di-glycerides, oleic, linoleic, and tallow), calcium (from bones), and flavor carriers or enhancers. Halal certification is required for these ingredients.

HALAL ANIMAL-DERIVED NON-HALAL INGREDIENTS

These are ingredients derived from halal animals but they are not halal and cannot be used as ingredients or processing aids in halal processed meats. These include blood or blood-based products such as blood serum proteins (albumin or Fibrimex) used as a cold-set binder for value-added processed meats.

PLANT-DERIVED HALAL INGREDIENTS

A variety of plant-derived ingredients are used in processed meats to serve various purposes including as gelling agents (carrageenan, alginates), binders/fillers and bulking agents (e.g., soy proteins, alginates, starches, gums and colloids), fat sources (vegetable oil, lecithin, mono and di-glycerides)), fat mimics (konjac flour, xanthan gums), antioxidants (tocopherol, rosemary), encapsulating agents (agar-agar, guar gum), release and anti-stick agents (vegetable oils, starches), and acidulants. Extracts and essences of spices and herbs (e.g., garlic and pepper resins) are acceptable as long as the extracting medium is halal. Halal certification is required to accompany these ingredients.

MINERAL/PETROLEUM-DERIVED HALAL INGREDIENTS

Substances derived from minerals such as sodium chloride, nitrate salts, and phosphates are acceptable to be used at predetermined safe levels in halal processed meats. Halal certification may be required. Petroleum-based products are considered acceptable.

OTHER ADDITIVES

Fat-soluble antioxidants such as BHA, BHT, DMP, catechins, and quercetin, and water-soluble additives such as citric, ascorbic, and phosphoric acids are acceptable for use in halal processed meats.

In this chapter, only the general methods, technologies, and HCP in the manufacture of the broad categories of processed meats in Table 9.1 are discussed to provide guidance on the relevant halal issues regarding these products.

CATEGORIES, PROCESSING, AND HCP FOR HALAL PROCESSED MEATS

HALAL PROCESSED MEATS CATEGORIES

For the purpose of this chapter, processed meats are grouped into 18 broad categories based on the level of comminution and preservation technologies/methods involved in the primary and secondary stages of the product manufacture (Table 9.2) (Farouk, 2011). In this chapter, only the general methods, technologies, and HCP in the manufacture of the broad categories of processed meats in Table 9.2 are discussed to provide guidance on the relevant halal issues regarding these products.

HALAL PROCESSED MEATS MANUFACTURE AND HCPs

The general flow diagram for the manufacture of pre-cooked sausages (Figure 9.1), pot roast, jerky, and canned luncheon meat (Figure 9.2), salami or pepperoni, and breakfast beef or related products (Figure 9.3) were summarized from Lucy and Kiteley (1982), Pearson and Gillett (1999), Salvage (1999), Hoogenkamp (2001), Farouk (2011), and Farouk and Bekhit (2013). These references provide the relevant steps in the manufacture of these processed products and the critical controls points during their manufacture. However, the halal status of the finished products also require adherence to the relevant HCP. These HCPs are determined using the following broad considerations:

1. The importance of making sure that all incoming raw materials and ingredients are halal compliant and from acceptable suppliers and certifiers
2. Ensuring the proper storage of the received materials and their traceability
3. Determining the ease with which halal ingredients and processing aids can be contaminated with non-halal or hazardous biological, chemical, or physical materials
4. Determining the ease with which process specifications including time, temperature, and weights can be missed, abused, or neglected
5. Ensuring the whole process flow eliminates any chance of cross contamination from equipment, water supply, air, and other sources
6. Ensuring that finished and raw ingredients are separated during preparation and storage
7. Ensuring the cleaning process is adequate for general hygiene and to avoid cross contamination.

Halal Processed Meat Requirements

TABLE 9.2
Categories and Examples of Processed Meats and Their General Food Safety (Wholesome) Requirements

Primary Processing Step	Secondary Processing Step	Product Examples and their Halal and Wholesome Assurance and Control Points/Considerations (HCP)
Whole-tissue	Non-cured	Pre-cooked roasts, steaks, delicatessen, rotisserie chicken Cooking process must be sufficient to render product microbiologically safe for intended purpose. A 6D process for destruction of *L. monocytogenes* is sufficient to inactivate all other vegetative forms of pathogens. Validated alternatives may be proposed. Cooling processes must ensure no growth of *Clostridium botulinum*, and growth of *C. perfringens* is limited to 1 $\log_{10}$ (accounting for lag phase)
	Cured/cured smoked	Pastrami, corned silverside, beef bacon, breakfast beef Cooking/smoking to achieve same outcome as non-cured. Products must meet microbiological limits in food standards for product. Addition of salt, nitrate/nitrite and/or adjuncts to preserve color and prevent spoilage
	Non-fermented semi-dry/dried	Jerky, biltong, beef prosciutto/bresaola Reduction in water activity (A_w) to a level that will ensure inactivation of targeted vegetative pathogens and/or viable spores. Typically, dry meats have an A_w of <0.6, while semi-dry products have an A_w of 0.6–0.85
	Fermented semi-dry/dried	Pastarma Raw material to be protected from contamination. Reduction in pH (and potentially also Eh) through fermentation using a starter culture. Reduction in water activity (A_w) to a level that will ensure inactivation of targeted vegetative pathogens and/or viable spores
	Canned/pouched	Corned beef, canned sliced beef Cooking process must be sufficient to render product microbiologically safe for intended purpose. A 12D process for destruction of *C. botulinum* is sufficient to inactivate all other vegetative and spore-forming pathogens; where heat tolerance of product permits, higher/longer temperature/time processes, these are recommended
	Specialty items/cocktails	Breaded/crumbed steaks, sandwiches containing meat, smoked tongue Operator to establish measurable food safety objectives, i.e., microbiological limits, time/temperature combination or performance criteria

(Continued)

TABLE 9.2 (CONTINUED)
Categories and Examples of Processed Meats and Their General Food Safety (Wholesome) Requirements

Primary Processing Step	Secondary Processing Step	Product Examples and their Halal and Wholesome Assurance and Control Points/Considerations (HCP)
Coarsely comminuted/diced/sliced/pulled	Non-cured	Restructured roasts, deli meats, hamburgers, patties, stews, pulled beef, kebabs
		Same food safety considerations as for whole-tissue non-cured products but more important because contamination is distributed throughout the product rather than only on the surface
	Cured/cured smoked	Sucuk, kielbasa
		Same food safety considerations as for whole-tissue cured products
	Non-fermented semi-dry/dried	Dry salami, dry chorizo, meat floss
		Same food safety considerations as for whole-tissue category
	Fermented semi-dry/dried	Salami, chorizo, pepperoni
		Same food safety considerations as for whole-tissue, non-cured products. Additionally, uncooked comminuted fermented meats (UCFM) must meet the requirements of the UCFM Standard.
		Cooked comminuted fermented meats (CCFM) must meet the regulatory microbiological limits for cooked/cured meat products in, for example, the Food Standards Code, Standard 1.6.1 (New Zealand). [Relevance to the US—FSIS regulations.]
	Canned	Luncheon meat, meatballs
		Same food safety considerations as for whole-tissue canned products
	Specialty items/cocktails	Aspics, meat loaves, restructured steaks, sandwich steaks, crumbed/breaded products, microwaveable breakfast sandwiches
		Operator to establish measurable food safety objectives, i.e., microbiological limits, time/temperature combination or performance criteria

(Continued)

TABLE 9.2 (CONTINUED)
Categories and Examples of Processed Meats and Their General Food Safety (Wholesome) Requirements

Primary Processing Step	Secondary Processing Step	Product Examples and their Halal and Wholesome Assurance and Control Points/Considerations (HCP)
Finely comminuted/emulsion	Non-cured	Cooked fresh sausages
		Same food safety considerations as for whole-tissue and coarsely ground
	Cured/cured smoked	Pre-cooked sausages, bologna, frankfurters (hot dogs)
		Same food safety considerations as for whole-tissue and coarsely ground
	Non-fermented semi-dry/dried	Mortadella
		Same food safety considerations as for whole-tissue and coarsely ground
	Fermented semi-dry/dried	Summer sausages, pepperoni
		Same food safety considerations as for whole-tissue and coarsely ground
	Canned	Canned luncheon meat, canned sausages
		Same food safety considerations as for whole-tissue and coarsely ground
	Specialty items/cocktails	Meat spreads, meat loaves, jellied roasts; pâté
		Same food safety considerations as for whole-tissue and coarsely ground

(Additional Information in Farouk and Mills, 2016)

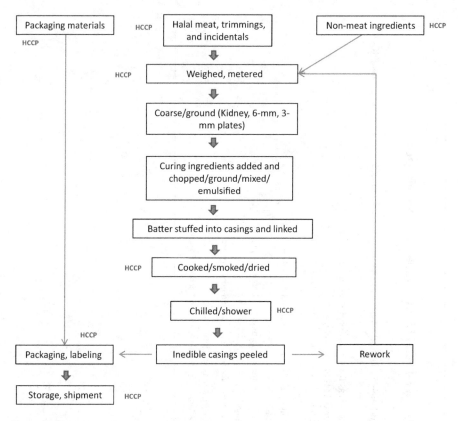

FIGURE 9.1 Process flow-chart for the manufacture of pre-cooked sausages and halal critical control points (HCCPs).

The main HCPs are shown in Figure 9.1. Most of these points are similar for most processed meats, with some variations as seen in Figures 9.2 and 9.3. The most common of these HCPs for processed meats and the associated preventative, monitoring, and corrective actions relevant to halal meat processing are included in Table 9.3.

WHOLESOME ASPECTS OF HALAL PROCESSED MEATS

In establishing the HCP for processed meats, some of the trends in consumer demands and the response of the meat/food industries to those demands should be considered carefully to ensure the wholesome aspects of processed meats are not compromised. Some of the recent consumer trends include the demand for the following: (1) Natural—the demand for natural foods is growing and manufacturers of processed meats are responding by avoiding the use of artificial colors, flavors, and preservatives (Messenger, 2007; Newswire.com, 2010); (2) Reduced/no salt—sodium reduction in processed meats is a growing trend (Pszczola, 2010; Tarver, 2010); (3) Organic/country of origin—processed deli meats produced from free-range and locally based producers that are free from hormones, antibiotics, artificial

Halal Processed Meat Requirements

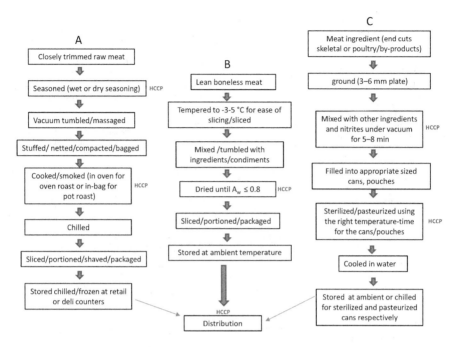

FIGURE 9.2 Generalized flow-charts for the manufacture of (A) pot roasts and other non-cured pre-cooked meats that are sliced for retail at deli; (B) jerky and (C) canned luncheon meat. Additional halal critical control points (HCCPs) to ones in Figure 9.1.

additives, and are considered "authentic," "real," and free of negative ingredients are growing in popularity (Crews, 2011; Sloan, 2015); (4) Less processed—consumers are seeking less processed foods with fewer ingredients (Browne, 2011); and (5) Convenience—a review of ready-to-eat convenience meats by Farouk (2011) concluded that the trend for convenience revolves around smaller product portions for single consumers (Figure 9.2).

Some of the ingredients that meat processors are using to meet the trends for naturalness, salt reduction, and minimum processing include the use of natural cures such as sea salt, vegetable juice powders, and concentrates such as celery juice powder, beet juice powder, celery juice concentrate, and carrot juice concentrate (O'Donnell, 2009; Pellegrini, 2009; Sindelar et al., 2007). All of these materials contain high levels of sodium nitrate so product safety can be maintained. Potassium chloride, autolyzed yeast extracts and hydrolyzed vegetable proteins are some of the most common substitutes for salt. The texture and appearance of pre-cooked ready-to-eat (RTE) meats are being improved by coating, battering, breading, crumbing, or encrusting using ingredient blends tailored to give a certain finish to the product. Many of the processed meats that are stored chilled during merchandising could be relatively mildly processed and unpreserved because of the consumer demand for natural, organic, and preservative-free. Cured meats with claims such as "all natural" and "no added nitrites/nitrates" are in the market posing another challenge to the wholesome aspects of the products. In addition to processors not achieving sufficient

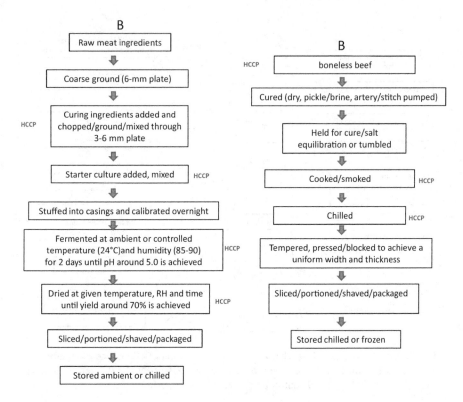

FIGURE 9.3 Generalized flow-chart for the manufacture of (A) salami, pepperoni, and other semi-dry/dried fermented sausages; (B) breakfast beef and related products. Additional halal critical control points (HCCPs) to ones in Figure 9.1.

level of control, which could lead to products that are improperly processed; other issues may also arise due to poor tracking/recording of raw materials origins, not being able to track finished processed RTE products' physical location, and temperature abuse along the supply chain (Pellegrini, 2008). These challenges require additional wholesome measures, apart from simple reliance on the chill chain if the risk of food poisoning and overall reduction in wholesomeness is to be minimized.

The major challenge in terms of being "wholesome" for some of these trendy processed meats products where natural ingredients are used, the salt content is reduced, and/or are minimally processed, is being able to meet the need of consumers for these desired attributes and still make the products more (or at least equally) wholesome than their counterparts (O'Donnell, 2009).

PROCESSED MEATS WHOLESOME-RELATED INCIDENTS AND REGULATIONS

The majority of the wholesome risks associated with RTE processed meats are due to microorganisms. A range of pathogenic bacteria have been shown to be associated with RTE meats, including *Listeria monocytogenes, Salmonella spp., Enterohaemorrhagic Escherichia coli (EHEC), Campylobacter spp., Staphylococcus*

Halal Processed Meat Requirements

TABLE 9.3
Common Halal and Wholesome Assurance Control Points (HCP) for Processed Meats

HCP	Process Stage	Halal and Wholesome Risks	Preventative Measures	Monitoring	Corrective Action (CA)
1	Meat and ingredient procurement	Noncompliant meats, ingredients, contamination, pathogens, foreign bodies, animal welfare abuse	Agree on written specifications and list of approved suppliers	Routine supplier audits, certificates, QA records	Reject non-compliant ingredients or from unapproved suppliers
2	Meat and ingredients intake and storage	Cross contamination with non-halal, spoilage, physical contamination with insects, metals, and other substances	Inspect all deliveries, clear labelling and separation of ingredients, and strict temperature control	Check QA records, temperatures, physical separation of halal and non-halal ingredients	Enforce storage separation, reject spoiled or contaminated materials
3	Packaging and processing aids procurement	Noncompliant materials, contamination	Agree on specifications and list of approved materials and suppliers	Routine supplier and materials audits, certificates	As in HCP CA 1
4	Product formulary	Noncompliance to specifications and unauthorized substitutions of ingredients	Agree on written specifications and ingredients	Routine checks of QA records, weighing and measuring equipment	As in HCP CA 1
5	Blending, mixing, chopping	Prolonged holding at wrong temperatures leading to increased pathogen risk, contamination with non-halal and foreign matter	Good manufacturing practices (GMP) and effective hygiene practices	Visual inspection by operators, and QA and swabs to monitor hygiene	Hold products back to last clean-up, dispose of contaminated products, and re-evaluate procedures
6	Cooking, drying, fermentation	Time-temperature, pH, A_w, and process noncompliance. Raw-unprocessed-processed cross contamination of pathogens	Strict process controls and separation of products and various stages of processing	QA to check process controls and verify their calibrations and product separations	Hold products to determine whether to re-work or dispose or reject as non-halal

(Continued)

TABLE 9.3 (CONTINUED)
Common Halal and Wholesome Assurance Control Points (HCP) for Processed Meats

HCP	Process Stage	Halal and Wholesome Risks	Preventative Measures	Monitoring	Corrective Action (CA)
7	Cooling	Biological and non-halal materials contamination	Time-temperature and cooling medium control	QA and inspectors to check controls and process specifications	Hold and dispose of or reject non-conforming products
8	Slicing, portioning	Biological recontamination	Proper sanitization of equipment and control of room temperature	Routine check by QA and inspectors of sanitization program	As in HCP CA 7
9	Packaging, labelling	Migration of packaging material, label ambiguity/misinformation	Agree on packaging materials, and type and label specifications	Routine supplier audits, QA records, labelling	Hold to repackage/label or reject
10	Storage/distribution	Growth of spoilage/pathogenic bacteria, mixing halal and non-halal products	Hygiene and control of storage and distribution temperatures/separation	Check storage and distribution hygiene and temperature	No product should be despatched until all the hygiene and temperature requirements are met
11	Cleaning processes	Growth of pathogens, contamination with detergents and sanitizers, poor hygiene	GMP; agree on cleaning and sanitization processes, materials, and schedules	QA, routine visual inspection, and swaps for hygiene and biological hazards	Reject contaminated products, and enforce hygiene and GMP

Source: Drawn from Figures 9.1 through 9.3. Additional details in Essien (2003).

aureus, *Clostridium perfringens*, and *Yersinia enterocolitica*. *Clostridium botulinum*, while rarely reported (due to an adequate lethal processing of the product) also remains a significant risk if a process fails. Infections due to toxoplasma have not been considered in most countries' risk assessment activities but has been associated with RTE meats in the U.S. (Crerar et al., 2011; Gilbert et al., 2009; Sumner et al., 2005).

HALAL AND WHOLESOME PROCESSED MEATS CERTIFICATION, ACCREDITATION, AND AUDITING

The globalization of ingredients procurement, international trade in finished foods/goods, and the heightened consumer awareness and demand for food safety, and the many food scandals and food-related disease outbreaks have together resulted in the rise of regulatory bodies to ensure that foods are wholesome including processed meats. These regulatory bodies have a wide range of responsibilities and powers, and include governmental and nongovernmental organizations. For halal processed meats, there can be up to five layers of assurances depending on the product and the country where it is produced:

1. Processor internal halal and quality control teams
2. Producing/exporting country specific regulations
3. Halal food certifiers
4. Importing country competent authorities/halal auditing teams
5. International standards organizations.

Each of these layers of assurances play an important role in ensuring the wholesomeness of processed meats and in assuring the consumers of the halal status and/or hygiene and safety of the products.

The process involved in the certification of a processed meat product by a halal certifying body are essentially the same across organizations and is exemplified by the processed meat certification process shown in Figure 9.4. The main steps include a processor applying to certify a processed meat product as halal, then establishing contact very early during the product and process development with a halal certifying body, and then the certification process will go through some of the steps shown in Figure 9.4. These steps may vary with certification bodies and the rigorousness of their process (Farouk, 2013).

The responsibilities for making sure the whole process is thorough and painless, that processed meats are certified, and the type of documentation required are shown in Figure 9.5. Most of the responsibilities falls on the processors to form the right teams to develop the HCP, and to test and make sure that they work for certification and maintenance of the certification. They should also be easy so that a halal culture can be established in the organization.

In many countries, governmental bodies often provide oversight and regulate the wholesome aspects of processed meats to protect the health and safety of all consumers. This does not however mean that halal certifiers should neglect this aspect even though the primary responsibility is with the government. Some of

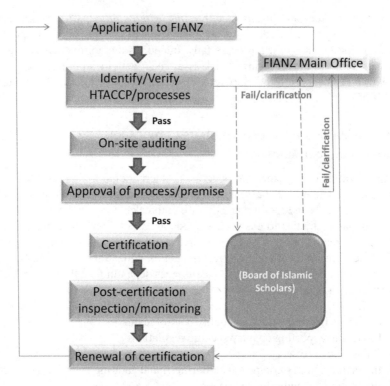

FIGURE 9.4 Federation of Islamic Associations of New Zealand (FIANZ) Halal Certification Flow-Chart for processed meats. HCCP = halal critical control points.

FIGURE 9.5 Halal critical control points plans responsibilities processor = Staff responsible for in-plant quality control, logistics, and various forms of evaluations.

TABLE 9.4
Microbiological Limits that Apply to RTE Meats (Food Safety Standards, Standard 1.6.1)

RTE Meats	Pathogen		Limits		
Packaged cooked	Coagulase-positive staphylococci/g	5	1	10^2	10^3
Cured/salted meat	Listeria monocytogenes/25g	5	0	0	
	Salmonella/25g	5	0	0	
Packaged heat-treated meat paste and packaged heat-treated pâté	Listeria monocytogenes/25g	5	0	0	
	Salmonella/25g	5	0	0	
All comminuted fermented meat that has not been cooked during the production process	Coagulase-positive staphylococci/g	5	1	10^3	10^4
	Escherichia coli/g/25g	5	1	3.6	9.2
	Salmonella/25g	5	0	0	

the relevant regulations in selected countries where large amounts of halal processed meats are produced are summarized in the following paragraphs (Farouk and Mills, 2016).

New Zealand and Australia: In New Zealand, the production of RTE meats is legislated under the Animal Products Regulations (2000), and the Food Standards Code (2002), which includes microbiological limits for food (Section 1.6.1; Table 9.4) and standards for Meat and Meat Products (Section 2.2.1).

U.S. and Canada: In the U.S., cooked meat products are regulated by the Food Safety and Inspection Service (FSIS) of the Department of Agriculture, either though Performance Standards or Compliance Guidelines. Performance Standards for the production of processed meat and poultry products were published in the Federal Register in 2001 (9 CFR Parts 301, 303, and a number of compliance guidelines have since been issued for the control of Listeria (in May 2006) and more recently, salmonella in small establishments (in April 2011). Compliance focuses heavily on control of the lethality step, verifying the effectiveness of the sanitation program, and post-process handling. In Canada, according to the Canadian Food Inspection Agency (CFIA) processed meats production facilities must implement food contact surface testing for Listeria.

European Union: In the EU, RTE processed meats are controlled under regulation 852/2004 (which sets out general and specific hygiene requirements, HACCP, and registration) and regulation 2073/2005 (which establishes the microbiological criteria for foodstuffs). According to those requirements, companies have to use the principles of Good Hygienic Practices (GHP), Good Manufacturing Practices (GMP), and the HACCP system. The latter regulation contains a specific criterion for control of *L. monocytogenes*, which requires the premise to answer the following (Dufour, 2011): (1) Is the food RTE? (2) Is the food for infants or special medical purposes? (3) Will the food allow growth of *L. monocytogenes* based on intrinsic factors, for example, pH and/or water activity? (4) Does the product have a shelf-life of less than five days? (5) Is there scientific evidence of the absence of growth? A tolerance of up to 100 cfu/g for *L. monocytogenes* products that could support the growth of

L. monocytogenes are recommended to sample the processing areas and equipment for *L. monocytogenes* as part of their sampling scheme.

PROCESSED MEATS ACCREDITATION BY INTERNATIONAL STANDARDS ORGANIZATIONS (ISO)

Halal certified processed meats are now traded globally. EU products are traded under required uniform regulations. The precondition for trading under uniform conditions is that every product, officially accepted in one country, becomes accepted in the markets of other states, without repeated testing, inspection, or certification (Popović and Popović, 2014). The conformity of halal processed meats to the requirements of importing countries is done by halal certifying bodies with some government oversight in many Western countries like New Zealand and Australia. The competence of the halal certifying bodies in most exporting countries is currently being monitored and audited by the competent authorities from some key importing countries such as JAKIM in Malaysia, MUI in Indonesia, and GSO of GCC countries. Figure 9.5 show Halal critical control points plan responsibilities, staff responsible for in-plant quality control, logistics, and various forms of evaluations.

Because of a lack of internationally recognized halal standards for certifying products and services, there are no "third party" halal verification bodies akin to ISO to impartially verify/assess the competence of the multitudes of halal certifying bodies or even the many competent authorities from the importing countries. The absence of an International Halal Standards Organization (IHSO) led to the competent authorities in UAE (UAE.S 2055–2:2014) recently demanding that their halal certifiers meet the relevant general, auditing, reporting, product analyses, and certification requirements specified in ISO 19011; ISO/IEC 17021:2011; ISO/IEC 17025; and ISO/IEC 17065:2012. Halal processed meats processors should take note of this requirement of the UAE for potential domino effects by other halal meat importing countries.

REFERENCES

Browne, D. (2011). Convenience meals and processed meats. *Prepared Foods*, 180(3), 51–58.
Crerar, S. K., Castle, M., Hassel, S., and Schumacher, D. (2011). Recent experiences with Listeria monocytogenes in New Zealand and development of a food control risk-based strategy. *Food Control*, 22(9), 1510–1512.
Crews, J. (2011). Unveiling ideas. New food products highlight quality, convenience and flexibility. *Meat and Poultry*, 105–107.
Dufour, C. (2011). Application of EC regulation no. 2073/2005 regarding Listeria monocytogenes in ready-to-eat foods in retail and catering sectors in Europe. *Food Control*, 22(9), 1491–1494.
Farouk, M.M. (1997). Meeting the halal criteria. *Food Technology in NZ*, 32–34.
Farouk, M.M. (2010). Restructured whole-tissue meats. *Handbook of Meat Processing*, Ed: Toldrá; F, 399–421. Hoboken, NJ: Blackwell Publishing.
Farouk, M.M. (2011). Improving the quality of restructured and convenience meat products. *Processed Meats*, Eds: Kerry, J. P. and Kerry, J. F, 450–477. Cambridge, UK: Woodhead Publisher.

Farouk, M.M. (2013). Advances in the industrial production of halal and kosher red meat. *Meat Science*, 95(4), 805–820.
Farouk, M.M. and Bekhit, A.E.D. (2013). Processed camel meat (Chapter 12). *Camel Meat and Meat Products*, 186–204. Wallingford, UK: CABI Publishing.
Farouk, M.M. and Mills, J. (2017). Processing of ready to eat meat products (accepted book chapter). In *Advances in Meat Processing*, Ed: Bekhit, A.E.D., 447–486. Boca Raton, FL: Taylor & Francis.
Farouk, M.M., Al-Mazeedi, H.M., Sabow, A.B., Bekhit, A.E.D., Adeyemi, K.D., Sazili, A.Q., and Ghani, A. (2014). Halal and Kosher slaughter methods and meat quality: A review. *Meat Science*, 98(3), 505–519.
Farouk, M. M., Regenstein, J. M., Pirie, M. R., Najm, R., Bekhit, A. E. D., and Knowles, S. O. (2015). Spiritual aspects of meat and nutritional security: Perspectives and responsibilities of the Abrahamic faiths. *Food Research International*, 76, 882–895.
Gilbert, S., Lake, R., Hudson, A. and Cressey, P. (2009). *Risk Profile: Listeria Monocytogenes in Processed Ready-to-eat Meats*. Christchurch, NZ: Christchurch Science Centre.
Hoogenkamp, H. W. (2001). *Soy Protein and Meat Formulations*. St. Louis, MO: Protein Technologies International.
Hussain, M.M. and Sakr, A.H. (1984). *Islamic dietary laws and practices*, 2nd edn. Chicago, IL: Islamic Food and Nutrition Council of America.
Lucy, L. and Kiteley, T.D. (1982). *Meats, Poultry, Fish, Shellfish*. Westport, CN: AVI Publishing.
Messenger, J. (2007). Deli-cious trends. IDDBA's Carol Christison talks about deli trends with M&P. *Meat and Poultry*, 28–30. Available online: www.meatpoultry.com.
Newswire.com. (2010). Innovation in ready meals. Market drivers, NPD and alternative sales channels, 1–107. Business Insights Ltd, July 1. Available online: www.newswire.com/innovations-in-ready-meals-market/46156
O'Donnell, C.D. (2009). Formulation trends in processed meats and alternatives. *Prepared Foods*, 115–120. Available online: www.PreparedFoods.com.
Pearson, A.M. and Gillett, T.A. (1999). *Processed Meats*, 3rd edn. Gaithersburg, MD: Aspen Publishers Inc.
Pellegrini, M. (2008). Batter and breading function vs flavor. *The National Provisioner*, 48–56.
Pellegrini, M. (2009). Prepared/ready-to-eat foods 101. *The National Provisioner*, 22–33.
Popović, P. and Popović, D. (2014). Implementation of new international standards for certification and inspection bodies. *Journal of Applied Engineering Science*, 12(3), 187–196.
Pszczola, D.E. (2010). Tackling meaty issues in a lean economy. *Food Technology*, 64(4), 49–58.
Salvage, B. (1999). Pre-cooked meats. *Meat Marketing and Technology*, 46–51.28.
Sindelar, J.J., Cordray, J.C., Sebranek, J.G., Love, J.A., and Ahn, D.U. (2007). Effects of varying levels of vegetable juice powder and incubation time on color, residual nitrate and nitrite, pigment, pH, and trained sensory attributes of ready-to-eat uncured ham. *Journal of Food Science*, 72(6), S388–S395.
Sloan, A.E. (2015). 10 top food trends. *Food Technology*, 22–40. Available online: www.ift.org.
Sumner, J., Ross, T., Jenson, I. and Pointon, A. (2005). A risk microbiological profile of the Australian red meat industry: Risk ratings of hazard–product pairings. *International Journal of Food Microbiology*, 105(2), 221–232.
Tarver, T. (2010). Desalting the food grid. *Food Technology*, 64(8), 45–50.

10 Halal Production Requirements
For Fish and Seafood

Mian N. Riaz, Rafiuddin Shaik, and Munir M. Chaudry

CONTENTS

Requirements for Slaughtering or Killing Fish and Seafood 147
General Guidelines for Processing Fish and Seafood 147
Further Processed Products 148
Food Ingredients and Flavors 149
Imitation Seafood Products 149
General Halal Control Points for the Production of Seafood Products 149
    HCP1 Sorting by Category 149
    HCP2 Ingredients and Packaging 149
    HCP3 Equipment 149
References 154

Fish and seafood refers to all non-plant life from natural bodies of water as well as human-made fish farms. Fish and seafood are reliable source of halal food, but there are a number of differing opinions on the halal or haram status of fish and seafood (Regenstein et al., 2003) The Quran states:

> To hunt and to eat the fish of the sea is made lawful for you, a provision for you and for seafarers; but to hunt on land is forbidden you so long as ye are on the pilgrimage. Be mindful of your duty to Allah, unto Whom ye will be gathered.
>
> **Chapter V, Verse 96**

> And He it is Who hath constrained the sea to be of service that ye eat fresh meat from thence, and bring forth from thence ornaments which ye wear. And thou seest the ships ploughing it that ye (mankind) may seek of His bounty, and that haply ye may give thanks.
>
> **Chapter XVI, Verse 14**

> And two seas are not alike; this, fresh, sweet, good to drink, this (other) bitter, salt. And from them both ye eat fresh meat and derive the ornament that ye wear. And thou seest the ship cleaving them with its prow that ye may seek of His bounty, and that haply ye may give thanks.
>
> **Chapter XXXV, Verse 12**

These verses state that it is lawful to fish for food. In fact, God has given the bounty of the seas to human beings so they may partake of it and benefit by what has been provided. In addition, a number of hadiths (traditions of the Prophet Muhammad, PBUH) also address the subject of seafood (Al-Quaderi, 2002).

It is stated in a hadith that a group of the Prophet Muhammad (PBUH) companions were on a journey when they ran out of food on a journey. They came upon a huge sea creature, often referred to as a huge fish or whale, washed up on the shore. They debated whether it was permissible to eat from it because it was already dead, but finally decided that their need for food exempted them should there be any sin in it. After returning home and informing the Prophet (PBUH), they were told it was a blessing provided to them by God. Three points of jurisprudence were thereby established:

- It is permissible to eat whale even though it is not considered a true fish because it is a mammal. Similarly, animals that wholly live in water (not water and land) are permitted for food.
- There is no requirement to slaughter sea animals unlike the requirements for land animals, even if they are mammals. They do have to be killed humanely, generally by leaving them out of the water to let them die their natural death.
- Unlike land animals, it is permitted to eat dead sea animals. However, they must not show any visible signs of deterioration and spoilage.

Islamic scholars have studied the question of which seafood is permitted and which is prohibited to be eaten by Muslims. Some of the scholars believe that only live catches are halal. They believe that if the object is found dead, it comes under the restriction of prohibiting the consumption of dead land animals. The majority of scholars opine that seafood is exempt from this restriction, and use the tradition about the dead whale to justify their opinion (Regenstein et al., 2003).

As to the species of sea creatures that are permitted, all scholars have agreed that fish with scales are halal. Some believe that only fish with scales are halal and other creatures are not. This group believes that lobster, shrimp, octopus, eels, and so on, are not permitted. Some have opined that anything that can only live in water is halal, whereas creatures that can live in and out of the water are haram. The latter include turtles, frogs, and alligators (Ahmed, 2008). To help guide readers the following more detailed categorization may be helpful:

- *Category one*: Includes fish with fins and removable scales. All fish with scales have fins. This group includes most of the traditional fish species. This category is accepted by all Muslim consumers. List of such fish have been developed by the Jewish community, which uses this as the criteria for fish they accept.
- *Category two*: Includes fish or fishlike animals that may have fins but not removable scales. Some of these may breathe oxygen from the air rather

than water, even if they live in water all the time. Examples are catfish, shark, swordfish, sturgeon, eel, monkfish, cusk, and blowfish. This category is acceptable to the majority of Muslim consumers, but not all. Some may consider them makrooh (disliked or detested).
- *Category three*: Comprises several unrelated species, mobile or not, of various shapes and sizes, that cannot survive without being in water. These are generally either mollusks or crustaceans, including clams (these can live outside and may live in the sand during low tide), mussels (these also may be exposed during tidal changes), lobsters, shrimp, oysters, octopus, scallops, and squid. This group also includes marine mammals that live totally in the sea such as whales and dolphins. The majority of Muslim consumers eat them; however, others consider them either haram or makrooh. Shrimp seems to be in a special category: some only eat them but not the rest of category.
- *Category four*: Includes many of the animals generally falling under the definition of seafood. They live in and around water most of their life cycle, but are capable of living outside water because they can breathe air. These are generally not considered halal although some Islamic scholars are of the opinion that they are from the seas because they live in and around water. These include crabs, snails, turtles, alligators, and frogs.

REQUIREMENTS FOR SLAUGHTERING OR KILLING FISH AND SEAFOOD

Animals from the water are not required to be killed in any religiously specified manner as practiced for land animals. However, fish and seafood should be prepared in a manner that the animals do not suffer excessively. They should not be skinned or scaled while still alive, for example, as practiced by some Eastern countries (Wan Hassan and Awang, 2009).

The cultured aquatic animals should be fed healthy balanced diet, using clean ingredients. Because of the high cost of fish meal, some growers may use rendered animal meals (animal by-products, feather meal, and porcine blood meal) as the protein source in the feeds. Such ingredients may render these aquacultured fish as makrooh or haram (Qureshi, 2016).

Newer, hopefully more humane methods of killing fish are being developed. Percussive stunning either by hand or with appropriate equipment is the preferred method. Allowing animals to die in air or using a slush chilling bath, although traditionally used, have animal welfare concerns. The Muslim community supports efforts to improve the killing of fish and seafood.

GENERAL GUIDELINES FOR PROCESSING FISH AND SEAFOOD

General guidelines for processing fish and seafood consist of maintaining the identity of the product and not using any prohibited ingredients during processing. The guidelines also include not using equipment that has been used for haram products.

FURTHER PROCESSED PRODUCTS

Further processed seafood products must not contain any haram or doubtful ingredients. Several nutritional supplements are also made from fish and shellfish, such as fish oils high in omega-3 and omega-6 fatty acids; chondroitin sulfate and glucosamine for joint support; fish gelatin/collagen/isinglass for use in pharmaceuticals and in cosmetics; and fish glue. These products must be from acceptable raw materials based on the Muslim consumer's requirements but also require that all processing equipment, ingredients, and packaging materials be halal.

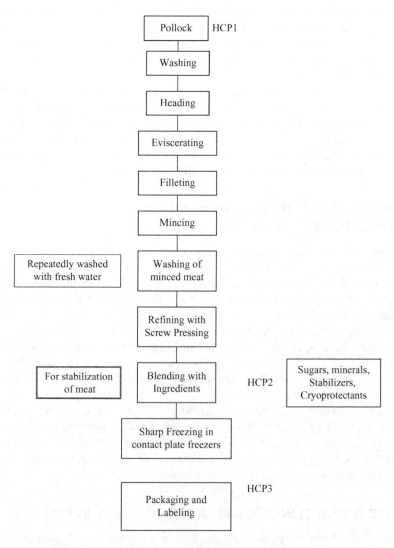

FIGURE 10.1 Halal control points for processing of surimi made from pollock.

FOOD INGREDIENTS AND FLAVORS

Extracted flavors and other ingredients from fish and seafood are used in non-seafood products. These again must be consistent with the Muslim consumer's requirements. When exporting, it is advisable to learn the market requirements in a particular country.

IMITATION SEAFOOD PRODUCTS

Imitation seafood products are generally made using surimi. Surimi is usually made from category 1 fish although category 2 fish may also be used. They are flavored with natural and artificial flavors and texturized using different stabilizing agents. Thus, the flavors in particular may be from category 3 or 4, which have more limited acceptance in the Muslim community. There are some such products that do not use any category 2, 3, or 4 ingredients for those Muslims who only accept category 1.

GENERAL HALAL CONTROL POINTS FOR THE PRODUCTION OF SEAFOOD PRODUCTS

Figure 10.1 shows halal control points for processing of surimi made from pollock.

HCP1 Sorting by Category

The initial catch must be sorted so that all products kept together are in one category.

HCP2 Ingredients and Packaging

All ingredients and packaging must be checked that they are halal. Preferably with a well-recognized halal certificate.

HCP3 Equipment

All equipment used for seafood processing must be clean and free of any haram ingredients. Thus, product in any category should ideally be prepared on equipment that had none of the higher categories, for example, equipment for category 1 production must be thoroughly cleaned if category 2, 3, or 4 animals were processed. Figure 10.2 shows halal control points for processing of fish balls. Figure 10.3 shows halal control points for processing of fish fingers. Figure 10.4 shows halal control points for processing of fish sausages. Figure 10.5 shows halal control points for processing of smoked fish and Figure 10.6 shows halal control points for processing of fish oil.

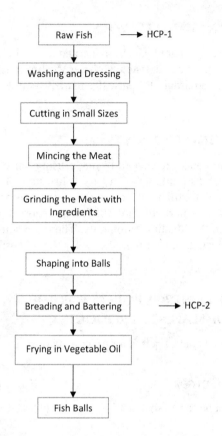

FIGURE 10.2 Halal control points for processing of fish balls.

Halal Production Requirements

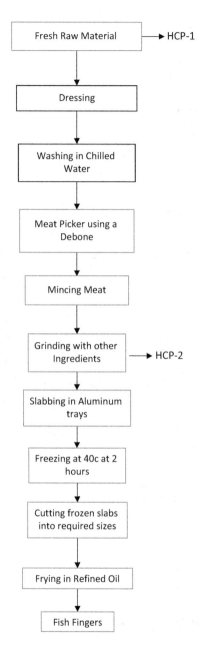

FIGURE 10.3 Halal control points for processing of fish fingers.

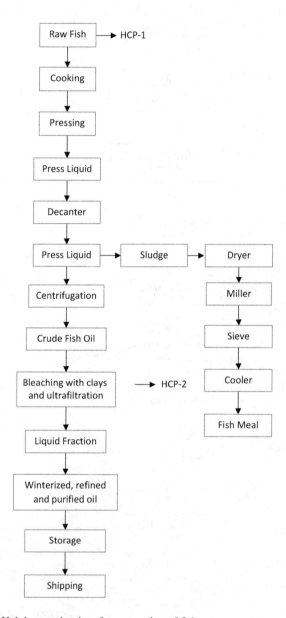

FIGURE 10.4 Halal control points for processing of fish sausages.

Halal Production Requirements

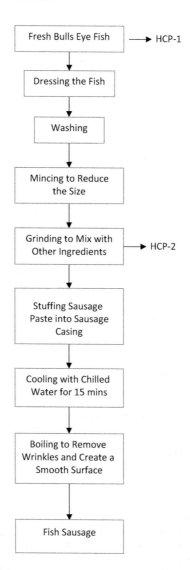

FIGURE 10.5 Halal control points for processing of smoked fish.

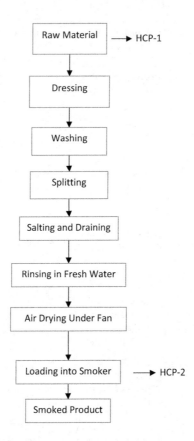

FIGURE 10.6 Halal control points for processing of fish oil.

REFERENCES

Ahmed, A. (2008). Marketing of halal meat in the United Kingdom: Supermarkets versus local shops. *British Food Journal*, 110(7), 655–670.

Al-Quaderi, S.J. (2002). Seafood: what's halal, what's not? *Halal Digest*, September, available online: www.ifanca.org.

Qureshi, F.A. (2016). Fish and Seafood requirements in Halal food Production. https://www.linkedin.com/pulse/fish-seafood-requirements-in-halal-food-production-ep-8-qureshi/ Accessed on June 25, 2018.

Regenstein, J.M., Chaudry, M.M., and Regenstein, C.E. (2003). The kosher and halal food laws. *Comprehensive Reviews in Food Science and Food Safety*, 2(3), 111–127.

Wan Hassan, W.M. and Awang, K.W. (2009). Halal food in New Zealand restaurants: An exploratory study. *International Journal of Economics and Management*, 3(2), 385–402.

11 Halal Production Requirements for Dairy Products

Mian N. Riaz and Munir M. Chaudry

CONTENTS

Milk in the Quran 156
    Milk: Whole, Low-Fat, Skim, and Flavored 156
Cream, Half and Half, and Butter 156
Dry Milk Powder and Nonfat Dry Milk Powder 156
Cheeses 156
    Anticaking Agents 157
    Preservatives 157
    Whey, Whey Protein Concentrate, Whey Protein Isolate, and Lactose 158
Cultured Milk, Sour Cream, and Yogurt 158
Ice Cream and Frozen Desserts 158
Flavors and Enzyme-Modified Products 159
Halal Control Points in Cheese Making 159
    HCP1: Raw Milk 159
    HCP2: Additions of Ingredients (Enzymes, Cultures, and Colors) 159
    HCP3: Aging 159
    HCP4: Packaging 160
References 162

The dairy products industry produces a vast number of products ranging from fresh milk to ice creams and frozen desserts. Dairy is one of the oldest food industries. Cheese, which has been produced for about 5000 years, is one of the classical fabricated foods in the human diet (Fox et al., 2000). A wealth of information is available on all aspects of milk and dairy processing; however, the processors generally do not take into account the needs of the Muslim consumer. It is appropriate to mention that producing halal products is very similar to producing kosher products, although halal requirements are a bit simpler. Therefore, most kosher dairy products can easily be halal certified ("dual certified") and some that are not kosher can still be halal. For a comparative analysis between kosher and halal, the reader is encouraged to consult Chapter 25.

    There are two main types of dairy cattle in the world: the buffalo in South Asia and parts of Africa, and the cow in the rest of the world. There are other minor milk sources such as goats, sheep, mares, and camels. In this chapter, milk means cow's milk, although the processing of milk from other kinds of animals, particularly buffalo, is similar to cow's milk.

MILK IN THE QURAN

Milk is one of the recommended foods for Muslims. It is considered pure and palatable for drinkers (Pickthall, 1994):

> And lo! In the cattle there is a lesson for you. We give you to drink of that which is in their bellies, from between the refuse and the blood, pure milk palatable to the drinkers.
>
> **Chapter XVI, Verse 66**

Milk is almost a complete food. It provides nourishment, minerals, vitamins, and protein. There are so many different types of dairy products that the halal issues vary from simple to complex.

MILK: WHOLE, LOW-FAT, SKIM, AND FLAVORED

In the U.S., milk is generally fortified with vitamins A and D. Some vitamin A comes from shark oil. These may present some halal issues. (Please see Chapter 10 on fish and seafood.) To make these vitamins soluble in milk, they are mixed with or standardized with emulsifiers such as polysorbates. Other functional ingredients can also be added to increase the stability and shelf life of milk. Polysorbates are fatty chemicals that can be made from vegetable oils or from animal fats. For producing halal milk, these emulsifiers and other functional ingredients must be from halal sources such as plant oils. Some dairies use gelatin as a thickener when making chocolate and other flavored milks. This is another critical ingredient that must be halal for halal certification (Karim and Bhat, 2008; Shariff and Lah, 2014).

CREAM, HALF AND HALF, AND BUTTER

Mono- and di-glycerides are sometimes added to these products to prevent the fat phase from separating from the water phase. Both animal- and vegetable-derived mono-glycerides are available, so must be checked for its halal status.

DRY MILK POWDER AND NONFAT DRY MILK POWDER

These are heat-processed, dehydrated milk powders. Normally no other ingredients are added to them. However, the spray drier must not have run any haram products. Properly cleaning a spray drier is very difficult, so it is recommended that the equipment be dedicated for only halal production. These are products, when kosher, are therefore acceptable for Muslims.

CHEESES

There are many different types of cheeses, and they are processed using different methods and different ingredients. Cottage cheese, for example, may be made by

Halal Production Requirements for Dairy Products

curdling milk with acid, which makes it a halal-suitable process. Other cheeses such as mozzarella, cheddar, and colby are made use milk-curdling enzymes and bacterial cultures. Bacterial cultures are generally halal, as long as the media they are grown in are halal, but enzymes can come from many different sources, as explained in Chapter 13. One must make sure that enzymes are halal suitable. Some cheeses are ripened or aged by using bacterial cultures, molds, or enzymes. Cheeses processed with enzymes are more complex and might contain some objectionable ingredients. Transgenically produced enzymes are not only permitted but are preferred for use in the production of halal foods. For example, bovine rennet produced from calves that have not been slaughtered according to Muslim requirements is not acceptable to most Muslims, whereas chymosin (the main enzymes found in rennet) produced microbially through transcription from the bovine chymosin genes is universally accepted by Muslims, as long as the standardizing ingredients and media in which they were raised contain no haram ingredients nor has the fermenter been used previously with haram ingredients without receiving a halal-approved cleaning (Cowan et al., 2001). (Note: Because most of these products are almost always kosher, these requirements are routinely met. However, in the absence of a kosher certification, a more comprehensive halal inspection will be necessary.)

Enzymes are a major area of concern for halal cheese production. Several enzymes are obtained from pigs, which are haram. Some enzymes are also derived from calves or other permitted animals, but if these animals are not halal slaughtered the enzymes are not acceptable for halal cheese production (Al-Mazeedi et al., 2013). For enzyme manufacturers there is an opportunity to capture the halal market by making enzymes such as lipase through genetic engineering. However, if the gene is transferred from a pig, it will not be acceptable (Khattak et al., 2011; Riaz, 2000a). An interesting question is if one sequenced the pig gene and then produced a synthetic gene that is halal, would it be accepted? Muslim jurists have not ruled on this at the time this book was being completed, although a few Muslim writers have suggested that it, too, would be unacceptable.

ANTICAKING AGENTS

Shredded cheese might contain anticaking agents such as animal or vegetable stearates. These ingredients should be from halal sources.

PRESERVATIVES

All preservatives and mold inhibitors should be from halal sources. The preservatives can be proprietary mixtures of natamycin, sodium benzoates, calcium propionates, and other compounds, which may contain emulsifiers from animal sources, and so can be a concern for Muslim consumers.

The above ingredients, especially enzymes, not only affect the status of cheeses but also have a major impact on cheese by-products such as whey. Because whey contains a considerable number of enzymes, Muslim consumers are concerned about foods containing whey (Riaz, 2000b), that is, it needs to have appropriate supervision. At this time, food grade whey in the U.S. is almost always kosher and at this time there is no information to suggest that such whey is not also halal (Chaudry et al., 2000).

WHEY, WHEY PROTEIN CONCENTRATE, WHEY PROTEIN ISOLATE, AND LACTOSE

These are by-products of the cheese making process used in a myriad of food products ranging from baked goods to frozen desserts. These ingredients are further processed into powders from liquid whey. Usually, nothing further is added to whey after it is drained to make cheese; therefore, if the cheese manufactured is halal, then liquid whey and ingredients from whey, such as whey powder, whey protein concentrate, whey protein isolate, and lactose, are also halal, as long as the drying equipment is halal.

CULTURED MILK, SOUR CREAM, AND YOGURT

These are compounded, cultured, and further processed milk products. Ingredients such as gelatin, emulsifiers, flavorings, stabilizers, and colors can be added to these products for various functional properties. Gelatin is one of the most widely used ingredients in yogurt. Halal gelatin is now available, as are other halal-texturizing ingredients, including pectin, carrageenan, and modified starches, which are suitable as gelatin replacements (Anir et al., 2008; Chaudry et al., 2000; Karim and Bhat 2008).

ICE CREAM AND FROZEN DESSERTS

Ice cream and frozen desserts are complex food systems requiring dozens of different ingredients to manufacture them. There are several possibilities for doubtful ingredients being incorporated into ice cream and frozen desserts, but the three ingredients that present the greatest difficulties are gelatin, flavors, and emulsifiers.

To make natural vanilla ice cream, a company must use natural vanilla flavor, which by its "standard of identity" must contain at least 35% alcohol. Even when this alcohol is diluted down to its use level, the final ice cream may contain from 0.2% to 0.5% alcohol. Other flavors, which are uniquely liquor flavors, such as rum, might contain even higher amounts of alcohol. One must decrease the amount of alcohol in these products to the minimum level needed to provide the technical effects desired from that particular ingredient. The final alcohol content of the finished products should be lowered to less than 0.1% according to the standard of IFANCA. Manufacturers will have to check with other certifiers to determine their standard. A dried vanilla has been prepared and it might substitute for all or some of the alcohol containing vanilla (Cullen-Innis, 2005).

Halal gelatin and halal marshmallows are now available for companies that wish to formulate ice cream products with those ingredients (Karim and Bhat, 2008; Shah and Yusuf, 2014). Flavors that have a connotation related to alcoholic beverage can be formulated from nonalcoholic extracts of natural flavors or by the use of synthetic ingredients, such as rum flavor. However, Muslim consumers may have some hesitation in using such products.

FLAVORS AND ENZYME-MODIFIED PRODUCTS

These days many flavors such as butter flavor are extracted from dairy ingredients by concentrating the flavor compounds and further intensifying the flavor with modifications through enzymatic reactions. Manufacturers should make sure that enzymes used for these flavors are halal suitable (Al-Mazeedi et al., 2013). Currently, some cheeses are becoming available where animal flesh, rather than an extractive are incorporated into the cheese. Such animal ingredients must be from halal-slaughtered animals and maintain as halal (Jahangir et al., 2016).

HALAL CONTROL POINTS IN CHEESE MAKING

Cheese making generally is a dedicated process for each type of cheese and there are very few chances of cross contamination. It also is a fixed process, requiring few changes, if any, during the year. Figures 11.1 and 11.2 give the halal control points (HCP) for dairy technology.

HCP1: Raw Milk

In the U.S., all milk is cow's milk unless specified otherwise; for example, cheese is also made from sheep and goat's milk. Other sources of milk may be used as long as they come from a halal animal and are maintained as halal. In the U.S., cow's milk is highly regulated and so it is probably a safe assumption that it is really cow's milk and that it is safe and wholesome if part of the commercial milk supply. Artisan operations not part of the Class A milk program in the U.S. probably need to be checked.

HCP2: Additions of Ingredients (Enzymes, Cultures, and Colors)

This is the most critical point in cheese manufacture because enzymes used can be from haram or halal animals as well as from microorganisms. For universal acceptance, enzymes should be of microbial or plant origin and other ingredients should be free from doubtful ingredients. All ingredients need to be checked that they are not from animals that are not halal or that were not slaughtered halal. The use of these ingredients should be properly documented.

One must consider not only the primary ingredient, but also potential processing aids that might be added to the main ingredient to standardize or stabilize it.

HCP3: Aging

After the cheese is processed and ready to be stored for aging or ripening, mold and bacteria can be applied to its surface to be incorporated internally. Preservative such as sorbates, propionates, or natamycin can also be applied where cheese is ripened without added cultures to inhibit mold growth. All these specialized chemicals must conform to halal guidelines.

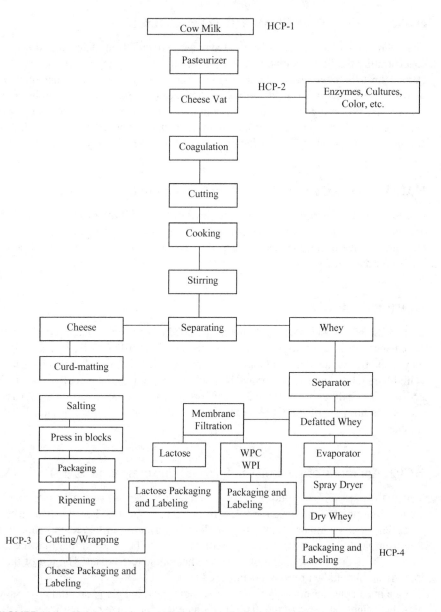

FIGURE 11.1 Halal control points for cheese and whey processing.

HCP4: Packaging

Finally, packaging must be clean and halal acceptable. Labels should have clear halal markings. It is advisable to use the description microbial enzymes if microbial enzymes are used so that Muslim consumers are reassured.

Halal Production Requirements for Dairy Products

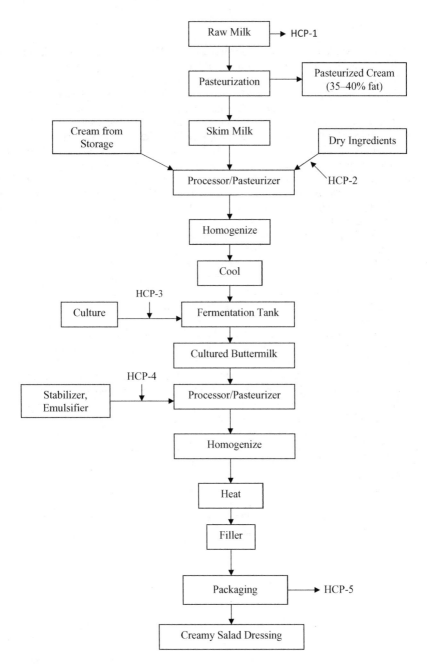

FIGURE 11.2 Halal control points (HCP) in creamy salad dressing.

REFERENCES

Al-Mazeedi, H. M., Regenstein, J. M., and Riaz, M. N. (2013). The issue of undeclared ingredients in halal and kosher food production: A focus on processing aids. *Comprehensive Reviews in Food Science and Food Safety*, 12(2), 228–233.

Anir, N. A., Nizam, M. N. M. H., and Masliyana, A. (2008). RFID tag for Halal food tracking in Malaysia: Users perceptions and opportunities. In *WSEAS International Conference. Proceedings. Mathematics and Computers in Science and Engineering*, Istanbul, Turkey (No. 7). World Scientific and Engineering Academy and Society.

Chaudry, M. M., Jackson, M. A., Hussaini, M. M., and Riaz, M. N. (2000). *Halal Industrial Production Standards*. Deerfield, Illinois: J&M Food Products Company, 1–15.

Cowan, C., Meehan, H., McIntyre, B., and Cronin, T. (2001). *Food Market Studies in; Meat Packaging, Nutritional Meat Products, Speciality Cheeses, Extruded Meats*. Dublin: Teagasc. https://t-stor.teagasc.ie/handle/11019/125. Accessed June 25, 2018.

Cullen-Innis, E. (2005). An introduction to kosher and halal issues in process and reaction flavors, Chapter 4, pp. 56–67; Eds: Weerasinghe, D and Sucan M. ACS Symposium Series, Vol. 905, American Chemical Society.

Fox, P. F., Guinee, T. P., Cogan, T. M., and McSweeney, P. L. H. (2000). *Fundamentals of Cheese Making*, 2nd edn. Gaithersburg, MD: Aspen Publishers.

Jahangir, M., Mehmood, Z., Bashir, Q., Mehboob, F., and Ali, K. (2016). Halal status of ingredients after physicochemical alteration (Istihalah). *Trends in Food Science and Technology*, 47, 78–81.

Karim, A. A. and Bhat, R. (2008). Gelatin alternatives for the food industry: Recent developments, challenges and prospects. *Trends in Food Science and Technology*, 19(12), 644–656.

Khattak, J. Z. K., Mir, A., Anwar, Z., Abbas, G., Khattak, H. Z. K., and Ismatullah, H. (2011). Concept of halal food and biotechnology. *Advance Journal of Food Science and Technology*, 3(5), 385–389.

Pickthall, M. M. (1994). *Arabic Text and English Rendering of the Glorious Quran*. Chicago, IL: Kazi Publications, Library of Islam.

Riaz, M. N. (2000a). What is halal (press article), *Dairy Foods*, 101(4), 36.

Riaz, M. N. (2000b). How cheese manufacturers can benefit from producing cheese for the halal market. *Cheese Market News*, 20(18), 4, 12.

Shah, H. and Yusof, F. (2014). Gelatin as an ingredient in food and pharmaceutical products: An Islamic perspective. *Advances in Environmental Biology*, 8, 774–780.

Shariff, S. M. and Lah, N. A. A. (2014). Halal certification on chocolate products: A case study. *Procedia-Social and Behavioral Sciences*, 121, 104–112.

12 Halal Production Requirements for Cereals and Confectionaries

Mian N. Riaz and Munir M. Chaudry

CONTENTS

Product Types 164
    Breakfast Cereals 164
    Bread 164
    Cakes, Cookies, and Pastries 165
    Doughnuts and other Fried Goods 165
    Chewing Gum 165
    Marshmallows 165
Halal Control Points (Hcp) for Flour- And Starch-Based Products 165
References 166

Cereal-based products include a large number of staple food products such as bread, breakfast cereals, cakes, candies, doughnuts, cookies, pastries, and chewing gum. The processing and composition of these products vary a great deal. The major ingredients used in this group of products are flour, sugar, and shortening. There are possibly hundreds of other ingredients used in cereal products, depending on the nature of each product (Riaz, 1996).

Some of the commonly used ingredients by the cereal and confectionary industries whose halal status is questionable include the following.

- *Gelatin*: May be used as a glaze component on doughnuts and strudels as well as some types of cake and pastry (Nuruddin, 2007). As discussed in Chapter 14, two types of gelatin are suitable for halal: (1) mammalian gelatin from halal-slaughtered animals, and (2) fish gelatin from fish consistent with the needs of a Muslim consumer. Those from fish with fins and removable scales are accepted by all Muslims. (All halal fish gelatins are mostly from category 1 fish although some may be from category 2, see Chapter 10.)
- *Mono- and di-glycerides*: These are emulsifiers quite widely used in the bakery and confectionary industries, and, to a lesser extent, in candy products. Although mono- and di-glycerides can be made from any fat or oil,

the only acceptable sources for halal foods are vegetable mono- and diglycerides (Riaz, 1998) or those produced from halal-slaughtered animals.
- *Other emulsifiers*: Many different emulsifiers are available. These need to have a reliable halal certification to avoid those from unacceptable animal sources. Cream liquor generally contains varying amounts of alcohol and must be avoided in halal production (Riaz, 1997).
- *Pan greases and release agents*: Same rules apply as for "other emulsifiers."
- *Food coatings*: Same rules apply as for "other emulsifiers."
- *L-cysteine*: This amino acid may be used in some baked products to modify the texture of the batters and breading. Halal L-cysteine must be from either a vegetarian source, made synthetically under halal acceptable conditions, or from bird feathers either prior to slaughter or from birds slaughtered according to halal requirements. L-cysteine from human hair is also available but generally not accepted as halal, because it is considered offensive to one's psyche, hence makrooh.

PRODUCT TYPES

Some of the key points in different product types such as breakfast cereals, bread, cookies and pastry, doughnuts and other fried goods, chewing gum, and marshmallows are discussed next.

BREAKFAST CEREALS

Most of the breakfast cereals are rather simple formulations containing pure grain-based ingredients mixed in with sugar, salt, and a few other minor ingredients such as colors and flavors. However, many are also fortified with vitamins and minerals. All of these minor ingredients may have halal issues and require halal certification. Some breakfast cereals may also contain gelatin-based marshmallows, which must be presumed to not be halal unless the product is certified by a reliable halal agency (Sakr, 1999).

BREAD

There is a large variety of breads. Many of the trace ingredients may have halal concerns and do need to be checked. As baked products there may also be issues with pan greases and release agents.

Many breads are made with yeast, which not only generates carbon dioxide but also produces alcohol. Although alcohol is one of the haram ingredients, the purpose of the yeast in making bread is very different from its purpose in brewing alcoholic drinks, that is, the carbon dioxide is what is sought rather than the alcohol. Hence, there is no concern with the presence of any residual alcohol in the bread after baking.

Cakes, Cookies, and Pastries

Again, the emphasis needs to be on checking all of the minor ingredients for proper halal status.

Doughnuts and other Fried Goods

The foremost requirement is that the frying oil should be vegetable oil or from a halal-slaughtered animal source. The processing of the oil should be checked. According to U.S. regulations, pure oils can contain up to 0.1% of ingredients that do not have to be labeled. Thus, the oil requires certification. Most vegetable oil is kosher certified and distributed separately from animal oils. Thus, it can generally be considered to be halal. As always, minor ingredients need to be carefully checked (Nasaruddin et al., 2012).

Chewing Gum

Gum-base and chewing gum might be composed of generally recognized as safe (GRAS) ingredients according to U.S. FDA regulations. Again, all of these ingredients need to be checked for their halal status including stearates and gelatin (Uddin, 1994), which are known to be of concern.

Marshmallows

Marshmallows are primarily sugar, gelatin, and flavors. A special kind of gelatin, usually pork, is required to make the best fluffy, non-sticky marshmallows. One can make halal marshmallows with halal-certified mammalian or fish gelatin. All other ingredients also need to be checked. In most plants, marshmallows are rolled in corn starch so that they do not stick to each other. It is very important that the equipment be thoroughly cleaned and old starch replaced with fresh starch before producing halal marshmallows. This is an expensive proposition as the starch is used over and over again. So, ideally, marshmallows will be produced in a plant dedicated to halal production.

HALAL CONTROL POINTS (HCP) FOR FLOUR- AND STARCH-BASED PRODUCTS

Although the process of bread making is quite simple, there are few points to consider in the production of bread (Figure 12.1).

HCP1: All major and minor ingredients used in manufacturing of bread.
HCP2: Release agent and pan grease if used should be halal suitable.
HCP3: Packaging materials used should also be halal suitable.

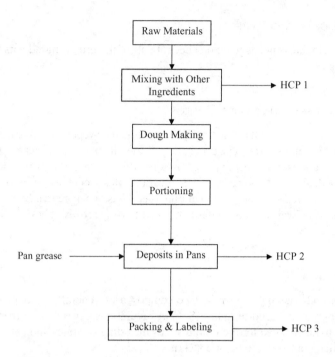

FIGURE 12.1 Flow-chart of bread making with respect to halal critical points (HCP).

REFERENCES

Nasaruddin, R. R., Fuad, F., Mel, M., Jaswir, I., and Hamid, H. A. (2012). The importance of a standardized Islamic manufacturing practice (IMP) for food and pharmaceutical productions. *Advances in Natural and Applied Sciences*, 6(5), 588–596.

Nuruddin, N. A. (2007). Islamic Manufacturing Process (IMP). Retrieved from Institute of Islamic Understanding Malaysia. Available online at: www. ikim. gov. my.

Riaz, M. N. (1996). Hailing halal. *Prepared Foods*, 165(12), 53–54.

Riaz, M. N. 1997. Alcohol: The myths and realities. In *A Handbook of Halaal and Haraam Products*, Vol. 2, Uddin, Z., Ed. Richmond Hill, NY: Center for American Muslim Research and Information, 16–30.

Riaz, M. N. (1998). Halal food: An insight into a growing food industry segment. *International Food Marketing and Technology*, 12(6), 6–9.

Sakr, A.H. (1999). *Gelatin, Foundation for Islamic Knowledge*. Lombard, IL: Islamic Food and Nutrition Council of America (Chicago, IL).

Uddin, Z. (1994). *A Handbook of Halaal and Haraam Products*. Richmond Hill, NY: Publication Center for American Muslim Research and Information.

13 Enzymes in Halal Food Production

Mian N. Riaz and Munir M. Chaudry

CONTENTS

Uses of Enzymes in Food 168
Classification of Enzymes... 169
Bioengineered Enzymes.. 169
Halal Control Points in Enzyme Production..................................... 170
Halal Control Points for a Conventional Process for Extracting Enzymes
From Animal Organs... 171
    HCP1: Animal Organs... 171
    HCP2: Preparation of Tissues for Extraction 172
    HCP3: Use of Acceptable Release Agents 172
    HCP4: Ingredients for Standardization 172
    HCP5: Packaging and Labeling 172
Halal Control Points for Enzyme Production in a Conventional Fermentation
Process 172
    HCP1: Raw Materials for Growth Media 172
    HCP2: Origin of Culture 173
    HCP3: Processing Aids 174
    HCP4: Additives in the Media 174
    HCP5: Standardization Ingredients.. 174
    HCP6: Packaging and Labeling 174
To Label or Not to Label Enzymes 174
References... 174

Enzymes play a vital function in the regulation and performance of living cells. They speed up reactions and catalyze specific reactions while producing few unintended by-products. The food industry has taken advantage of these properties to produce enzymes that can reduce food production costs, manufacturing time, and waste production, and improve taste, color, and texture (Birch et al., 2012; Oort, 2009). The advent of bioengineering has enabled the enzyme industry to produce microbial enzymes that are derived initially from animal or vegetable enzymes. Bioengineered enzymes are cheaper to produce and easier to control for purity (Dodge, 2010). The use of microbial enzymes is more compatible with halal food production because it eliminates the use of animal-derived enzymes (Al-Mazeedi et al., 2013; Khattak et al., 2011). However, animal-derived enzymes are still used in the food industry, particularly in the dairy industry. Unless the animal from which the enzymes are derived had been slaughtered halal, these enzymes would be unacceptable.

USES OF ENZYMES IN FOOD

Enzymes have been used for many centuries. There is evidence of the use of enzymes in cheese making as mentioned in Greek epic poems dating back to 800 BCE (Ashie, 2003). Today, enzymes are used for many purposes and in many industries. Major food industry uses are for cheese making, baking, fruit and vegetable processing, and food ingredient production. Enzymes are used in the production of sugar, processing of starch, hydrolysis of proteins, and modification of fats and oils. Other food uses include brewing and winemaking, which are not relevant for halal (Reed, 2012). The first major food industry use of enzymes began in the 1960s, when the enzyme glucoamylase was developed to allow starches to break down to glucose (Olsen, 2000a). Until then, glucose production involved acid hydrolysis of the starch. While the acid hydrolysis method produces the desired product, use of the enzymatic process reduces production costs, waste, and undesirable by-products. Today, almost all glucose is produced using enzymes. High-fructose corn syrup, which is found in most soft drinks, is also produced using an enzyme.

In the baking industry, enzymes help to improve the quality, freshness, and shelf life of breads and baked goods. They convert sugars to alcohol and carbon dioxide causing the dough to rise; strengthen the gluten network resulting in greater flexibility and machinability; and modify triglycerides resulting in larger loaf volumes (Reed, 2012).

In cheese making, enzymes help the milk to coagulate, the first step in making cheese. In the dairy industry, both microbial and animal enzymes are used. Enzymes are also used to accelerate cheese ripening and to reduce the allergic properties of dairy products. Chymosin is the enzyme used for coagulation, lipase is used for ripening, and lactase is used to improve digestibility for those who may be lactose intolerant. The source of animal enzymes is a concern for halal consumers. Cheese and whey (the yellowish liquid remaining after the cheese curds are formed) produced by using animal enzymes are haram if the source is haram animals (Ermis, 2017; Man and Sazili, 2010). Cheese and whey produced using animal enzymes from halal animals not slaughtered according to Islamic requirements are doubtful, because only some Muslim consumers accept these products. Because whey has become a popular food ingredient, mainly because of its proteins, the use of animal-derived enzymes to produce whey poses a problem for the halal consumer (Chaudry, 2002).

Protein hydrolysis is another application for enzymes. Animal and vegetable proteins are hydrolyzed using enzymes to improve the functionality and nutritional values of proteins (Man and Sazili, 2010). Proteins are used in foods as emulsifiers; hydration agents; to control viscosity; as gelling agents; and to improve cohesion, texture, and solubility. Production of these proteins by chemical reactions is undesirable because it requires severe conditions and produces many undesirable by-products, which are difficult or expensive to remove. Enzymes are faster, allow production under milder conditions, and produce fewer undesirable by-products. Enzymes can be tailored to catalyze a specific reaction, with very little by-product formation. Enzymes can be used to produce meat extracts from scrap-bone residues that can be used in soups, sauces, broths, and other applications (Ermis, 2017).

In the juice industry, enzymes increase yields and improve color and aroma. Enzymes are used to clarify juices and extract essential oils from citrus peels (Oort, 2009). Enzymes can also improve the texture of fruit pieces used in food products, such as fruit-flavored yogurt (Reed, 2012). Enzymes can be used in the extraction of vegetable oil. Conventionally, high oil containing materials such as rapeseeds, coconuts, sunflower seeds, palm kernels, and olives are first pressed to extract the oil. The remaining oil is then extracted by using an organic solvent. In this second phase, enzymes can be used instead to allow the oil to be extracted into a water solution, avoiding the need for an organic solvent, making it a more environmentally friendly method of producing oils and avoiding consumer concerns about the use of organic solvents. However, this method is not yet in wide use. Enzymes are also used to modify oils to improve nutritional value or to produce lubricants and cosmetic ingredients (Reed, 2012; Ermis, 2017). (Note: All chemical compounds contain carbon are considered to be organic compounds. The term "organic" as used to designate food products is based on a set of standards governed in the U.S. by the Agricultural Marketing Service of the U.S. Department of Agriculture.)

CLASSIFICATION OF ENZYMES

The International Union of Biochemists has developed an identification system to classify enzymes. Naming of enzymes involves a numerical classification, a long systematic name, and a short, easy-to-use name. For example, the enzyme that catalyzes the conversion of lactose (milk sugar) to galactose and glucose is classified as EC 3.2.1.23, has the systematic name b-d-galactoside galactohydrolase, and the common name lactase.

Enzymes are classified into six categories (Ako and Nip, 2012; Fernandes, 2010; Olsen, 2000b):

- *Oxidoreductases*: Catalyze oxidation reactions, such as the conversion of alcohol.
- *Transferases*: Enzymes that catalyze the transfer of a group of atoms, referred to as a radical, from one molecule to another, such as the transfer of amino groups.
- *Hydrolases*: Catalyze the reaction of a chemical with water. This usually breaks up large molecules into smaller ones by adding water at each break point, such as the hydrolysis of proteins.
- *Lyases*: Catalyze reactions producing or resulting in double bonds, for example, conversion of sugars.
- *Isomerases*: Catalyze the transfer of groups on the same molecule, resulting in a new structure for the molecule.
- *Ligases*: Catalyze the joining of molecules to form larger molecules.

BIOENGINEERED ENZYMES

The identification and understanding of how DNA functions has led to gene splicing and the development of bioengineered enzymes. Many enzymes are now produced by bioengineering, with a variety of methods employed for the growth and production of these enzymes (Khattak et al., 2011; Søndergaard et al., 2005). One method of production is submerged fermentation. In this process, selected microorganisms (bacteria and fungi) are grown in closed vessels in the presence of liquid nutrients and oxygen. As the microorganisms break down the nutrients, they release enzymes into the solution. The nutrients are normally sterilized foodstuff such as cornstarch, sugars, and soybean grits (Wingard et al., 2014). On the other hand, some such processes use animal by-products as part of the nutrient mix, which unless from a halal-slaughtered animal and properly prepared would render the products of such a production haram (Fischer, 2015). The process can be operated continuously or in a batch mode. The temperature of the vessels and oxygen consumption and pH are carefully controlled to optimize enzyme production.

Recovering the enzymes, referred to as harvesting, is done in a number of steps. First, the solids, referred to as biomass, are removed by filtration or centrifugation. The enzyme remains in the solution, called the broth. The broth is evaporated to concentrate the enzymes. The enzymes can be further purified by ion exchange, before further processing into powders, liquids, or granules. The residual biomass is often stabilized using a lime treatment and used as fertilizer. Table 13.1 gives a sampling of the major sources of enzymes.

HALAL CONTROL POINTS IN ENZYME PRODUCTION

Enzymes are harvested or extracted from natural biological sources such as animal tissues, plant materials, or microbial sources. Enzymes can also be produced

TABLE 13.1
A Sampling of Major Sources of Enzymes

Source	Type	Activity	Uses
Bacteria	Bacillus	Protease	Meat, beverages
Bacteria	Streptomyces	Isomerase	Beverages, starch
Fungi	Aspergillus	Protease	Cheese
Fungi	Mucor	Lipase	Cheese, fat
Yeast	Saccharomyces	Invertase	Cocoa
Yeast	Kluyveromyces	Chymosin/renin	Cheese
Plants	Barley/malt	Amylase	Bakery, sugar
Plants	Papaya	Papain/protease	Bakery, beverages
Animal	Bovine liver[a]	Catalase	Beverages, dairy
Animal	Ruminants[a]	Rennin/protease	Cheese
Animal	Pig/cattle stomach[a]	Pepsin/protease	Cheese/cereals

Source: From Mathewson, P. R. (1998). Major biological sources of enzymes (Appendix C). In *Enzymes*. St. Paul, MN: Eagan Press, pp. 93–95. With Permission.

[a] Examples of enzymes that are of particular concern in halal food production.

Enzymes in Halal Food Production

through a fermentation process. Requirements for each process are quite different and unique to each product. Therefore, halal control points (HCPs) have been determined although they may need to be adjusted for the uniqueness of the particular processes (OIC/SMIIC, 2011).

HALAL CONTROL POINTS FOR A CONVENTIONAL PROCESS FOR EXTRACTING ENZYMES FROM ANIMAL ORGANS

There are five HCPs: use of halal organs; clean equipment; acceptable release agents; approved standardization ingredients; and proper packing and labeling (Figure 13.1).

HCP1: ANIMAL ORGANS

In commercial practice, enzymes are extracted from many animal organs of various species such as porcine, bovine, or ovine. Enzymes extracted from pigs' organs are not acceptable by any halal consumer or halal regulatory group, so these must not be used individually or in combination with other animal organs. Organs should be from halal-slaughtered animals to be universally acceptable. Such organs might have to be harvested from regions where Muslims do the animal slaughter. Some

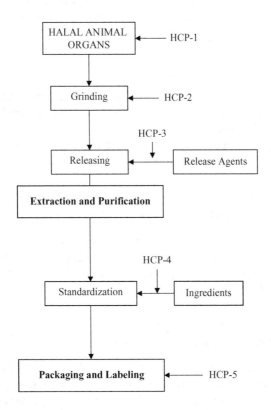

FIGURE 13.1 Conventional process for extracting enzymes from animal organs.

countries might accept enzymes from halal animals even though the animals are not slaughtered by Muslims (Ermis, 2017; Shafii and Wan Siti Khadijah, 2012).

HCP2: Preparation of Tissues for Extraction

Because a majority of the enzymes of animal origin are from non-halal sources, use of common equipment presents a concern. Before using the equipment for extracting halal enzymes, it must be thoroughly cleaned to avoid contamination from previous runs (Al Mazeedi et al., 2013; Man and Sazili, 2010).

HCP3: Use of Acceptable Release Agents

The enzymes might not be present in the tissues in a soluble form and may therefore need to be released or made soluble to increase the yield. Chemicals used for this purpose must be suitable for halal production.

HCP4: Ingredients for Standardization

Besides salt and water, several other ingredients can be used to adjust the enzyme's strength. Preservatives and emulsifiers are also used to enhance or increase shelf life. All standardizing ingredients used should be suitable for halal production (Ermis, 2017).

HCP5: Packaging and Labeling

Finally, labels should be marked properly, including halal markings to identify the product correctly. If the enzymes are from animals slaughtered by Muslims (preferably using the dhabh procedure), they should be identified as such rather than labeling them simply as bovine and so on (Ermis, 2017). Packaging materials and labels themselves should conform to the guidelines discussed in Chapter 23.

HALAL CONTROL POINTS FOR ENZYME PRODUCTION IN A CONVENTIONAL FERMENTATION PROCESS

There are six critical control points in this process: raw materials; origin of cultures; acceptable processing aids; growth media; approved standardization ingredients; and packaging and labeling. Many of the process steps are similar to those shown in Figure 13.2, two major differences being the origin of the cultures and the growth media.

HCP1: Raw Materials for Growth Media

Growth media was not considered to be a major halal issue until seven employees of P.T. Ajinomoto were arrested in Indonesia for violating the country's law for manufacturing halal foods. In the growth media, they used a soy peptone that was manufactured using porcine enzymes (Roderick, 2001).

Enzymes in Halal Food Production

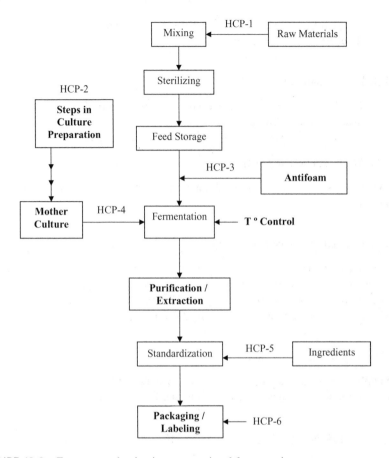

FIGURE 13.2 Enzyme production in a conventional fermentation process.

The first control point in the production of enzymes by fermentation is the use of acceptable media. Any raw materials even of vegetable origin should not have been modified or processed by using porcine or other non-halal enzymes or materials (Olempska-Beer et al., 2006).

HCP2: Origin of Culture

Microbial cultures, yeast, algae, or bacteria can be indigenous or can be genetically engineered. All indigenous sources of culture are acceptable in commercial practice as long as the culture media is halal. However, if the bacteria or other microorganisms have been modified through biotechnology, the source of the genetic material used for the original gene becomes important (Man and Sazili, 2010). Genetic material from halal species of animals and all plant sources are generally acceptable. The food safety of such a genetically modified organism is a major concern but it is normally the responsibility of the government food regulatory agencies such as the U.S. Food and Drug Administration (FDA), the U.S. Department of Agriculture (USDA), and the Environmental Protection Agency (EPA) in the U.S. and not of the religious regulatory agencies. As a general rule, all genetic materials from haram animals should be avoided (Ermis, 2017).

HCP3: Processing Aids

Processing aids such as antifoam agents should be free of prohibited materials, especially derivatives from pork fat.

HCP4: Additives in the Media

All materials used in the growth media and in the preparation of the mother culture should be halal.

HCP5: Standardization Ingredients

Preservatives, emulsifiers, and other standardizing materials must be from acceptable sources. Alcohol may sometimes be used as a preservative to protect the enzyme activity. Non-beverage ethyl alcohol is generally acceptable if below 0.5% by volume in the final enzyme preparation (Ab Talib and Johan, 2012).

HCP6: Packaging and Labeling

Enzymes should be packed in acceptable containers and labeled properly with halal markings. For suitable packaging materials, see Chapter 23.

TO LABEL OR NOT TO LABEL ENZYMES

Enzymes are usually used as processing aids and functional catalysts. They are usually inactivated in the final product and are not included on the labels. For example, in fruit juice processing, enzymes are deactivated during pasteurization and the active form might not be detected in the finished product, hence, they may not need to be labeled. In cheese and bakery products, however, enzymes might remain active in the final product and then should be listed on the finished product label (Mannie, 2000).

Muslim consumers as well as halal authorities in many countries are concerned about the presence and source of enzymes. It is better to list not only the descriptions of the enzymes but also their sources, even when the products are marked halal.

REFERENCES

Ab Talib, M. S. and Johan, M. R. M. (2012). Issues in halal packaging: A conceptual paper. *International Business and Management*, 5(2), 94–98.

Ako, H. and Nip, W. K. (2012). Enzyme classification and nomenclature. *Food Biochemistry and Food Processing*, 109–124.

Al-Mazeedi, H. M., Regenstein, J. M., and Riaz, M. N. (2013). The issue of undeclared ingredients in halal and kosher food production: A focus on processing aids. *Comprehensive Reviews in Food Science and Food Safety*, 12(2), 228–233.

Ashie, I. N. (2003). Bioprocess engineering of enzymes. *Food Technology*, 57(1), 44–51.
Birch, G. G., Blakebrough, N., and Parker, K. J. (2012). *Enzymes and Food Processing*. Berlin, Germany: Springer Science and Business Media.
Chaudry, M. M. (2002). Enzymes catalysts for life. *Halal Consum*, 4, 5–7.
Dodge, T. (2009). Production of industrial enzymes. In *Enzymes in Food Technology*. Hoboken, NJ: Blackwell Publishing, 44–58.
Ermis, E. (2017). Halal status of enzymes used in food industry. *Trends in Food Science and Technology*, 64, 69–73.
Fernandes, P. (2010). Enzymes in food processing: A condensed overview on strategies for better biocatalysts. *Enzyme Research*. Available online: http://dx.doi.org/10.4061/2010/862537.
Fischer, J. (2015). Keeping enzymes kosher: Sacred and secular biotech production. *EMBO Reports*, e201540529.
Khattak, J. Z. K., Mir, A., Anwar, Z., Abbas, G., Khattak, H. Z. K., and Ismatullah, H. (2011). Concept of halal food and biotechnology. *Advance Journal of Food Science and Technology*, 3(5), 385–389.
Man, Y. B. C. and Sazili, A. Q. (2010). Food production from the halal perspective. In *Handbook of Poultry Science and Technology*. Hoboken, NJ: John Wiley & Sons, 183–215.
Mannie, E. (2000). Active enzymes, *Prepared Foods*, 169(10), 63–66, 68.
Mathewson, P. R. (1998). Major biological sources of enzymes (Appendix C). In *Enzymes*. St. Paul, MN: Eagan Press.
OIC/SMIIC. (2011). General guidelines on halal food (with the references of CODEX, ISO 22000, ISO 22005 þ Islamic Fiqh Rules). Turkey, Istanbul: SMIIC.
Olempska-Beer, Z. S., Merker, R. I., Ditto, M. D., and DiNovi, M. J. (2006). Food-processing enzymes from recombinant microorganisms—a review. *Regulatory Toxicology and Pharmacology*, 45(2), 144–158.
Olsen, H. S. (2000a). *The Nature of Enzymes, in Enzymes at Work*. Bagsvaerd, Denmark: Novozymes A/S. Available online: https://www.novozymes.com/-/media/Project/Novozymes/.../Enzymes_at_work.pdf.
Olsen, H. S. (2000b). *Enzyme Applications in the Food Industry, in Enzymes at Work*. Bagsvaerd, Denmark: Novozymes A/S. Available online: https://www.novozymes.com/-/media/Project/Novozymes/.../Enzymes_at_work.pdf.
Oort, M. V. (2009). Enzymes in food technology-introduction. In *Enzymes in Food Technology*, 2nd edn, 1–17.
Reed, G. (Ed.). (2012). *Enzymes in Food Processing*. New York: Elsevier.
Roderick, D. (2001). Hold the pork, please, *Time Asia*, 157(3), January 22. Available online: www.time.com/time/world/article/0,8599,2055065,00.html.
Shafii, Z. and Wan Siti Khadijah, W. M. N. (2012). Halal traceability framework for halal food production. *World Applied Sciences Journal*, 17, 1–5.
Søndergaard, H. A., Grunert, K. G., and Scholderer, J. (2005). Consumer attitudes to enzymes in food production. *Trends in Food Science and Technology*, 16(10), 466–474.
Wingard, L. B., Katchalski-Katzir, E., and Goldstein, L. (Eds.). (2014). *Immobilized Enzyme Principles: Applied Biochemistry and Bioengineering* Vol. 1. New York: Elsevier.

14 Gelatin in Halal Food Production

Mian N. Riaz and Munir M. Chaudry

CONTENTS

Status of Gelatin in Islam 177
Sources of Gelatin 178
Production of Halal Gelatin 178
Preparation of Gelatin 178
Vegetable Substitutes for Gelatin 179
Some Products That Use Halal Gelatin 180
    Gelatin in Foods 180
Halal Gelatin in Pharmaceuticals And Cosmetics 181
    Medicinal, Dietetic, and Therapeutic Uses 181
Control Points in Halal Gelatin Production 181
    HCP1: Raw Materials 181
    HCP2: Equipment 182
    HCP3: Chemicals and Ingredients 182
References 183

Gelatin is used in many food products, including jellies, ice cream, confectionery, cookies, and cakes. It is also used in non-food products, including medical products and in veterinary applications (Schrieber and Gareis, 2007). Gelatin can be from halal or haram sources. Common sources of gelatin are pigskin, cattle hides, cattle bones, and, less frequently, fish skins and poultry skins. In general, a product label does not indicate the source of the gelatin (Chaudry, 1994; Cheng et al., 2012; Hermanto and Fatimah, 2013; Jaswir et al., 2009), so halal consumers normally avoid products containing gelatin unless they are certified halal. As Muslim countries have increased imports of food products, there has been growing awareness of the problem gelatin presents to Muslim consumers (Hermanto and Fatimah, 2013; Widyaninggar et al., 2012).

Malaysia, Indonesia, and several other Muslim countries now require that imported as well domestic products containing gelatin be produced with halal gelatin. Several gelatin manufacturers in Europe, India, and Pakistan produce halal gelatin.

STATUS OF GELATIN IN ISLAM

Gelatin is an animal by-product, the partially hydrolyzed collagen tissue from various animal parts. Its halal status depends on the nature of raw materials used in its manufacture (Widyaninggar et al., 2012). Most gelatin is one of two types: (1) Type A gelatin is exclusively made from pork skins, hence haram for Muslims to use. (2) Type B gelatin is made either from cattle and calf skins or from demineralized cattle bones. Cattle and calf skins used in gelatin manufacture are usually from animals slaughtered by non-Muslims. Whether this type of gelatin is permitted or prohibited for Muslims is controversial. However, gelatin made from the bones of halal-slaughtered cattle is available (Cheng et al., 2012). Fish-skin gelatin is halal as long as it is free from contamination from other sources and is made from a fish species that is accepted by the Muslims who use the product (Chapter 10) (Gómez et al., 2009; Karim and Bhat, 2009). A food processor understands that a non-specific gelatin is highly questionable regarding its source, and must be suspected of containing pork gelatin. So its use is very strongly discouraged for Muslims (Hermanto and Fatimah, 2013; Sakr, 1999; Widyaninggar et al., 2012).

SOURCES OF GELATIN

For gelatin from cattle skins, cattle bones, poultry skins, or other permitted animals to be halal, the animals have to be slaughtered according to Islamic requirements, as explained in Chapter 3. Again, the slaughter of the animals must be acceptable to the targeted Muslim consumers.

In modern slaughterhouses, bones are sold to rendering companies, which turn fresh bones into dry bone chip used for gelatin (Hermanto and Fatimah, 2013; Jaswir et al., 2009). However, in many Asian and African countries, bones are discarded as waste and are subject to a natural degreasing process by microbes and other organisms. Without going into details of the degreasing process, it suffices to say that when collecting and selecting the bones for food-grade or pharmaceutical-grade halal gelatin, bones must be examined and segregated into bones from halal species (if acceptable even if not slaughtered halal) and those from non-halal species, which cannot be used. Bones from animals that have died without being properly slaughtered or that were used for religious ceremonies are also prohibited. So bones collected from cattle that died naturally are not permitted (Cheng et al., 2012).

PRODUCTION OF HALAL GELATIN

Gelatin is derived from collagen, an insoluble fibrous protein that occurs in vertebrates and is the principal constituent of connective tissues and bones. Gelatin is recovered from collagen by hydrolysis. There are several varieties of gelatin, the composition of which depends on the source of collagen and the hydrolytic treatment used.

PREPARATION OF GELATIN

The principal raw materials used in halal gelatin production currently are cattle bones and cattle hides. Non-collagen substances such as minerals (in the case of bones) and fats and proteins (in the case of hides) are removed by various treatments to prepare collagen for extraction (Cheng et al., 2012).

- *Bones*: Fresh bones, also called green bones, from the halal-slaughtered cattle are cleaned, degreased, dried, sorted, and crushed to a particle size of ca. 1 to 2 cm. The pieces of bone are then treated with dilute hydrochloric acid to remove mineral salts. The resulting sponge-like material is called ossein.
- *Hides*: Cattle hides from halal-slaughtered animals are received from the trimming operations of leather production. These are of concern because of the chemicals used. Therefore, hides dedicated for gelatin are preferred. The hide pieces are usually de-haired chemically with a lime and sulfide solution, followed by a mechanical loosening.

For the production of halal gelatin, both ossein and cattle hide pieces are subjected to lengthy treatment with an alkali, usually lime and water, at ambient temperature. Depending on previous treatment, the nature of the material, size of the pieces, and exact temperature, and liming usually takes 8 to 12 weeks. The material is then thoroughly washed with cold water to remove excess lime, its pH adjusted with acid, and extracted with hot water to recover the soluble gelatin (Tasara et al., 2005).

Dilute halal gelatin solutions from the various hot water extractions are filtered, deionized, and concentrated using cross-flow membrane filtration or vacuum evaporation, or both. The vacuum evaporator can be cleaned to permit it to be used for halal production while the membrane filters generally cannot be cleaned sufficiently to be used for both halal and non-halal production. Halal gelatin solutions are then chilled and either cut into ribbons or extruded as noodles, and the gelled material is deposited as a bed onto an endless, open-weave stainless steel belt. The belt is passed through a drying chamber. If this equipment has been used for non-halal gelatin, it will require an extensive and complicated cleaning. After drying, it is broken into pieces that are ground to the required particle size. Again the grinder requires cleaning if it was previously used for non-halal purposes (Tasara et al., 2005).

Where halal gelatin is not available, food manufacturers can use plant substitutes. However, currently gelatin is the only material that melts below body temperature and is reversible, that is, can be melted and gelled more than once at low temperatures, so the these substitutes are not a complete replacement. A process of gelation processing is shown in Figure 14.1

VEGETABLE SUBSTITUTES FOR GELATIN

- *Agar*: Also called agar-agar, gelose, Chinese isinglass, Japanese isinglass, Bengal isinglass, or Ceylon isinglass.
- *Carrageenan*: A polysaccharide extracted from red seaweed.
- *Pectin*: A polysaccharide substance present in cell walls of all plants.
- *Xanthan gum*: A polysaccharide gum produced by bacteria. The bacterial medium must be halal for the product to be halal.
- *Modified corn starch*.
- *Cellulose gum*.

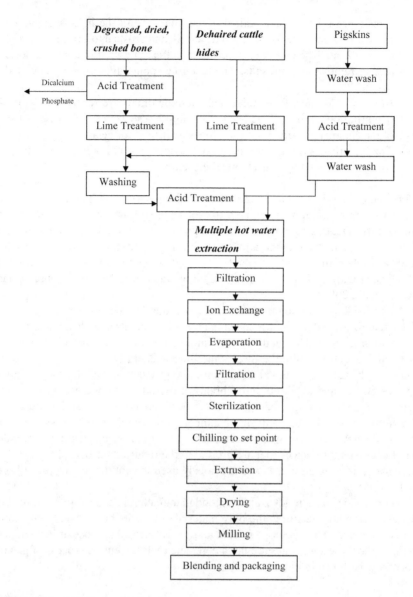

FIGURE 14.1 Gelatin production process.

SOME PRODUCTS THAT USE HALAL GELATIN

Gelatin in Foods

Halal gelatin can be used for gelatin desserts; dairy products such as yogurt, sour cream, and cottage cheese; and other dairy and imitation dairy foods. It is also widely used in frozen desserts such as ice cream, cream pies, and cheesecakes. Gelatin is the primary ingredient in marshmallows. It is used in some other confections. In the meat industry, gelatin is used in luncheon meats, jellied beef, and corned beef loaves. Gelatin is also used as a processing aid in the food industry for the clarification of cider and fruit juices (Al-Mazeedi et al., 2013; Blech, 2008; Regenstein 2012; Regenstein et al., 2003).

HALAL GELATIN IN PHARMACEUTICALS AND COSMETICS

The major use of halal gelatin in the pharmaceutical industry is in the manufacture of capsules. Both soft and two-piece hard capsules as well as enteric capsules contain gelatin as the main ingredient.

Halal gelatin is also used in tablets, lozenges, and cough drops. It has been used for the external application of drugs to treat various skin disorders and as an adhesive to hold bandages and dressings together. Other pharmaceutical uses of gelatin include as a glycerinated base for suppositories and a carrier for certain dietary supplements.

Gelatin is frequently used in creams and wave-set lotions, and may be the protein used in "protein" shampoos and hair conditioners. The use of halal gelatin in any of these products will increase the market for these products in Muslim countries.

MEDICINAL, DIETETIC, AND THERAPEUTIC USES

Gelatin in various forms may be used by medical professionals as absorbable sponges to arrest hemorrhage and as a dusting powder for surgical gloves. Gelatin can be used in open wounds. Gelatin is an excellent dietetic and therapeutic agent for the prevention of obesity. Low-sugar gelatin desserts require more calories to digest than they contain. Gelatin (when properly supplemented) has been used as a protein food in malnutrition and infant feeding. Other therapeutic uses include treatment of digestive disorders, peptic ulcers, muscular disorders, and brittle fingernails. Gelatin is commonly used as a plasma extender for the treatment of shock. However, although not desirable, when these non-halal products are used in life-saving applications they are acceptable. But under less compelling circumstances, Muslim consumers may reject the product or treatment. Thus, pharmaceutical and drug industries should start using halal gelatin rather than just gelatin to cater to the Muslim market. Halal gelatin, which has the same functional properties as regular gelatin and might only be slightly more expensive, opens up many new markets.

CONTROL POINTS IN HALAL GELATIN PRODUCTION

The most important issue in producing halal gelatin is obtaining the proper raw materials. Because pigskins cannot be used for halal gelatin production, plants that manufacture pigskin gelatin should not be considered for halal production. Halal raw materials, both bones and hides, are a limiting factor; hence, gelatin companies have to work upstream with their suppliers through tanneries and bone mills, all the way to halal slaughterhouses. Figure 14.2 shows halal control points (HCPs) in gelation processing.

HCP1: RAW MATERIALS

All sources, hides, and bone chips should be approved and constantly monitored. Gelatin factories normally receive pieces of hide and bone chips. Gelatin

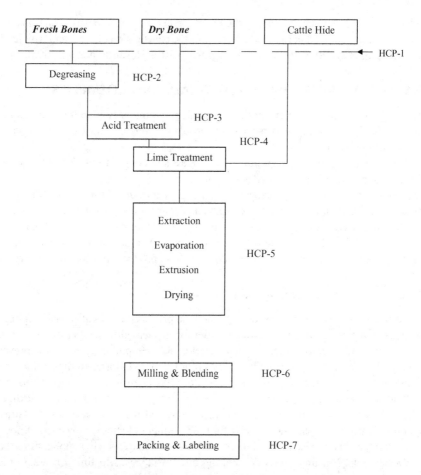

FIGURE 14.2 Gelatin production with respect to halal critical points (HCPs).

manufacturers must execute controls at their supplier's plants to make sure raw materials are properly segregated throughout the entire process.

HCP2: Equipment

All equipment used in the production of halal gelatin must either be dedicated equipment or cleaned consistent with the halal requirements of the halal certifying agency. If cleaning is used, it must be fully documented.

HCP3: Chemicals and Ingredients

Although most of the chemicals and ingredients used in gelatin production are non-organic chemicals, the source and production of all of them must be determined to be sure they are made under conditions that do not compromise their halal status. This also covers all packaging material.

REFERENCES

Al-Mazeedi, H. M., Regenstein, J. M., and Riaz, M. N. (2013). The issue of undeclared ingredients in halal and kosher food production: A focus on processing aids. *Comprehensive Reviews in Food Science and Food Safety*, 12(2), 228–233.

Blech Z. (2008). The story of gelatin. In: *Kosher Food Production*, 2nd edn. Ed. Blech, Z. Hoboken, NJ: Wiley Blackwell, 317–322.

Chaudry, M. M. (1994). Is Kosher Gelatin Really Halal? *Islam Perspectif.* XI (1), 6, 13–28.

Cheng, X. L., Wei, F., Xiao, X. Y., Zhao, Y. Y., Shi, Y., Liu, W., and Lin, R. C. (2012). Identification of five gelatins by ultra-performance liquid chromatography/time-of-flight mass spectrometry (UPLC/Q-TOF-MS) using principal component analysis. *Journal of Pharmaceutical and Biomedical Analysis*, 62, 191–195.

Gómez-Estaca, J., Montero, P., Fernández-Martín, F., and Gómez-Guillén, M. C. (2009). Physico-chemical and film-forming properties of bovine-hide and tuna-skin gelatin: A comparative study. *Journal of Food Engineering*, 90(4), 480–486.

Hermanto, S. and Fatimah, W. (2013). Differentiation of bovine and porcine gelatin based on spectroscopic and electrophoretic analysis. *Journal of Food and Pharmaceutical Sciences*, 1(3), 68–73.

Jaswir, I., Mirghani, M. E. S., Hassan, T., and Yaakob, C. M. (2009). Extraction and characterization of gelatin from different marine fish species in Malaysia. *International Food Research Journal*, 16, 381–389.

Karim, A. A. and Bhat, R. (2009). Fish gelatin: Properties, challenges, and prospects as an alternative to mammalian gelatins. *Food Hydrocolloids*, 23(3), 563–576.

Regenstein, J. M., Chaudry, M. M., and Regenstein, C. E. (2003). The kosher and halal food laws. *Comprehensive Reviews in Food Science and Food Safety*, 2(3), 111–127.

Regenstein, J. M. (2012). The politics of religious slaughter-how science can be misused. In *65th Annual Reciprocal Meat Conference at North Dakota State University in Fargo, ND*.

Sakr, A. H. (1999). *Gelatin.* Lombard, IL: Foundation for Islamic knowledge, and; Chicago, IL: Islamic Food and Nutrition Council of America, 13–28.

Schrieber, R. and Gareis, H. (2007). *Gelatine Handbook: Theory and Industrial Practice.* Hoboken, NJ: Wiley.

Tasara, T., Schumacher, S., and Stephan, R. (2005). Conventional and real-time PCR–based approaches for molecular detection and quantitation of bovine species material in edible gelatin. *Journal of Food Protection*, 68(11), 2420–2426.

Widyaninggar, A., Triyana, K., and Rohman, A. (2012). Differentiation between porcine and bovine gelatin in capsule shells based on amino acid profiles and principal component analysis. *Indonesian Journal of Pharmacy*, 23(2), 104–109.

15 Flavors, Flavorings, and Essences in Halal Food

*Eric Butrym, Laura LaCourse,
Mian N. Riaz, and Munir M. Chaudry*

CONTENTS

Simple Mixtures 188
    Concentrated Liquid Flavors 188
    Dilutions 188
Processed Flavoring Materials 189
    Extraction 189
    Reaction Flavors 190
Engineered Flavor "Systems" 191
    Spray-Dried Flavors 191
    Emulsions 191
    Other Encapsulation Techniques and Flavor Delivery Systems 192
    Synthetic Chemicals 192
    Oleoresins and Essential Oils 192
    Fruit Juices and Concentrates 192
    Sugars and Starches 193
    Amino Acids 193
    Lipids and Fatty Acids 193
Areas of Special Concern for Halal 194
    Biotechnology 194
    Animal-Derived Materials 195
    Alcohol 196
Production Processes 197
    Raw Material Receipt 197
    Equipment Segregation 197
    Cross-Contamination, HACCP Controls 198
    Cleaning and Storage 198
References 198

Flavoring materials are among the most complex components of modern food products; they are made from a large number and variety of ingredients from different sources. The complexity of flavoring materials makes them particularly interesting with regard to regulatory compliance and other requirements, such as halal status. It is important to note that more than 90% of raw materials used in the manufacture of flavors pose no risk whatsoever to the halal status of the finished products because their origin is either synthetic or from acceptable plant sources. This is a welcome simplification, as there are over 4000 ingredients in general use in the flavors industry, and most suppliers of these raw materials provide no halal certification (Uhl, 2000).

In most markets, the terms "flavors" or "flavoring" can include many materials as long as they collectively act to alter the sensory perception of the finished food product as perceived in the mouth and nasopharynx, with the *exception* of altering the basic tastes of sweet, salty, sour, bitter, or umami. Thus, an additive whose purpose may be to lower sugar content by boosting the perception of sweetness may be used and declared as a flavoring material only in concentrations that do not by themselves taste sweet. At levels above the threshold at which sweetness is observed from the material by itself, it must be declared as a food additive or sweetener. As another example, citric acid is used as an acidulant (without which many fruit flavors would taste flat or cloying) in soft drinks and must be declared as such, as at almost any effective level, the material by itself will impart a sour taste. It cannot therefore be represented in ingredient statements under the generic category of "flavor" because by itself it alters one of the five basic taste modalities. Some ingredients in flavors have more than one function and it may be left to the discretion of the manufacturer whether to declare them under the category of "flavors" or as individual additives (U.S. FDA, 2016).

The complexity inherent in flavoring preparations has led to the acceptance of a wide range of materials under the broad category of "flavors," especially for the purposes of consumer labeling. Although the reasons for this are pragmatic (consider the effects of listing an additional 20 or more ingredients—which the flavor manufacturer wants to keep confidential even from the food companies they serve—on every food label), the practice can raise questions about flavorings as places where undeclared additives of unknown or dubious origin may be introduced into food products. To address these concerns, most regulatory organizations such as the European Council and the U.S. Food and Drug Administration (FDA) have succinct definitions that determine what may or may not be included under the term "flavor": U.S.: 21 CFR 501.22, 21 CFR 101.22; EU: EC 2334/2008; and so on. The halal status and related characteristics of non-flavoring food additives are covered in other chapters of this book; the following pages will therefore treat several classes of flavor products and those raw materials commonly declared as flavors or flavoring, in addition to some important intermediates used in their manufacture.

Flavor products are available in a wide variety of different forms, from simple mixtures compounded from a few liquid ingredients to highly engineered "flavor systems" that may include sophisticated encapsulation technology. The form taken by any particular flavoring material is dictated by several factors but is heavily influenced by the nature of the application (a carbonated soft drink and a breakfast cereal, for instance, can have vastly different requirements, including solubility and dissolution, heat stability, and compatibility with other food ingredients, to name only a few). Shelf-life, ease of handling, and cost in use are other factors that may determine the form in which a flavoring product is distributed and used.

Halal food producers and consumers benefit greatly from the information flavor manufacturers collect to meet product quality and regulatory requirements. Raw materials used in the manufacture of flavors must be scrutinized closely for assurance that the producer maintains the ability to consistently make and appropriately market and label their products. To supply a product that meets customer requirements for natural or organic status, for example, information must be maintained on each raw material in the product's formula, as a single synthetic ingredient or sufficient quantity of uncertified components can affect the status of the finished goods. The need to identify and track such diverse information on every raw material allows the tracking of the halal status of any individual ingredient to be integrated as a routine process.

How then does one begin to evaluate the halal status of products containing dozens of ingredients (from a palette of literally thousands)? The answer is to focus closely on those materials that pose the highest risk. Halal rules for flavorings can be generalized to four principles, with some minor exceptions:

1. Animal products must be halal either intrinsically (honey, dairy products) or from acceptable species slaughtered according to Islamic law. Vegetable materials are halal except if modified to include genetic material from animal sources.
2. Ethanol and other intoxicants are not halal. Some allowances must be made for ethanol that occurs naturally or in proportions that cannot result in intoxication, but products derived from the manufacture of alcoholic beverages are forbidden.
3. Fermentation agents including microbes and enzymes must be derived and maintained according to halal principles—that is, origin and genetic modification should be monitored, and growth media must be free of non-halal ingredients or additives.
4. Production equipment and environments must be managed to preclude any possibility of contamination of halal products with haram materials.

The first three of these requirements can be adequately managed through control of sourcing and formulation information. Collecting data from suppliers about ingredient origin, halal certification, and other aspects, such as specifics of genetic modification or manufacturing details, will allow the automatic screening of product formulas for non-compliant ingredients as described previously. Monitoring of production environments usually requires the establishment and enforcement of manufacturing procedures to ensure the integrity of halal flavors. This will be discussed in depth later in this chapter.

A final corollary to these principles is that any gray area or question arising in relation to the halal status of ingredients or finished products can only be resolved through the decision of a competent halal authority.

An unavoidable characteristic of the global business of food production is that certification authorities will have differing standards and approaches to halal manufacturing. Successful outcomes are the product of excellent relationships with certifying bodies as well as a fundamental commitment to maintaining the integrity of halal programs, including policies governing the entire value chain from raw material purchasing and formula creation to the management of inventory, production, and shipping processes.

It is entirely fitting, even intuitive, that food should hold a central place in most cultures, and that those who provide such necessities to the general population are held to high standards of integrity. This is reflected in the detailed legislation dealing with food and its production in most parts of the world; it is also reflected in the suspicion with which corporations and other profit-driven entities are sometimes regarded with respect to providing products for consumption. These circumstances combine to represent a significant risk to businesses engaged in the production of food and food ingredients. Inasmuch as halal production is undertaken to provide opportunities to access markets that would otherwise be closed (i.e., a commercial motivation), the costs and sacrifices in flexibility involved must be fully understood. A lapse in commitment or procedure in the manufacture of halal products may naturally lead to questions of compliance in other areas, and by itself can be sufficient to severely damage a brand. If the necessary resources and attention to program design and maintenance cannot be met, it is far wiser to forego business opportunities in the halal sphere than risk noncompliance. The information that follows has been accumulated over many years of managing the halal aspects of flavor manufacturing in many different contexts and circumstances.

SIMPLE MIXTURES

CONCENTRATED LIQUID FLAVORS

Perhaps the most common form flavoring materials take is that of concentrated liquids. The reasons for this are many, including greatly simplified handling and manufacturing, ready availability of raw materials in liquid form, general (though not universal) lack of solubility problems, and the more practical commercial advantages related to shipping and cost in use. The halal status of liquid flavoring materials, as that of all other flavor preparations, depends not only on the status of the raw materials used in the formulations, but also on other aspects peripheral to the ingredients themselves, and specifically on the production equipment used to manufacture, handle, and store them. The primary simplifying characteristic of liquid flavors with respect to halal is that the manufacturing formula can be used to determine the status of the raw materials in a straightforward way. That is, in most cases the finished product is a simple mixture of ingredients, and there are usually very limited circumstances (covered below) in which interactions between the raw materials can affect the halal status. Additionally, a liquid flavor formula that has been certified as halal can itself be used as a component of any number of flavoring materials (of almost any form), and its halal status can be verified by certificate instead of scrutinizing each raw material in its formula.

DILUTIONS

Diluted liquid flavors provide a case to illustrate the value of certifying concentrated flavoring materials. To establish the halal status of a dilution, one need only to provide the certificate for the concentrated form and assure that the dilution solvent is acceptable for use in halal products at the level considered appropriate by the certifying authority. The obvious complicating case is the use of ethanol as a dilution solvent (covered in more detail below), but there are other solvents in common use that can be derived from or adulterated with non-halal materials. Glycerin is a commonly cited example of a diluent that can be sourced from animal fats, although the synthetic version and even natural versions from vegetable sources are more widely available and less expensive. Again, careful attention to the collection and validation of information from raw material and solvent suppliers is critical to assuring the integrity of halal formulations.

PROCESSED FLAVORING MATERIALS

In contrast to simple mixtures, flavoring materials that undergo additional processing steps cannot necessarily be considered as mere combinations of the raw materials in their formulas.

EXTRACTION

Extraction is a widely used technique that selectively removes specific active or desirable flavoring components from more complex (usually a natural product) matrices by processing the starting material with a solvent. The key considerations for halal are the makeup and halal status of the base matrix as well as the extraction solvent. Common examples include:

1. *Natural extracts*: This process involves the removal of active flavor compounds from the biological matrix in which they naturally occur. This can be accomplished with a variety of solvents, although the manufacture of extracts for use in food (and flavoring) is limited to those with acceptable toxicity including water, carbon dioxide, and ethanol, among others. Natural vanilla extract has a specific standard of identity maintained by the U.S. FDA and must contain a minimum of 35% ethanol to meet the standard: U.S. FDA, 21CFR169.175. Halal certification authorities manage this requirement carefully, requiring information about the source of ethanol in the extract and verifying a level of use that will bring the alcohol content in consumer goods to an acceptable level before allowing the extract to be used in a product labeled as halal. The extract itself cannot be certified or labeled halal due to the high level of ethanol.
2. *Washed oils*: These are water and alcohol extractions of moderately polar flavor compounds from essential oils, which were in turn removed from their source via pressing (usually the case for citrus) or distillation and concentration to a homogenous oil phase. This technique uses water adjusted with some type of alcohol to provide an extraction solvent with the appropriate polarity. The solvent is combined with the essential oil (with which it is not miscible) and agitated to extract the desired polar compounds. The solvent is then decanted and may be filtered or standardized in subsequent

steps. The resulting product has the advantages of water solubility and relatively high concentrations of many of the most intense flavor compounds present in the original essential oil. If ethanol is used in the extraction solvent, specific information about the source and level will be considered by halal certifying bodies before allowing these extracts to be used in their halal foods.

REACTION FLAVORS

Flavoring preparations that specifically rely on chemical reactions between two or more raw materials to provide the desired flavor characteristics represent another level of complexity, as the composition of the finished product is quite intentionally different from any of the starting materials. However, the key to evaluating halal acceptability of these products again lies in examining their formulas. Two examples of this category of flavoring preparation bear special mention:

1. *Maillard reactions*: This term describes a complex chain of events, the details of which are beyond the scope of this chapter, except to mention that the first reaction step is between an amino acid and a reducing sugar molecule. In the context of halal acceptability, the focus will be on the source of the amino acids used in the reaction in addition to the other ingredients in the formula. It should be noted that although Maillard reactions can often be used to generate "savory" or "meaty" flavors, the tonality of the finished product does not necessarily provide an appropriate assessment of halal risk. Amino acids derived from animal sources can be involved in the creation of vegetable, cereal, or bakery flavors, and likewise many "meat" flavors can be created with amino acids strictly from vegetable sources. For a comprehensive treatment of this subject, see Nursten (2005).
2. *Esterification*: This reaction involves the creation of an ester compound by the controlled reaction of an organic acid with an alcohol. Esters occur naturally in many foods. Characteristically fruity and ethereal, they are commonly used in the manufacture of food flavors. The reaction itself occurs in aqueous solutions and can involve almost any alcohol-acid combination; it is enhanced with the proper control of pH and heat. Again, the primary concern for halal will be the source of ethanol if it is used as the alcohol in the reaction. Esters of ethanol are recognized by the word "Ethyl" at the beginning of the compound name (e.g., ethyl butyrate or ethyl propionate, denoting the combination of ethanol with butyric and propionic acids, respectively). Esters formed from other alcohols pose no risk for halal. It must be noted that the esterification reaction is reversible, and in the proper conditions, esters, including ethyl esters, can revert to their constituent alcohol and acid. It is therefore possible for a product that contains no ethyl alcohol from its formula to have small but measurable amounts of ethanol through this process. The concentrations of ethanol seen in these cases will rarely rise to a level of concern for halal.

ENGINEERED FLAVOR "SYSTEMS"

To meet specific needs of food producers and consumers, flavor manufacturers respond with innovation and new applications of technology. Whether the challenge is extending shelf-life, providing temperature stability for cooking or baking applications, or solving solubility problems, flavor producers routinely reach beyond the confines of flavor chemistry to address the evolving needs of their clients and global markets. The results of these efforts often take the form of products that perform some functional role in consumer goods in addition to delivering flavor. In assessing these materials for halal suitability, novel or unusual ingredients may be encountered that require special consideration. As the functionality of flavor systems may depend on unique raw materials that impart the functional properties desired, attention to these ingredients is a common principle across all technology platforms. Often, complex and expensive production equipment is required to manufacture these functional flavor products, and strict segregation policies should be followed to avoid inadvertent contamination of halal-dedicated machinery.

SPRAY-DRIED FLAVORS

Spray-drying is a mature technology used to dry liquid flavor materials to give a solid matrix, and the final product is a powder that may have a range of physical properties that largely depends on the nature of the matrix and particle size. Spray-drying equipment usually represents a large investment, and the drive to utilize capacity may lead manufacturers to run non-halal formulas on the same equipment that is used for halal production. Generally, this does not present a problem when operational procedures include appropriate cleaning steps in between products (as determined by the halal certifier); although, in most cases, the manufacturer's internal quality requirements are more than adequate to address halal concerns. The notable exception is the use of any materials of porcine origin; even carefully validated cleaning procedures are not sufficient to maintain halal acceptability for equipment that has been used to process pork or its derivatives. In the event of accidental contamination, the halal certifier may recommend some extraordinary measures to rehabilitate the equipment; however, extensive downtime and expense should be expected. Careful production planning and close consultation with certification authorities should minimize this eventuality. A toll-manufacturing strategy can be effectively applied when options for drying non-halal or pork products are limited. If non-halal-slaughtered beef gelatin were to be dried in a spray-drier—cleaning of spray-drier going to be impossible—the drier needs to be dedicated. Manufacturers are increasingly turning to external contract spray-drying operations to manage complex production issues, and many of these contract dryers have robust halal programs in place.

EMULSIONS

Emulsions are another time-tested technology used to achieve desired solubility properties, especially in beverages. High-shear mixing of aqueous and oil phases is used to achieve the complete suspension of small particles of one of the phases in the other. Stability—or rather, instability—of emulsions is a thermodynamic phenomenon and can be modified by the use of materials that act at the surfaces of the microscopic particles within the emulsion but it never can be fully stable. Many common emulsifiers such as esters of sorbitan (polysorbates) or glycerin (mono- and di-Glycerides) function because their molecular structure facilitates "bridging" aqueous and oil phases—a polar section of the molecule interacts well with the aqueous phase, and a lipophilic portion associates with the oil phase. These materials may be sourced partly or entirely from animals and appropriate validation of the origin of these or other surfactants and emulsion stabilization agents is needed for halal certification of products in which they are used.

OTHER ENCAPSULATION TECHNIQUES AND FLAVOR DELIVERY SYSTEMS

This subject is very dynamic given the rapid evolution of products for consumer use, which drives innovation in flavor technology. The principle concerns are the use of new ingredients that impart novel properties to the product, and any specific processing equipment or steps. Manufacturing equipment will not in itself pose a risk to the halal status unless it can be contaminated by *najis* (mainly pork) materials. Concerns related to ingredients in the area of innovation in flavor systems can be extended to the use of biotechnology, which is treated more completely below.

As previously mentioned, the focus of halal acceptability follows the entire value chain, from raw material receipt to manufacturing process. The following discusses a few raw material categories that are commonly used in flavor creation and possible concerns for halal suitability.

SYNTHETIC CHEMICALS

Synthetic chemicals are generally a low-risk category for halal as most synthetic materials are of petroleum origin and therefore halal. Production steps involving biotechnology or enzyme activity may be scrutinized by certifiers to assure consistency with halal rules.

OLEORESINS AND ESSENTIAL OILS

These materials are derived from vegetable sources and are not in themselves risky; however, attention should be paid to any solvents used in the extraction and residual levels in the material. Also, processing steps involving fermentation or enzyme activity should be evaluated to assure that all components meet halal standards.

FRUIT JUICES AND CONCENTRATES

These materials are among the most common natural flavoring ingredients and are by their nature halal unless contaminated or their naturally occurring ethanol is concentrated to high levels. Additives such as stabilizers or emulsifiers should be the subject of scrutiny from certifying bodies.

Sugars and Starches

The most prevalent concern for sugar processing is the filtration through bone char (charcoal made from animal bones) typically cattle or pig. If the bone char is of porcine origin, it is not permitted for use in halal-certified products. Cattle bones are accepted by most Muslim scholars because of the high heat reached in creating the bone char. Starch may be derived from corn, potato, tapioca, rice, wheat, or other cereal grains. Modified food starch is considered halal but the modification process should be recorded and approved by the certification authorities (as it may involve treatment with enzymes). Modified starches play a large role in manufacturing technologies as they are very useful in encapsulation applications, as well as texturing agents in finished consumer goods.

Amino Acids

Amino acids are widely used in the manufacture of flavors, especially for savory notes and enhancers of umami taste. They may be used as reactants in the aforementioned Maillard processes, and often take the form of crystalline salts of mineral acids (e.g., lysine HCl). With few exceptions they are derived from animal and plant proteins, of which they are constituent monomers (Jahangir et al., 2016). Amino acids can be extracted by hydrolyzing the natural proteins with or without the aid of enzymes. Proteins are, of course, ubiquitous in nature, and the biological source of any amino acid is vital in determination of its halal status. Amino acids derived from pork or other haram animals are not allowed in halal foods, and this prohibition extends to transgenic and biotechnologically derived ingredients (Regenstein et al., 2003).

Of specific note, the amino acid L-cysteine can be derived from human hair, animal hair, animal feathers, or synthetically. Cysteine from human hair is forbidden in halal foods, but material of feather origin may be acceptable if it comes from appropriately slaughtered halal animals. If it is made synthetically or from fermentation, it may also be considered halal, provided that all steps and ingredients are considered appropriate by an accepted halal authority (Regenstein et al., 2012).

Lipids and Fatty Acids

Lipids, including fats, fatty acids, waxes, and sterols among others, are widely used in flavors for both their sensory and physical properties. Many vitamins and specific fatty acids are also incorporated in foods for their nutritional value but are seldom confused in this context with "flavors." In flavor encapsulation systems, lipid derivatives such as lecithin and glycerol ester mixtures are often employed to stabilize the systems and aid the retention of flavoring ingredients. Most of these ingredients are available as natural materials from both plant and animal sources, in addition to synthetic versions. Animal (and biotechnology) sources of such materials will be suspect and must be reviewed by certification authorities for acceptance in finished products.

AREAS OF SPECIAL CONCERN FOR HALAL

BIOTECHNOLOGY

Biotechnology can be broadly defined as the use of biological processes to achieve an industrial goal. Biotechnology in some form has been employed in baking, brewing, and other traditional food preparation methods since the dawn of civilization. Techniques included within the field cover a wide range from the use of whole animal or plant organisms in what can be recognized as conventional agricultural activity, to advanced processes involving recombinant genetic material in these same organisms or in microbes. In the context of food and especially flavor production, these latter methods are increasingly being applied to achieve new properties and more efficient production routes for established materials (Nelson, 2001a,b). Materials of microbial origin should be assessed to assure that growth and culture media as well as the source environment are acceptable for halal. Ingredients from transgenic organisms will also be scrutinized by certification agencies to assure that halal standards are met, where they exist (Nelson and Bullock, 2001). In cases in which no precedent exists, or where halal status is broadly questioned, food and flavor manufacturers are advised to pursue other sources or face repeated challenges and delays in halal certification, as well as potential backlash from clients and consumers.

1. *Enzymes*: Enzymes are protein structures that are produced by biological systems and have the distinguishing characteristic of catalyzing specific chemical reactions, which is their purpose in living organisms. They can be extracted from living tissue and may retain their catalytic activity for use in food or flavor manufacturing processes, among others (Birch et al., 2012; Oort, 2009). As an example, cheese is made using rennet, an enzyme originally extracted from the digestive system of ruminant animals, to cause milk to coagulate into what eventually becomes the curd. Only non-animal (vegetable or fungus derived) rennet or rennet from an appropriately slaughtered halal animal may be used to make halal dairy products (Al-Mazeedi et al., 2013). The genetic provenance of the proteins should be thoroughly documented for review by halal certifiers to assure the acceptability of any enzymes from transgenic organisms. Enzymes from microbial sources must meet the same standards for carriers, growth media, nutrients, and so on, as microbes themselves (Khattak et al., 2011). (See *Fermentation*.)
2. *Fermentation*: Technically, fermentation is a generic term for microbiological or enzymatic activity, although it is often used beyond the scientific community to refer specifically to alcoholic fermentation, in which sugars are biologically converted to alcohol in a reduced-oxygen environment. It is important to understand that fermentation is widely used in traditional food preparation, where it is responsible for the development of distinct flavors without producing ethanol in any appreciable quantities—cheese making again serves as a typical example. When fermentation products are used in flavoring, the important factors are the sources of the microorganisms and other ingredients in the fermentation system. Bacteria or fungi from unclean

sources are considered haram, as are enzymes extracted from prohibited animals (e.g., porcine lipase). Some special cases may occur, such as in the production of vinegar, where bacteria are used to convert ethanol (the substrate) to acetic acid in a secondary fermentation. These rare cases should be managed in close consultation with an experienced certifying authority. When fermentation does result in the production of alcohol, the quantities permissible in the finished flavoring product are usually minute. The determination of the halal status for a product of fermentation usually requires complete disclosure to the halal certifier of the origin and constituents of every process step, including the source of transgenic material as well as any ingredients, substrates, reagents, nutrients, or other additives used in the creation, propagation, or extraction of the active enzyme or organism. Fermentation processes that are associated with the manufacture of alcoholic beverages are not considered halal, and by-products from such processes are likewise prohibited in halal food and flavors.

3. *GMO*: The acceptability of genetically modified organisms is generally a topic of much debate, and in the context of halal foods is no exception. For halal purposes, the areas of possible concern are the transfer of genes from animal to animal, insects to plants, and animal to plants. In these instances, it is possible to introduce the genes of a haram animal/insect into an otherwise acceptable ingredient source. While the details of this topic are beyond the scope of this chapter, it is worth noting that GMO status should be a data collection point for raw materials and this information must be available if requested by the halal certification body.

ANIMAL-DERIVED MATERIALS

There are many ways animal material can be introduced into flavorings. Processing aids, production methods, and even growth media in fermentation applications, each may contribute to "hidden" animal ingredients in raw materials. If any animal derivatives are present in the ingredients, they must be found acceptable by the halal certifying body before being included in a halal-certified product. Good record keeping and quality programs will aid in identifying raw materials that contain animal by-products. It is important that if the flavor requires using meat or poultry products that the animal has been religiously slaughtered according to Islamic law. Slaughtering records, or a validated halal certificate, must be maintained by the flavor manufacturer and approved by the certifying agency prior to use in halal production. A few other animal-derived materials that deserve note are:

1. *Castoreum, skatole, civet*: These are examples of (non-halal) animal-derived materials that were sometimes used in food manufacturing due to their sensory characteristics and "natural" classification but have been nearly universally replaced by cheaper synthetic versions or less expensive and more widely available natural ingredients. If they are still used, certification authorities will require product reformulation to remove them prior to certification.

2. *Fish and crustaceans*: Depending on the consumer market and the interpretation of Quranic law by the regional halal authority, some fish and crustaceans may be considered haram.
3. *Insects*: Again, this will depend on the interpretation of Quranic law by the halal authority. Some insects are considered permissible for use; carmine, a natural coloring agent, is a common example of an insect derivative that may be present in foods or flavors and which may or may not be considered acceptable for halal use.
4. *Meat flavoring*: As long as the product is entirely derived from halal sources (many "meat" flavors are, in fact, completely vegetable-based), meat flavorings can be approved as halal. Care should be exercised with meat flavoring that mimics bacon, ham, or pork in any way. The naming convention of the flavor is important so that a distinction can be made for those religiously observant (a haram name cannot be used for a product that requires halal certification). Of course, animal ingredients used in meat flavors must observe halal slaughter requirements.

Alcohol

The type of alcohol of concern in halal-certified products is ethanol or ethyl alcohol. In the context of halal, it is important to distinguish ethanol from other organic compounds that are classified chemically as alcohols, since the origin of the prohibition of ethanol is tradition in many cultures of use as *Khamr* or recreational intoxicant. Many flavoring materials are chemically classified as alcohols and pose no risk to the halal status of food but are often questioned (particularly by consumers) because of this association. Even the halal status of ethyl alcohol is not definitive, due to its natural presence in many materials that are promoted as beneficial in Islam, such as vinegar and fresh fruit juice (Riaz, 1997).

Ethanol from alcoholic beverages is strictly prohibited at any level in halal food. For ethyl alcohol to be permitted at any level, it must be naturally present in the raw material or produced as "industrial" ethanol (not part of a process specifically meant for beverage production such as fermentation of grapes for wine). When using industrial ethanol in flavorings, the halal certifier will often determine the status based on its origin because halal certificates are not available for ethyl alcohol. Ethanol that occurs naturally can be commonly found in derivatives or by-products of natural materials containing sugar, such as orange juice.

Ethanol is necessary in some production processes for efficiency (washed oils, etc.) and regulatory (e.g., vanilla) purposes. It has many functional applications, and can be used for extraction, as a processing aid, flavor carrier, solvent, or even as a disinfectant. Ethanol may be permissible in these capacities as long as the total amount of remaining ethanol in the finished flavoring (which is still an ingredient) is no more than 0.5%. Most certifying agencies globally are in agreement that this level is an acceptable cutoff for flavoring ingredients. It is important to note that the final consumer product will have a much lower acceptable level depending on the requirements of the final product halal certifier. In general, flavoring materials account for a very small portion of the total makeup of any consumer product (typically less than 1%, but this may vary widely depending on the application) (Halal Consumer Group, 2012).

PRODUCTION PROCESSES

As mentioned earlier in this chapter, the constraints related to manufacturing halal flavors are minimal and relatively easy to manage in a production facility as long as good quality assurance programs are already in place. Halal programs review the entire value chain from raw materials upon receipt through storage and manufacturing, equipment cleaning procedures, and finally labeling and shipping. The requirements for halal can be integrated into a HACCP, GMP, or food safety program, or can at least take advantage of the processes and information used in these programs. The following guidelines form the basis of a halal manufacturing program:

1. The finished product must be free of any derivative of animals that are forbidden, or other haram materials.
2. The finished product must not be prepared, processed, or manufactured using utensils or equipment that are considered haram by Islamic Laws.
3. During preparation, processing, and storage, the finished product should not come into contact with haram items.

RAW MATERIAL RECEIPT

A critical point of raw material review is to determine if any pork was used to manufacture/extract/obtain the raw material. As it can be such a costly mistake to use a pork-derived raw material, screening raw material suppliers is a vital component of the halal program. If a porcine derivative is present in any raw material, it is important that this information be clearly visible to manufacturing personnel, and these items must be strictly segregated and used with care. Any vessel, measuring instrument, or equipment that comes in contact with a pork-derived raw material cannot be used for the production of halal flavors. This strict segregation extends to reception, quality control evaluation, and to storage.

EQUIPMENT SEGREGATION

All equipment and vessels within the manufacturing facility should be dedicated to either pork contact or pork-free resources. It is an important distinction that items can be non-halal without being pork-derived. The clean-up after using pork is more stringent than that for other haram materials. While maintaining pork contact and pork-free resources can seem imposing when establishing a halal program, it is quite straightforward to manage in practice if properly implemented. Clearly labeling materials as pork-derived or not (whether by color coding or other means) will assist with proper resource management. Manufacturing non-halal items that do not contain porcine derivatives on pork-free equipment is permissible as long as approved cleaning standards are used between every batch. Using a systematic approach to regulating equipment enhances a manufacturer's ability to conform to and apply halal requirements at a global (or multi-plant) level.

Cross-Contamination, HACCP Controls

Proper cross-contamination protocols are vital to ensure that the integrity of the pork-free equipment remains intact. If GMP are followed and validated, it stands to reason that the facility will be in a good position to manufacture in accordance with halal principles. The Hazard Analysis Critical Control Point (HACCP) framework is a process control system that identifies where hazards might occur in the food production process and puts into place stringent actions that must be taken to prevent the identified hazards from occurring. The chance of a hazard occurring is reduced through strictly monitoring and controlling the identified control points at each step of the process. Allergen cross-contamination is considered a chemical hazard in the HACCP programs, and is controlled with the proper cleaning of equipment. If pork status is treated in a similar fashion, the risk of contamination should be adequately controlled for the purposes of halal certification (Bas et al., 2006, 2007).

Cleaning and Storage

The cleaning methods used during the manufacturing process must also meet the halal standard. If ethanol is used as a cleaning agent, any trace must be removed or it must be accounted for in the final calculations of ethanol in a product. Other cleaning agents should be reviewed to make sure they are from acceptable sources (be cautious of pig bristles on brushes). Cleaning according to GMP methods between every production run will be sufficient to re-qualify the equipment used in manufacturing flavors that may contain certain non-halal products other than pork. These standards of cleanliness are required to extend to the entire manufacturing property, meeting GMP standards, and remaining free from insect and other pest infestations.

Halal-certified flavors are a necessary part of consumer goods manufacturing. As flavors can be processed in a variety of ways and include a large number of ingredients in their production, halal certification provides assurance to food manufacturers (and ultimately the consumer) that all ingredients in their food are suitable for those following a halal diet. Implementing and following a halal program is a straightforward process that can be successful with the understanding of a few simple halal principles and a close working relationship with the certifying body.

REFERENCES

Al-Mazeedi, H. M., Regenstein, J. M., and Riaz, M. N. (2013). The issue of undeclared ingredients in halal and kosher food production: A focus on processing aids. *Comprehensive Reviews in Food Science and Food Safety*, 12(2), 228–233.

Baş, M., Ersun, A. Ş., and Kıvanç, G. (2006). Implementation of HACCP and prerequisite programs in food businesses in Turkey. *Food Control*, 17(2), 118–126.

Baş, M., Yüksel, M., and Çavuşoğlu, T. (2007). Difficulties and barriers for the implementing of HACCP and food safety systems in food businesses in Turkey. *Food Control*, 18(2), 124–130.

Birch, G. G., Blakebrough, N., and Parker, K. J. (2012). *Enzymes and Food Processing*. Berlin, Germany: Springer Science & Business Media.

Halal Consumer Group. (2012). Available online: www.muslimconsumergroup.com. Accessed on June 20, 2012.

Jahangir, M., Mehmood, Z., Bashir, Q., Mehboob, F., and Ali, K. (2016). Halal status of ingredients after physicochemical alteration (Istihalah). *Trends in Food Science and Technology*, 47, 78–81.

Khattak, J. Z. K., Mir, A., Anwar, Z., Abbas, G., Khattak, H. Z. K., and Ismatullah, H. (2011). Concept of halal food and biotechnology. *Advance Journal of Food Science and Technology*, 3(5), 385–389.

Nelson, G. C. (2001a). *Genetically Modified Organisms in Agriculture: Economics and Politics*. London: Academic Press.

Nelson, G. C. (2001b). Traits and techniques of GMOs, in *Genetically Modified Organisms in Agriculture*. Ed. Nelson, G. London: Academic Press, 7–13.

Nelson, G. C. and Bullock, D. (2001). The economics of technology adoption, in *Genetically Modified Organisms in Agriculture*. Ed. Nelson, G. London: Academic Press, 15–17.

Nursten, H. E. (2005). *The Maillard Reaction: Chemistry, Biochemistry and Implications*. UK: Royal Society of Chemistry.

Oort, M. V. (2009). *Enzymes in Food Technology-Introduction*, 2 edn, Hoboken, NJ: Blackwell Publishing, 1–17.

Regenstein, J. M. (2012, March). The politics of religious slaughter-how science can be misused. In *65th Annual Reciprocal Meat Conference* at North Dakota State University in Fargo, ND.

Riaz, M. N. (1997). Alcohol: The myths and realities, in *A Handbook of Halaal and Haraam Products*, Vol. 2, Uddin, Z., Ed. Richmond Hill, NY: Publication Center for American Muslim Research and Information, 16–30.

Regenstein, J. M., Chaudry, M. M., and Regenstein, C. E. (2003). The kosher and halal food laws. *Comprehensive Reviews in Food Science and Food Safety*, 2(3), 111–127.

Uhl, S. R. (2000). *Handbook of Spices, Seasonings, and Flavorings*. USA: CRC Press.

U.S. Food and Drug Administration. (2016). Available online: www.fda.gov/Food/IngredientsPackagingLabeling/FoodAdditivesIngredients/ucm228269.htm. Accessed on June 25, 2018.

U.S. Food and Drug Administration. (2015a). CFR—Code of federal regulations Title 21. *Code of Federal Regulations*, 21(6), 21cfr501.22. Available online: www.accessdata.fda.gov/scripts/cdrh/cfdocs/cfcfr/cfrsearch.cfm?fr=501.22. Accessed on June 25, 2018.

U.S. Food and Drug Administration. (2015b). CFR— Code of federal regulations title 21. *Code of Federal Regulations*, 21(2), 21CFR169.175. Available online: www.accessdata.fda.gov/scripts/cdrh/cfdocs/cfcfr/cfrsearch.cfm?fr=169.175. Accessed on June 25, 2018.

16 Alcohol in Halal Food Production

Roger Ottman, Mian N. Riaz, and Munir M. Chaudry

CONTENTS

Alcohol as a Beverage 202
Alcohol as a Flavorant in Cooking 202
Alcohol as an Industrial Chemical 203
Other Uses of Alcohol 203
Implications for Halal 204
Halal Certification of Items Containing Alcohol 205
References 206

The word alcohol derives from the Arabic word alkuhl, which refers to the fine powder usually made from antimony and used as an eye shadow. In chemistry, alcohol is the name of a family of chemicals containing the hydroxyl group attached to a carbon atom (C-OH). This includes ethanol (used in the liquor industry), isopropyl alcohol (used in the healthcare industry), cetyl alcohol (used in the cosmetics industry), sugar alcohols like mannitol, xylitol, and sorbital (used in the confectionary industry and in foods and pet food to provide sweetness), as well as other alcohols (Othman and Riaz, 2000). In language, Merriam-Webster defines alcohol as a clear liquid that has a strong smell, which is used in some medicines and other products, and is the substance in liquors (such as beer, wine, or whiskey) that can make a person drunk (Jolly, 2001). The word khamr used in the Quran is often considered to mean alcohol, although it has the broader meaning of any intoxicant. Intoxicants are substances that affect the brain and thought. Ethanol is one of the main intoxicants and the one most commonly found in foods and beverages (Halal Consumer Group, 2012). The use of the term alcohol in this chapter will mean ethanol (or ethyl alcohol).

Alcohol is a chemical that is common in nature. It can be found in overripe fruits, juices, and milk. Alcohol has many uses and applications. In ancient times, it was mainly consumed as a beverage. Historically, alcohol was made by fermentation from fruits such as grapes and dates. In modern times, alcohol is also made from grains such as rye, wheat, barley, and corn and by synthesis from hydrocarbons such as oil and gas (Al-Mazeedi et al., 2013).

In the U.S., products containing more than 7% alcohol by volume come under the jurisdiction of the Bureau of Alcohol, Tobacco, Firearms and Explosives (ATF). Products containing less than 7% alcohol come under the jurisdiction of the

Food and Drug Administration (FDA). According to the FDA (2000), if alcohol is part of the food composition or formula, then it must be included on the label as an ingredient. However, if alcohol is part of one of the ingredients in the formula, such as a flavor, it does not have to be listed separately on the ingredient label.

Alcohol is used in a number of applications, including:

1. As a beverage
2. As a flavorant in cooking
3. As an industrial chemical
4. Other

ALCOHOL AS A BEVERAGE

Alcoholic beverages are produced from a variety of fruits and grains, including grapes, dates, rye, wheat, and barley. Legally, they can contain between 0.5% and 80% ethanol by volume while pure industrial alcohol can contain 95% ethanol. There are three major classes of alcoholic beverages:

- Fermented beverages are made from agricultural products, including grains and fruits, and contain between 3% and 16% alcohol.
- Distilled or spirit beverages are made by distillation of fermented beverages. Distillation increases the alcohol content of these products up to 80%.
- Compound or fortified beverages are made by combining fermented or spirit beverages with flavoring substances. The alcohol content of these products can also be as high as 80%.

Alcoholic beverages may be consumed directly or added to foods, either as an ingredient during formulation or during cooking. When an alcoholic beverage is an added ingredient in a product, the ingredient label of the product must list the specific alcoholic beverage that has been added and the final amount of alcohol if it is greater than 0.5%. Examples of this would be liqueur-flavored chocolates, rum cakes, and meals containing wines, such as beef stroganoff in wine sauce.

ALCOHOL AS A FLAVORANT IN COOKING

Alcoholic beverages are sometimes added to food items during cooking to impart a distinctive flavor to the food item. Wine is the most common form of alcoholic beverage used in cooking, though beer and hard liquors are also used. Although it may seem that all of the added alcohol evaporates or burns off during cooking, it does not. The United States Department of Agriculture (USDA) prepared a table listing the amount of retained alcohol in foods cooked with alcohol. The retained alcohol varies depending on the cooking method (Al-Mazeedi et al., 2013; 3Halal Consumer Group, 2012). The following are some of the retained alcohol contents of foods prepared by different cooking methods (USDA Table of Nutrient Retention Factors, Release 6, 2000) (Table 16.1).

TABLE 16.1
Alcohol Retention in Cooking

Cooking Process	% Alcohol Retained
Added to boiling liquid and removed from the heat	85
Cooked over a flame	75
Added without heat and stored overnight	70
Baked for 25 minutes without stirring	45
Stirred into a mixture and baked or simmered for 15 minutes	40
Stirred into a mixture and baked or simmered for 30 minutes	35
Stirred into a mixture and baked or simmered for 1 hour	25
Stirred into a mixture and baked or simmered for 2 hours	10
Stirred into a mixture and baked or simmered for 2½ hours	5

ALCOHOL AS AN INDUSTRIAL CHEMICAL

Pure industrial alcohol can be up to 95% ethanol. As a raw material, the main use of alcohol is to convert it to acetic acid to make vinegar. Vinegar is used in salad dressings, mayonnaise, and other applications. Whereas the use of alcohol in alcoholic drinks is prohibited, converting it to acetic acid (vinegar) makes it halal as it is no longer an intoxicant. In fact, vinegar is one of the recommended condiments in Islam. As a solvent, alcohol is used to extract flavors from plant materials such as vanilla beans or oranges. Dilute ethanol is almost universally used for the extraction of vanilla beans. After the extraction, vanilla flavor, called natural vanilla flavoring, is standardized with alcohol. The FDA's standard of identity for vanilla flavoring requires it to contain a minimum of 35% alcohol by volume, otherwise it cannot be called natural vanilla flavoring.

Alcohol is also used to standardize other flavors. A standing container of flavor may not remain homogenous if not mixed with alcohol or a similar compound. The alcohol maintains the homogeneity as well as allowing the flavor to be standardized so it can have the same potency when used in a product. For example, to make an orange-flavored carbonated beverage, orange flavor is added to the water. Orange flavor is an oil obtained from orange skins. As an oil, it does not dissolve in water, but it does dissolve in alcohol. When standardized in alcohol, the orange flavor will dissolve and disperse evenly in the water, creating the orange-flavored carbonated drink. The drink will maintain the consistent orange flavor over the expected shelf life of the product.

OTHER USES OF ALCOHOL

Alcohol is also used in pharmaceuticals, cosmetics, and topical products. It is frequently present in cough syrups and mouthwash, although alcohol-free products are now available. In perfumes, the use of SD alcohol is common. SD alcohol is ethanol that has been denatured, meaning substances have been added to it to make it undesirable or even toxic for consumption. The denaturing substances are very difficult to remove from the mixture, so denatured alcohol cannot be used in food items or drinks.

IMPLICATIONS FOR HALAL

Prior to the revelation of the Quran, alcoholic beverages were consumed by the Arabs in Arabia. When the Quran was revealed, and people began embracing Islam, they continued to consume alcoholic beverages, just as they had previously. Alcoholic beverages had not been specifically prohibited yet (Riaz, 1997). The prohibition occurred in three stages. The first revelation was a response to questions regarding the consumption of alcohol:

> They ask you about wine and gambling. Say, "In them is great sin and [yet, some] benefit for people. But their sin is greater than their benefit." And they ask you what they should spend. Say, "The excess [beyond needs]." Thus Allah makes clear to you the verses [of revelation] that you might give thought.
>
> **Chapter II, Verse 219**

In this revelation, we are told that there are benefits to be gained from alcohol but that indulging in drinking it is a sin. While this was enough for some to refrain from drinking alcohol, it was not a clear prohibition. The next revelation placed some limits on drinking:

> O you who have believed, do not approach prayer while you are intoxicated until you know what you are saying or in a state of janabah, except those passing through [a place of prayer], until you have washed [your whole body]. And if you are ill or on a journey or one of you comes from the place of relieving himself or you have contacted women and find no water, then seek clean earth and wipe over your faces and your hands [with it]. Indeed, Allah is ever Pardoning and Forgiving.
>
> **Chapter IV, Verse 43**

In this revelation we are told not to offer prayers while intoxicated. One must be able to focus when in prayer and being under the influence of alcohol does not allow one to focus, so it was prohibited to pray while intoxicated. Again, this did not clearly prohibit drinking. So the next verses were revealed:

> O you who have believed, indeed, intoxicants, gambling, [sacrificing on] stone alters [to other than Allah], and divining arrows are but defilement from the work of Satan, so avoid it that you may be successful. Satan only wants to cause between you animosity and hatred through intoxicants and gambling and to avert you from the remembrance of Allah and from prayer. So will you not desist?
>
> **Chapter V, Verses 90-91**

These final two verses made it absolutely clear that alcoholic beverages should be avoided completely. To further clarify, there are a number of Hadiths that clearly state that every intoxicant is prohibited and that anything that intoxicates in large quantities are also prohibited in small quantities.

It is important to note that there two major opinions on the prohibition of alcohol: (a) that the prohibition applies to all alcohol that is not naturally occurring regardless of the quantity or reason for its presence or (b) that is applies only to intoxicating beverages. The first opinion considers there is zero tolerance for alcohol regardless of whether it is ingested or applied topically. This means that non-intoxicating items containing alcohol like cakes, ice cream, and carbonated drinks; mouthwash; as well as all alcoholic beverages, including non-alcoholic drinks with any trace of alcohol would be prohibited. It does not include naturally occurring items with non-intoxicating levels of alcohol such as ripe fruits and milk. The second opinion considers only alcoholic beverages and other items that can cause intoxication if a reasonable quantity is consumed to be prohibited. It does not consider the use of alcohol for non-intoxicating purposes such as flavor extraction and standardization to be prohibited if the amount of residual alcohol in a product is below a low threshold. It does consider even traces of an alcoholic beverage or alcohol derived from such a beverage to be prohibited, even if the item it has been added to is not intoxicating. Under this opinion, a cake with a drop of rum would be prohibited but a cake with some residual alcohol introduced by an added flavoring compound would be acceptable if the alcohol level in the cake was below the threshold limit and the alcohol in the flavoring was pure industrial alcohol and not from an alcoholic beverage.

Different certifying bodies and government regulators have placed limits on the amount of residual alcohol that is acceptable in a product. These opinions would support any efforts to eliminate the residual alcohol but understand the challenges that would present to the food industry at the present time.

HALAL CERTIFICATION OF ITEMS CONTAINING ALCOHOL

1. Alcoholic beverages and any ingestible or topical product containing even a trace of alcoholic beverage cannot be certified halal. This includes all the alcoholic beverages, any food product containing an alcoholic beverage, any item cooked with an alcoholic beverage, and any topical product with an added alcoholic beverage. Examples are beer, wine, whiskey, rum, tequila, beer battered fish, shampoo containing beer, rum cake, liqueur containing candy, and so on.
2. Non-alcoholic beer and similar products are made from their alcoholic counterparts by removing the alcohol. In the U.S., if the final product contains less than 0.5% alcohol, it can be labeled as non-alcoholic. As these likely contain some alcohol, they would not be certifiable under the zero tolerance opinion but might be certifiable under the second opinion. However, many would not certify any product made to resemble an alcoholic beverage. That includes non-alcoholic beer packaged to look similar to its alcoholic counterpart. Fresh products that have been processed only to remove contaminants and to package them like fruits, vegetables, and milk that may contain some naturally occurring alcohol are halal under both opinions. They may need halal certification to assure consumers nothing

unacceptable has occurred during processing. Fruit juices that are filtered or clarified will require halal certification to ensure the filtering and clarifying media are acceptable.

Based on organoleptic tests to detect alcohol by sight, smell, and taste, the Islamic Food and Nutrition Council of America has adopted the following guideline for certifying ingredients and products containing residual alcohol:

1. The alcohol used must be industrial alcohol, not from an alcoholic beverage.
2. The alcohol must be used for a processing requirement, not for a flavor or other reason.
3. For ingredients, such as flavors, the residual amount of alcohol should not be more than 0.5%. This means vanilla flavoring, which must legally contain 35% alcohol will not be halal certified. If such vanilla flavor is dried and is below 0.5%, it can be considered for certification.
4. For food products, the residual amount of alcohol should not be more than 0.1%. This is the total amount of residual alcohol and includes any naturally occurring alcohol plus any added by way of an ingredient containing residual alcohol, like a flavor. Vanilla flavoring may be used, even though it is not halal certified, as long as the final edible product does not contain more than 0.1% residual alcohol.

Some certification bodies may have more lenient or stricter guidelines than these. The food industry should consult its customer companies or halal-approval agencies for their exact standard.

Addition of any amount of fermented alcoholic drinks such as beer, wine, or liquor to any food product or drink renders the product haram. However, if the essence is extracted from these products and alcohol is reduced to a negligible amount, most halal-certifying agencies and importing countries accept the use of such essences in food products. Consultation with proper authorities or end users can clarify this issue.

REFERENCES

Al-Mazeedi, H. M., Regenstein, J. M., and Riaz, M. N. (2013). The issue of undeclared ingredients in halal and kosher food production: A focus on processing aids. *Comprehensive Reviews in Food Science and Food Safety*, 12(2), 228–233.

FDA. (2000). 21 CFR 169.3(c). Silver Spring, MD: US Food and Drug Administration.

Halal Consumer Group. (2012). Available online: http://www.muslimconsumergroup.com. Accessed on June 20, 2012.

Jolly, L. (2001). The commercial viability of fuel ethanol from sugar cane. *Sugar Cane International*, Feb, 6–14.

Othman, R. and Riaz, M. N. (2000). Alcohol: A drink/a chemical. *Halal Consumer*, Fall Issue (1), 17–21.

Riaz, M. N. (1997). Alcohol: The myths and realities, in *A Handbook of Halaal and Haraam Products*, Vol. 2, Uddin, Z., Ed. Richmond Hill, NY: Publication Center for American Muslim Research and Information, 16–30.

17 Food Additives and Processing Aids

Mian N. Riaz and Munir M. Chaudry

CONTENTS

The Food Additive Market 209
Control Points for Halal Food Additives and Processing Aids 209
References 211

Food additives are substances added to foods for many purposes including preserving flavor or improving taste, texture, and appearance. The term food additive applies broadly to chemicals, both natural and synthetic, that are added to food, either intentionally or indirectly, to facilitate processing, extend shelf life, improve or maintain nutritional value, or enhance the food's organoleptic qualities. Some products would not be possible to produce without additives. However, some of the food additives also have serious problems with respect to their halal status. For the food industry to serve the halal market properly, it is very important that they determine the halal status (suitability) of these additives. Unfortunately, some of the common additives are derived from sources that are not halal, for example, pigs and other haram animals and animals that were not slaughtered as halal. Even if the food additive is listed in the ingredients statement, the source of the additive is usually not mentioned. Furthermore, food additives do not need to appear on the ingredients label statement when used as carriers, processing aids, and anticaking agents. Because of how they are processed, some foods may become contaminated with unintentional food additives that are not halal such as non-halal food grade equipment lubricants. To make sure all aspects of the food's production is halal, food companies need to be able to assure halal consumers that all food additives they use are halal. This requires that a food company work closely with their halal certifier and that the halal certifier is knowledgeable about and checks the halal status of all materials that are used in or come in contact with a food product (Zweig, 2013).

Food additives have been around for centuries to make foods more desirable, appear more attractive to eat, stay fresh longer, healthier, and even prevent disease. There are many types of food additives available with different uses. Food additives must be tested both for efficacy and safety. There are approximately 3000 food additives that have been formally studied and classified into different groups under the category of food additives that are being used by the food industry and are maintained in the U.S. by the Food and Drug Administration (FDA) in its food additive database (Kansas State Research and Extension, 2010). The main groupings of food additives include acids, acidity regulators, anticaking agents, antifoaming agents, antioxidants, bulking agents, food colorings, color retention agents, emulsifiers, flavors, flavor enhancers, flour treatment agents, glazing agents, humectants, tracer gasses, preservatives, sequestrants, stabilizers, sweeteners, surface active agents, and thickeners. There are some other categories, but the above groupings include the majority of the food additives that are commonly used.

The food additives that affect nutritional quality are primarily vitamins and minerals. In some foods, these may be added to enrich the food or replace nutrients that may have been lost during processing. In other foods, vitamins and minerals may be added for fortification to supplement nutrients that may often be lacking in human diets (Branen and Haggerty, 2002; Zweig, 2013). For example, vitamins and minerals are added to many foods including flour, cereal, margarine, and milk. This helps to make up for vitamins or minerals that may be low or lacking in an individual's diet. However, the fortification of foods in the U.S. is regulated and not all foods can be fortified.

Preservatives or antimicrobial substances are used to prevent bacterial and fungal growth in foods. These additives can delay spoilage or extend the shelf life of the finished product (Branen and Haggerty, 2002). Antioxidants are additives that can also extend the shelf life of foods by delaying rancidity or lipid oxidation. For example, antioxidants help baked goods preserve their flavor by preventing the fats and oils from becoming rancid. They also keep fresh fruits from turning brown when exposed to the air.

Additives that maintain product quality may also ensure food product safety for the consumer. For example, acids may be added to prevent the growth of microorganisms that cause spoilage and may also prevent the growth of microorganisms that can cause foodborne illness. Food additives may be used as a processing or preparation aid, that is, to help during processing while not changing the character of the finished product. Thus, in the U.S. and other countries, they may not require identification within the ingredient statement.

Leavening agents that release acids when they are heated react with baking soda to help biscuits, cakes, and other baked goods rise. Some food additives are used to enhance the flavor or color of foods to make them more appealing to the consumer. Flavoring chemicals may be used to magnify the original taste or aroma of food ingredients or to restore flavors lost during processing. Natural and artificial coloring substances are added to increase the visual appeal of foods, to distinguish flavors of foods, to increase the intensity of naturally occurring color or to restore color lost during processing (Branen and Haggerty, 2002).

However, most food flavorings are classified in a category called generally recognized as safe (GRAS) in the U.S. The standard for such a classification is that there is sufficient data in the public domain to establish the safety and efficacy by a panel of experts. Thus, over 4000 flavor compounds are recognized as GRAS through the GRAS panels of the Flavor and Extract Manufacturers Association and the list of accepted compounds is published regularly in Food Technology magazine rather than in the Code of Federal Regulations. The process of getting materials classified as GRAS has been changing, and the FDA website (www.cfsan.fda.gov) will have the latest information on the GRAS process in the U.S.

Food additives may be natural, nature identical, or synthetic. Natural additives are substances found naturally in a foodstuff and are extracted from that food so that it can be used in another food. For example, beet root juice with its bright purple color can be used to color other foods such as sweets. However, if the beet root juice is further processed to isolate the color compounds, then this material becomes an artificial color when used with any product other than beets. Nature identical additives are man-made copies of substances that occur naturally. For example, benzoic acid is a substance that is found in nature, for example, cranberries, but is made synthetically at a much lower cost. It is widely used as a preservative. Artificial additives are substances made synthetically that are not found in nature. An example is azodicarbonamide, a flour improver, which is used to help bread dough hold together.

THE FOOD ADDITIVE MARKET

According to new market data from Leatherhead Food Research (2014), there is a growing demand for functional foods and beverages that help consumers improve their diet, health, and well-being. This is driving the growth in the USD 24.5 billion global food additives market in 2010 (Leatherhead Food Research, 2014). According to the Institute of Food Technologist (IFT), the global food additives market exceeded USD 33 billion in 2015 (www.IFT.org).

Consumers' demand for healthy and natural ingredients are key factors driving food additive companies to develop a host of new additives including emulsifiers, hydrocolloids, sweeteners, vitamins and minerals, soya ingredients, probiotics, prebiotics, and plant stanol esters. The consumer's attitude toward natural (and clean label) food and beverages is also forcing additive suppliers to develop ingredients from natural sources for the flavor and color categories as well. Because of increased material and energy cost, enzymes, acidulants, and hydrocolloids are having the highest growth rates in the food additives sector because of their ability to reduce costs, while preservatives and sweeteners are seeing the lowest growth rates as many manufacturers are moving away from using these generally artificial additives and ingredients, although the demand for natural preservatives and sweeteners is increasing. The demand for food additives also continues to increase in the emerging economies such as China and India. This has led many ingredient suppliers to establish distribution and/or production bases in these parts of the world. Chinese producers have assumed an increasingly significant presence in such key sectors of the ingredients market as acidulants and vitamins. These new sources of ingredients from the emerging economics require more effort to assure the halal integrity of these materials.

CONTROL POINTS FOR HALAL FOOD ADDITIVES AND PROCESSING AIDS

The source material from which an ingredient is obtained and the details of the processing methods will determine the acceptability of these compounds for use. Among the undeclared processing aids used in the food industry are antifoaming and release agents, extraction aids, bleaching compounds, and anti-clumping, filtering, and clarifying agents.

Processing aids may help reduce the potential contamination of foods during production as well as aid in the removal of impurities. Thus, they may play an important role in food safety. Regulatory agencies such as the FDA and USDA have approved a number of processing aids for use during food processing. De-coloring agents, fruit and vegetable washes, and dough conditioners are approved by the FDA, while pH control compounds, chemical agents to control bacteria in chill water, and antimicrobial agents to reduce pathogens in meat and poultry are approved by the FSIS of the USDA (Chaudry et al., 2000).

Enzymes may also be considered processing aids in some food products, such as during cheese production. The source or origin of the enzyme(s) and the media used for the growth of microorganisms from which many of the enzymes are derived will determine the acceptability of these enzymes for use in the production of halal food products (Ermis, 2017; Riaz, 2000). To determine that the enzyme, even as a processing aid, meets the needs of the halal consumer requires that the halal certifying agency carefully and thoroughly investigate the detailed process of the enzyme's production.

Even if the processing aid is no longer found in the finished product, the product may still be haram. The presence of a non-halal processing aid should result in the rejection of the finished product by the halal certification agency. In fact, contact with any non-halal additive or processing aid with halal ingredients will result in a product that should not be acceptable for use in the halal market (Anir et al., 2008).

The raw materials used for manufacturing packaging materials, their processing, and any coatings used with the packaging materials are of importance with foods for the halal market. It is important to note that the packaging used for halal foods is as important as the other ingredients that are used for halal foods. There are additives in packaging materials for improvement of its functionality and some of those are not acceptable for halal products, such as compounds from materials of animal origin. Even something as innocuous as the steel used to make the typical "tin" can or 55-gallon drums, which are widely used in the food industry, may have been coated with a protective coating that had in the past been made containing a pork ingredient. In the U.S., the American steel industry understands this concern and has eliminated these materials. But again, as the food industry globalizes, the issue needs to be addressed in other countries. Plastics may contain chemicals that are meant to or may accidentally migrate to the food contact surfaces, that is, where the packaging touches the food product, making packaging material unacceptable for halal products. Glycerol mono-stearate and glycerol mono-oleate are used as releasing agents or lubricants for metal and plastic containers and for some films to keep the food product from "sticking" to the packaging. These may be of animal origin, so the use of these compounds must be checked to be sure they are acceptable, that is, from plant sources or from halal-slaughtered animals.

For the halal market, it is necessary to have a paper trail of the auditing of ingredient production by a recognized and acceptable halal certifying agency for every ingredient. If this is not possible, then the certifying agency may consider a notarized affidavit for the ingredient from the ingredient supplier stating that the additive, processing aid, or packaging materials do not contain any component of animal origin (Hashim, 2011; Riaz, 2004) or only animal products obtained from halal animals without slaughter, for example, milk, eggs, fish, and honey. However, the

Food Additives and Processing Aids

halal certifying agency should be using its global contacts to independently verify such information. Information about the packaging and transport of each ingredient starting on the farm should also be ideally obtained to assure that a product that is initially halal remains so (Noordin et al., 2014).

Going back to the steel plate discussion. The U.S. Steel Corporation asked and obtained affidavits from the suppliers of the grease coating material that the product was free from animal-derived materials. However, it turned out that on further investigation, the suppliers of the grease had a subcomponent supplier that was using the porcine (pig)-derived ingredient much to the surprise of both the grease manufacturer and U.S. Steel. So again vigilance and thorough investigation at many levels is needed (Riaz and Chaudry, 2004).

In general, food production facilities have their own computer programs to keep track of the inventory that will be used for halal certified products. This list and the actual inventory need to be spot checked by the halal supervision agency for accuracy. Assignment of a unique letter code, number code, alphanumeric code, or color code for halal ingredients, especially those found on the premises that may also be non-halal, including additives and processing aids as well as packaging materials for halal production will assist in monitoring and tracing of ingredients during production, shipping, and distribution. Ideally, a company would not be allowed to have the identical ingredient in both a halal and non-halal form within a factory. Despite all coding systems, human nature is such that if a person needs an ingredient, they may at that time not pay attention to the coding system.

A flow diagram showing the point of addition of any processing aid that is not declared on the food label or a halal-sensitive ingredient is suggested to remind the halal supervisors to check that the right material is used.

At a minimum, segregation of halal and non-halal additives, processing aids, and packaging materials in receiving, storage, preparation, and scaling areas is recommended. It is important for a food company to work closely with its halal certifying agency to be sure that all ingredients and packaging are suitable for halal products (Chaudry et al., 2000).

REFERENCES

Anir, N. A., Nizam, M. N. M. H., and Masliyana, A. (2008). RFID tag for Halal food tracking in Malaysia: Users perceptions and opportunities. In *WSEAS International Conference. Proceedings. Mathematics and Computers in Science and Engineering* (No. 7) Istanbul, Turkey: World Scientific and Engineering Academy and Society.

Branen, A. L., Davidson, P. M., Salminen, S., and Thorngate, J. H. III (2002). *Introduction to Food Additives, in Food Additives*, 2nd edn. New York, NY: Marcel Dekker, 1–9.

Chaudry, M. M., Jackson, M. A., Hussaini, M. M., and Riaz, M. N. (2000). *Halal Industrial Production Standards*. Deerfield, IL: J&M Food Products Company, 1–15.

Ermis, E. (2017). Halal status of enzymes used in food industry. *Trends in Food Science and Technology*, 64, 69–73.

Hashim, D. D. (2011). Introduction to the Global Halal Industry and Its Services. http://www.asidcom.org/IMG/pdf/L1-_Darhim_Dali_Hashsim.pdf. Accessed on June 25, 2018.

IFT. (2015). Global food additives market to exceed 33 B by 2015. Available online: www.ift.org/food-technology/daily-news/2010/june/08/global-food-additives-market-to-exceed-33-b-by-2015.aspx. Accessed on June 25, 2018.

Kansas State Research and Extension. (2010). The role of food additives. *Nutrition News*, www.ksre.ksu.edu/humannutrition1/nutritionnews/foodaddroles.htm (accessed on January 11, 2012).

Leatherhead Food Research. (2014). *The Global Food Additives Market*, 6th edn. UK. http://www.foodsciencematters.com/wp-content/uploads/2014/10/Food-Additives-White-Paper-Leatherhead-Food-Research-2014.pdf. Accessed on June 25, 2018.

Noordin, N., Noor, N. L. M., and Samicho, Z. (2014). Strategic approach to halal certification system: An ecosystem perspective. *Procedia-Social and Behavioral Sciences*, 121, 79–95.

Riaz, M. N. (2000). How cheese manufacturers can benefit from producing cheese for the Halal market. *Cheese Market News* 20, 4–18.

Riaz, M. N. (2004). Halal foods gaining popularity as wholesome and quality products. *Middle East Food* 21(5): 17–19.

Riaz, M. N. and Chaudry, M. M. (2004). *Labeling, Packaging and Coating for Halal Food, in Halal Food Production*. Boca Raton, FL: CRC Press, 129–135.

Zweig, G. (Ed.). (2013). *Principles, Methods, and General Applications: Analytical Methods for Pesticides, Plant Growth Regulators, and Food Additives*, Vol. 1. New York: Elsevier.

18 Food Ingredients in Halal Food Production

Mian N. Riaz and Munir M. Chaudry

CONTENTS

Bacon Bits	214
Amino Acids	214
Civet Oil	214
Liquor	214
Liquor and Wine Extracts	214
Fusel Oil Derivatives	215
Encapsulation Materials	215
Thickening, Gelling, Texturizing, and Stabilizing Agents	215
Production of Vinegar from Alcohol	215
Minor Ingredients	215
Manufacturing of Food Products	216
Spices and Seasoning Blends	216
Condiments, Dressings, and Sauces	216
Batters, Breadings, and Breadcrumbs	216
Flavorings	216
Cheese Flavors	217
Meat Flavors	217
Smoke Flavors and Grill Flavors	217
Colorants	217
Curing Agents	217
Fruit and Vegetable Coatings	217
Dairy Ingredients	218
Summary	218
References	218

This chapter discusses various types of single and compound ingredients, including spices, seasonings, condiments, sauces, dressings, batters, and breading, as well as curing agents, coatings, and flavorings. Thousands of ingredients are used in food and related industries. Most are categorized as GRAS or food additives, although a few are covered by other legal categories such as "prior-sanctioned" chemicals, that is, used before 1958, when the current law came into being (Al-Mazeedi et al., 2013; Kamaruddin et al., 2012). Some of the critical ingredients and processes from the halal perspective are listed and explained.

BACON BITS

Natural bacon bits are manufactured from real bacon made from pork. Artificial bacon bits are made with plant proteins, generally soy protein, and colored and flavored with other ingredients (Kunert and Herreid, 2002). Although natural bacon bits are not acceptable for halal products, one may use artificial bacon bits as long as the colors, flavors, and other incidental ingredients, as well as the production equipment, are halal approved.

AMINO ACIDS

Many amino acids are used for various technical functions. They are either made synthetically (generally microbially or from starting materials of plant origin) or extracted from natural proteins. One of the most common and questionable amino acids is L-cysteine, which is used in pizza crust, doughnuts, and batters (Jahangir et al., 2016). Natural cysteine is generally obtained from human hair, animal wool, and duck feathers. These days, L-cysteine is also made using sugar as the starting material. L-cysteine from human or animal wool, unless sheep are sheared when they are alive, is not acceptable as halal. However, it is acceptable if made from duck feathers, especially if the ducks have been slaughtered in an Islamic manner. Synthetic L-cysteine is often sold as vegetable grade L-cysteine, which is halal acceptable as long as all the production requirements are properly carried out (Regenstein et al., 2003; Regenstein, 2012).

CIVET OIL

Civet oil is oil extracted from the glands of a cat-like animal called a civet. Civet oil is not accepted as halal (Goldberg, 2012; Jahangir et al., 2016).

LIQUOR

Liquors are alcoholic drinks commonly used in flavors either for taste or aesthetic appeal for people who enjoy drinking liquor. Use of liquor or any alcoholic drinks in preparation of flavorings or batters is not acceptable. One of the possible flavors for fried products or fried batter products is beer batter. Actual beer is used in the production of batter-coated fries, onion rings, or other fried appetizers. Although alcohol is flashed out during the frying process, beer batter still remains unacceptable as halal (Al-Mazeedi et al., 2013; Halal Consumer Group, 2012). This concept is discussed in detail in Chapter 16.

LIQUOR AND WINE EXTRACTS

The flavor industry has now created extracts of different types of wines and liquors to use strictly as flavoring agents. Such extracts may be used in formulating halal products as long as residual alcohol in the extracts is very low, generally less than 0.5%. One must consult the halal-certifying agency and the end user on whether there is a problem with the source (Al-Mazeedi et al., 2013).

FUSEL OIL DERIVATIVES

Some ingredients such as amyl alcohol or isoamyl alcohol are made from by-products of the alcohol industry. These are doubtful to use because of their origin from beverage alcohol and thus, halal agencies in some countries do not approve of the use of such ingredients. An example is tartaric acid from wine.

ENCAPSULATION MATERIALS

Several ingredients such as gelatin, cellulose, shellac (also called "lac resin"), and zein (from corn) are used to encapsulate food ingredients as well as food products. Shellac (derived as an excretion of the lac bug), cellulose, and zein are generally accepted in halal products if properly produced. Gelatin is acceptable only if it is from halal-slaughtered animals or acceptable fish (Al-Mazeedi et al., 2013; Blech, 2008; Regenstein, 2012).

THICKENING, GELLING, TEXTURIZING, AND STABILIZING AGENTS

Thickening and gelling agents have been used for centuries in foods and also pharmaceuticals, cosmetics, and personal care products. They are widely used as thickening, gelling, stabilizing, texturizing, and suspending agents to improve product eating quality, product appeal, consistency, and shelf life. They are also used with food liquids to improve spreadability, binding, hardening, creaminess, freeze thaw stability, crystal control, adhesion, and water binding. Flour, cornstarch, and pectin are some common examples. Most of these materials are derived from plants and seaweeds except for gelatin from animals and some like xanthan and gellan gum are microbial polysaccharides (Jahangir et al., 2016).

Modified starches made from corn, wheat, potato, and tapioca are used to replace other more expensive gums. Even if these materials are of vegetable origin, their production process could involve some steps making use of animal-derived material (e.g., enzymes) from non-halal sources (Al-Mazeedi et al., 2013).

PRODUCTION OF VINEGAR FROM ALCOHOL

Vinegar is more than 90% water so often vinegar is made on-site or nearby from ethanol, generally from grain or petrochemical ethanol. This practice is generally acceptable but should be monitored to maintain the minimal residual alcohol below 0.5%. Note that beverage ethanol is not permitted, for example, wine vinegar or cider vinegar (Kocher et al., 2006; Muslim-ibn-al-Ḥajjaj, 2008).

MINOR INGREDIENTS

Many different GRAS ingredients may be used in food manufacture. They often do not have a function in the final product. All these minor ingredients must be from halal sources and produced in a halal acceptable way as defined and accepted by the halal-certifying agency.

MANUFACTURING OF FOOD PRODUCTS

Manufacturers of some food products may purchase one or more blends from another supplier. This requires that the halal certifier either check both the "blending" operation as well as the original ingredients manufacturers, or obtain acceptable halal certificates for both the ingredients and the blending operation (Aida et al., 2005).

SPICES AND SEASONING BLENDS

Spices and seasonings are single botanical ingredients or a dry blend of many different ingredients. Seasoning manufacturers may use any of the food ingredients available to them whether they are of vegetable or animal origin.

There are two considerations in making halal seasoning blends. The first is the composition of the blends. All components should be halal suitable. Non-certified animal-based ingredients should not be used in halal blends. The second consideration is cross-contamination from the equipment. Halal blends should be manufactured on thoroughly cleaned equipment or in dedicated mixers. Minor ingredients such as an encapsulating agent, anti-dusting agent, and free-flow agent must also be halal suitable (Al-Mazeedi et al., 2013).

CONDIMENTS, DRESSINGS, AND SAUCES

These are generally pourable or spoonable liquid products. Some of them can contain bacon bits, gelatin, wine, or complex flavorings. For halal production, such non-halal ingredients must not be included in the formulations. The product should be made on clean equipment.

BATTERS, BREADINGS, AND BREADCRUMBS

Batters, breading, and breadcrumb manufacturing has evolved into a specialized manufacturing process.

L-cysteine is used to modify the texture of batter or breading coatings and must be halal acceptable. Alcoholic beverages are also prohibited.

FLAVORINGS

Flavors and flavorings are some of the most complex ingredients used in the food industry. Under U.S. regulations (FDA, 2002), individual components of a flavor need not be declared to customers. The flavor industry is exempt from revealing such information (which is considered to be proprietary) as long as the component ingredients of a flavor are either on an FDA or Flavor and Extract Manufacturers Association (FEMA) list.

Two groups of ingredients are of special concern to formulators of halal products: (1) unique flavoring agents such as civet oil and (2) ingredients of alcoholic origin. As a general guideline, ingredients of animal origin should be avoided in the development of flavorings unless those ingredients are halal certified (Aris et al., 2012;

Jahangir et al., 2016). It is permissible to use industrial ethanol for extracting flavors or dissolving them. However, the amount of alcohol should be reduced to less than 0.5% in the final flavoring product. Certain countries or customers require lower allowances or even the total absence of any ethanol for products brought into their countries. Some countries do not permit fuel oil derivatives. It is advisable for formulators to work with their client companies and certifying agencies to determine the exact requirements of a certain company or country (Jamaludin, 2013).

CHEESE FLAVORS

Dairy ingredients should be derived from processes that use either microbial enzymes or halal-certified animal enzymes.

MEAT FLAVORS

Meat and poultry ingredients should only be from animals slaughtered according to halal requirements.

SMOKE FLAVORS AND GRILL FLAVORS

Halal concerns include the use of animal oils as a base for smoke and grill flavors or the use of emulsifiers from animal sources.

COLORANTS

Colors have been used in foods since antiquity (Francis, 1999). Historically, colors have been used not only to make food look appealing but also as adulterants to hide defects. Colors can be synthetic such as FD&C-certified colors, where water-soluble ones are called dyes and oil-soluble ones are called lakes. Colors can also be natural or organic, such as fruit and vegetable extracts. Potential halal concern are materials such as shellac (lac resin), octopus ink, and squid ink. Inorganic colors are not a concern except like all ingredients there is a need to assure halal appropriate processing. Some of the other ingredients believed to be used in colors such as gelatin are halal sensitive (Al-Mazeedi et al., 2013).

CURING AGENTS

Curing agents, which are used for making sausage products, are specialized blends of salt, nitrites, and some other ingredients such as sodium ascorbate, sodium erythorbate, citric acid, and propylene glycol. They are generally halal suitable. However, one should make sure that they are made on clean equipment and questionable ingredients are not incorporated into them.

FRUIT AND VEGETABLE COATINGS

Animal-based ingredients must be avoided. Particularly upsetting to Muslim consumers is the use of any haram ingredients for the coating of fruits and vegetables sold in the fresh produce section. The Nutritional Labeling and Education Act of 1990 contains specific provisions that require the packer to identify the source of components in any such coating and put it on the outside carton. Supermarkets and other retailers are required by law to have an easily visible sign indicating that fruits and vegetables in their store might have been treated with such coatings. Although vegetable and mineral ingredients used for this purpose are halal, the coating formulators must avoid doubtful ingredients such as beef tallow and gelatin, or haram ingredients such as lard (Syahariza et al., 2005). Halal consumers are encouraged to look at the required sign in the supermarket, and if not found, ask the management to be in compliance with these U.S. regulations.

DAIRY INGREDIENTS

Ingredients such as dry milk powder, which are heat-processed, are suitable for halal manufacturing. Ingredients such as whey powders, lactose, whey protein isolates, and concentrates produced with the use of enzymes are questionable if the source of enzymes is unknown. To make these ingredients halal, producers must use microbial enzymes or enzymes from halal-slaughtered animal sources. Most certifying agencies require that dairy ingredients made with enzymes be halal certified (Al-Mazeedi et al., 2013; Regenstein, 2012). As many of these materials are spray-dried, it is also important that the spray-drier and other equipment used be halal compliant.

SUMMARY

This chapter covered some ingredients of concern to the Muslim consumer. A more detailed listing appears in the appendices. Use of E-numbers rather than chemical names is common in Europe and Asian countries. Appendix B lists food ingredients with their E-designation and halal, haram, or doubtful status, and Appendix C lists ingredients and their halal, haram, or doubtful status.

REFERENCES

Aida, A. A., Man, Y. C., Wong, C. M. V. L., Raha, A. R., and Son, R. (2005). Analysis of raw meats and fats of pigs using polymerase chain reaction for Halal authentication. *Meat Science*, 69(1), 47–52.

Al-Mazeedi, H. M., Regenstein, J. M., and Riaz, M. N. (2013). The issue of undeclared ingredients in halal and kosher food production: A focus on processing aids. *Comprehensive Reviews in Food Science and Food Safety*, 12(2), 228–233.

Aris, A. T., Nor, N. M., Febrianto, N. A., Harivaindaran, K. V., and Yang, T. A. (2012). Muslim attitude and awareness towards Istihalah. *Journal of Islamic Marketing*, 3(3), 244–254.

Blech Z. (2008). *The Story of Gelatin, Kosher Food Production*, 2nd edn. Hoboken, NJ: Wiley-Blackwell Publisher.

FDA. (2002) Foods; labelling of spices, flavorings, colorings and chemical preservatives. Accessed on February 28, 2018. www.gpo.gov/fdsys/granule/CFR-2012-title21-vol2/CFR-2012-title21-vol2-sec101-22. Accessed on June 25, 2018.

Francis, F. J. (1999). *Colorants*. St. Paul, MN: Eagle Press.

Goldberg Y. (2012). The primary ingredients in detergents. Available online: http://oukosher.org/index.php/common/article/the_primary_ingredients_in_detergents/Surfactants. Accessed on June 20, 2012.

Halal Consumer Group. (2012). Available online: http://www.muslimconsumergroup.com. Accessed on June 20, 2012.

Jahangir, M., Mehmood, Z., Bashir, Q., Mehboob, F., and Ali, K. (2016). Halal status of ingredients after physicochemical alteration (Istihalah). *Trends in Food Science and Technology*, 47, 78–81.

Jamaludin, M. A., (2013). Fiqh Istihalah: Integration of science and Islamic law. *Revelation and Science*, 2(2), 117–123.

Kamaruddin, R., Iberahim, H., and Shabudin, A. (2012). Willingness to pay for halal logistics: The lifestyle choice. *Procedia-Social and Behavioral Sciences*, 50, 722–729.

Kocher, G. S., Kalra, K. L., and Phutela, R. P. (2006). Comparative production of sugarcane vinegar by different immobilization techniques. *Journal of the Institute of Brewing*, 112(3), 264–266.

Kunert, G. F. and Herreid, R. M. (2002). Method for making crunchy bacon bits. *US Patent No. 6,391,355*. Washington, DC: US Patent and Trademark Office.

Muslim-ibn-al-Ḥajjaj, A. A.-Ḥ. (2008). Sahihmuslim (Urdu translation), Vol. 3, JamiaDarulaloomKorangi. Karachi, Pakistan: Islamic Organization, 1016.

Regenstein, J. M. (2012). The politics of religious slaughter-how science can be misused. In *65th Annual Reciprocal Meat Conference at North Dakota State University in Fargo, ND*.

Regenstein, J. M., Chaudry, M. M., and Regenstein, C. E. (2003). The kosher and halal food laws. *Comprehensive Reviews in Food Science and Food Safety*, 2(3), 111–127.

Syahariza, Z. A., Man, Y. C., Selamat, J., and Bakar, J. (2005). Detection of lard adulteration in cake formulation by Fourier transform infrared (FTIR) spectroscopy. *Food Chemistry*, 92(2), 365–371.

19 Halal Production Requirements for Nutritional Food Supplements

Mian N. Riaz and Munir M. Chaudry

CONTENTS

Ingredients to Watch 222
Types of Products..223
References..223

This chapter deals with dietary and nutritional supplements containing nutraceuticals, vitamins, minerals, and other nutritional products used to enhance and maintain health and wellness or prevent disease. The lines between pharmaceuticals, products that heal, and nutraceuticals, products that help maintain the well-being of a person, are merging. The purpose of this chapter is not to determine the effectiveness of these products but to reflect on their compositions and determine whether any of the components presents a problem for the Muslim consumer.

"Halal Pharmaceuticals" is soon going to be a multimillion dollar industry according to an estimate of the pharmaceutical business (Khan and Shaharuddin, 2015). Although in the Islamic tradition, one may consume a haram product as a medicine under compulsion, Muslim consumers generally avoid knowingly taking anything that is haram or doubtful. Some people may take prescription medicine in a gelatin capsule but not a multivitamin capsule (Al-Qaradawi, 2007). Gelatin capsules, unless certified halal or labeled bovine, are generally made of pork gelatin. Pork gelatin is considered haram by Muslim consumers (Aziz et al., 2014). Medicine that is used to cure disease and help overcome illness is considered exempt from halal food regulations.

The majority of pharmaceuticals contain ethanol and animal derivatives and hence cannot be certified as halal, so they need to be prepared using some alternative ingredients and methods according to Islamic law to make them halal (Sarriff, 2013). Prescription drugs generally do not have alternative products to replace a prescribed drug. If a drug is available in capsule form only, one is obligated to take it, whereas multivitamins are normally not taken to cure serious illness, but to improve one's health. Moreover, there are many alternative forms of multivitamins, such as tablets, liquids, and vegetable capsules, so one does not have to take vitamins in gelatin capsules. Many consumers try to purchase alcohol-free products such as cough syrups (Charles et al., 2011). They can also ask the pharmacist for tablets rather than gelatin capsules. The Malaysian Department of Health, for example, has determined that nutritional supplements are health and medication products and may not be grouped together with food products. It is against the local regulations to display halal markings on such products. Many Muslim consumers are very apprehensive of this regulation (Ramli et al., 2012). Indonesian authorities, on the other hand, are including not only foods but also drugs and cosmetics in their halal program, which is evident by the creation of an institute, the Assessment Institute for Foods, Drugs, and Cosmetics (AIFDC), under the guidance of the Religious Council of Indonesia, also called the Majelis Ulama Indonesia (MUI). AIFDC is responsible for assessing, evaluating, certifying, and monitoring establishments and products including foods, drugs, cosmetics, personal care products, and other consumables. AIFDC insists on using logos or proper halal markings on labels of certified halal products or ingredients. Moreover, the Food and Drug Administration (FDA) is the U.S. agency that supervise food safety, medications, cosmetics, veterinary and medical products, and so on. They have the responsibility to oversee the food safety, wholesomeness of human and veterinary drugs, sanitary conditions for the accomplishment of effective and hygienic biological and medical devices and cosmetics, and to maintain a check on electronic devices that emit radiation that can be hazardous to human health (Meaning of FDA, 2011). The GMP for manufacturing, processing, packing, or holding finished pharmaceutical products was first published in 1963 (Immel, 2001).

General guidelines for the production of nutritional supplements are similar to producing other food products. Nutritional food supplements, for the most part, are composed of botanicals and plant extracts. It is the animal-derived ingredients one has to avoid in formulating the supplements (Al-Akili, 1994).

INGREDIENTS TO WATCH

Databases of ingredients can often comprise thousands of entries. Companies might use several thousand different ingredients in a given time period. It is beyond the scope of this book to describe the halal status of every ingredient used in the industry. Only some of the ingredients with potential concern for halal are given here.

- *Flavors and colorants*: Might have hidden alcohol or ingredients of haram animal origin, such as civet oil, ambergris, and castoreum in the formulations (Uhl, 2000).
- *Beta-carotene*: Often formulated with gelatin in small quantities. Gelatin is used to encapsulate and protect its color and other characteristics. Some companies use fish gelatin for encapsulation, which makes the product halal as well kosher. Manufacturers also use halal bovine gelatin or plant gums to encapsulate b-carotene.
- *Gelatin*: Very commonly used to make capsules, both softgel and two-piece hard shell. Halal gelatin or cellulose or starch can be used instead of porcine gelatin or gelatin from non-halal-slaughtered animals that could be halal (Shah and Yusof, 2014).

- *Stearates*: Can be used as free-flow agents in powders or tableting aids in tablets. For halal products, manufacturers can use stearates from plant not animal sources (Shabana and Acunova, 2013).
- *Tweens*: Sometimes used for coating and polishing tablets. Vegetable-derived tweens rather than animal-derived ones should be used in halal products.
- *Glycerin*: Used in the manufacture of capsules and it may also be used in other products. Glycerin of plant origin is halal suitable for such applications (Grundy et al., 2007, 2008).

TYPES OF PRODUCTS

Nutritional food products come in many physical forms such as powders, liquids, tablets, one-piece capsules (soft shell), and two-piece capsules (hard shell). Nutraceutical ingredients can also be incorporated into foods.

- *Tablets*: Can be coated with gelatin (gel tabs) or with specialty lipids such as polysorbates. Halal-certified gelatin and lipids of plant origin should be used for halal tablets. Sugars and plant proteins can also be used as coating material for tablets in halal products.
- *Liquid supplements and drinks*: Many liquid formulations are standardized with ethyl alcohol as a preservative or solvent. Alternatives such as mixtures of propylene glycol and water can be used. The amount of alcohol in the finished product may not be more than 0.1% as discussed in Chapter 16.
- *Softgel capsules*: One-piece capsules used to be made exclusively from gelatin. They can now also be made using vegetable ingredients. Halal-certified bovine and fish gelatins are also available for this purpose. Besides the main ingredient, softgel capsules might also contain glycerin or fatty chemicals, which should be from plant sources for halal production (Nasaruddin et al., 2012).
- *Hardgel capsules*: Like softgels, two-piece hardgel capsules used to be made exclusively with gelatin. There are now vegetarian capsules especially for nutritional supplements. Glycerin and other ingredients can be used as processing aids. All such ingredients should be from vegetable or petroleum sources. Two-piece gelatin capsules, if used, should be halal-certified bovine gelatin or halal fish gelatin and other incidental ingredients should also be halal suitable.

To ensure almost universal acceptability, it is recommended that pharmaceutical products and nutritional and dietary food supplements be manufactured by avoiding all traces of slaughtered animal products, so that the product is acceptable for halal, kosher, and vegetarians or vegans.

REFERENCES

Al-Akili, M. (1994). *Black Seed, in Natural Healing with the Medicine of the Prophet.* Philadelphia, PA: Pearl Publishing House, 229–232 (translated and emended by Muhammad Al-Akili from Book of the Provisions of the Hereafter by Imam Ibn Qayyim Al-Jawziyya).

Al-Qaradawi, S. Y. (2007). *The Halal and Haram in the Private Life of a Muslim. The Lawful and Prohibited in Islam.* Lahore, Pakistan: Al-Falah Foundation, 2–379.

Aziz, N. A., Ibrahim, I., and Raof, N. A. (2014). The need for legal intervention within the halal pharmaceutical industry. *Procedia-Social and Behavioral Sciences*, 121, 124–132.

Charles, F. L., Lora, L. A., and Leonard L. L. (Eds.). (2011). *Drug Information Handbook: American Pharmacists Association* (APhA). Hudson, OH: Lexi-Comp Inc.

Lacy, F. C., Armstrong, L. L., Goldman M. P., and Lance L. L. (Eds.). (2011). *Drug Information Handbook: A Comprehensive Resource for All Clinicians and Healthcare Professionals (Lexi Comp's Drug Information Handbooks)*, 15th edn. Hudson, OH: Lexi-Comp Inc.

Grundy, H. H., Reece, P., Sykes, M. D., Clough, J. A., Audsley, N., and Stones, R. (2007). Screening method for the addition of bovine blood-based binding agents to food using liquid chromatography triple quadrupole mass spectrometry. *Rapid Communications in Mass Spectrometry*, 21(18), 2919–2925.

Grundy, H. H., Reece, P., Sykes, M. D., Clough, J. A., Audsley, N., and Stones, R. (2008). Method to screen for the addition of porcine blood-based binding products to foods using liquid chromatography/triple quadrupole mass spectrometry. *Rapid Communications in Mass Spectrometry*, 22(12), 2006–2008.

Immel, B. (2001). A brief history of the GMPs. *Compliance Leadership TM Series*, 1–8.

Khan, T. M. and Shaharuddin, S. (2015). Need for contents on halal medicines in pharmacy and medicine curriculum. *Archives of Pharmacy Practice*, 6(2), 38.

Meaning of FDA. (2011). Retrieved from http://coldflu.about.com/od/glossary/g/fda.htm.

Nasaruddin, R. R., Fuad, F., Mel, M., Jaswir, I., and Hamid, H. A. (2012). The importance of a standardized Islamic manufacturing practice (IMP) for food and pharmaceutical productions. *Advances in Natural and Applied Sciences*, 6(5), 588–596.

Uhl, S. R. (2000). *Handbook of Spices, Seasonings, and Flavorings.* Boca Raton, FL: CRC Press, 329.

Ramli, N., Salleh, F., and Azmi, S. N. (2012). Halal pharmaceuticals: A review on Malaysian standard, ms 2424: 2012 (p). *Journal of Arts and Humanities,* 1(1), 137.

Sarriff, A. (2013). Exploring the halal status of cardiovascular, endocrine, and respiratory group of medications. *The Malaysian Journal of Medical Sciences: MJMS*, 20(1), 69.

Shabana, K. and Acunova, E. (2013). Factor driving halal pharma market. *International Halal Pharmaceutical Summit 2013*, GTOWER Hotel, Kuala Lumpur, Malaysia. http://docplayer.net/35959527-Factors-driving-halal-pharma-market.html. Accessed on June 25, 2018.

Shah, H. and Yusof, F. (2014). Gelatin as an ingredient in food and pharmaceutical products: An Islamic perspective. *Advances in Environmental Biology*, 8, 774–780.

20 Biotechnology and GMO Ingredients in Halal Foods

Mian N. Riaz and Munir M. Chaudry

We have discussed halal laws in detail throughout this book. God requires us to eat halal foods (Pickthall, 1994):

> (Saying): Eat of the good things wherewith We have provided you, and transgress not in respect thereof lest My wrath come upon you; and he on whom My wrath cometh, he is lost indeed.
>
> **Chapter XX, Verse 81**

All things are considered good except the ones specifically prohibited, which are very few in number. This chapter will look at genetically modified organisms (GMO) and biotechnology using some of the basic principles from Al-Qaradawi (2007) as discussed in Chapter 2.

Everything is halal unless specifically prohibited. There is no specific mention of altered, modified, or genetically engineered foods and ingredients in the Quran or the traditions of Muhammad (PBUH), because these scientific developments are very recent. However, genetically modified or engineered products from prohibited animals are prohibited. For example, because pork is prohibited, by extension, any products made from genetically altered pigs are prohibited too (MS 1500, 2009).

- God is the only one who has the power to legislate for humans.
- A scientist can explain a new development, and a religious scholar can only try to interpret whether the development violates any of the tenets of Islam. Permitting haram and prohibiting halal is similar to shirk, meaning ascribing partnership with God. It would be most serious if GMO were clearly haram, and Muslim scholars interpreted them as halal. This certainly is not the case.
- Haram is usually associated with what is harmful and unhealthy.
- If it is determined beyond doubt that any of the foods or ingredients developed through genetic modifications are harmful and unhealthy, they will not be approved by governments and, therefore, they become haram.
- There is always a better replacement for something that is haram. There are better replacements for haram ingredients using biotechnology. Until the mid-1980s, porcine pepsin was used in some cheese manufacture. Since the introduction of GM chymosin, the use of pepsin as a replacement for calf

rennet has practically vanished. This is a big plus for the use of biotechnology to help increase the availability of prepared halal foods.
- To proclaim something halal that is not halal is also haram. Again, if GM food was clearly haram, scholars would have a huge issue with it. The items that are not halal are clearly mentioned in the Quran and the traditions.
- Good intentions do not make haram into halal. This applies to pigs and other haram animals, even if scientists try to make pig cleaner and disease-free or grow pig organs for food in the lab; such organs are still haram.
- Doubtful things should be avoided. This is perhaps the most significant guideline. Muslims are required to avoid doubtful things. There is a clear tradition of the Prophet avoiding doubtful things. If Muslim consumers feel that GM foods are doubtful, they must avoid them. Presently, doubtful GMO are the ones modified using genes from prohibited animals.

Biotechnology is an extension of plant and animal breeding and genetics, which have been practiced for decades, and, in some cases, for centuries. One example of animal breeding dates back to prehistoric times when a donkey and a mare were crossbred to produce a mule. The meat of a donkey is not accepted as halal food, and therefore neither is the meat of a mule. Plants have always been bred with closely related plants and animals with closely related animals (Papazyan and Surai, 2007). In modern times, genes for a specific protein product have been identified and scientists also learned how to take a gene from one species and move it to another species. Currently, genes from fish or insects or pigs can be introduced into plant species without affecting the appearance or taste, while making the plants more resistant to diseases or nutritionally better compared with the conventional products available (Nelson, 2001a,b). This modern technology was not available at the inception of Islam. Muslim scholars are, therefore, striving to come to an acceptable decision on some of the issues raised by these new technologies. At the inception of Islam, almost 14 centuries ago, Islamic dietary laws in most countries were the only regulations dealing with the safety and wholesomeness of food products because there were no governmental food safety regulations. Currently, food safety is the responsibility of the government agencies with help from organizations such as the United Nation's Food and Agriculture Organization, and the World Health Organization. Issues relating to the safety of GM foods are deferred to such agencies; so only those concerns related to the religious aspects of GMO need to be discussed (Nelson and Bullock, 2001). The underlying principle for halal acceptability is that food has to be halal and tayyab, meaning permissible, and wholesome or good. For example, two government agencies in Malaysia, the Institute Kefahaman Islam Malaysia (IKIM) and Jabatan Kemajuan Islam Malaysia (JAKIM), concur that GM food is halal as long as it is from halal sources using halal methods of production (Kurien, 2002).

Several additional points can be considered, such as the concept of change. Is there any change taking place in a gene transfer from a prohibited animal to a permitted animal? Does the gene change the character of the recipient animal or plant enough to make it prohibited? If not, then istihala (change of state) has not taken place. Most GM products and ingredients fall within this concept. Is the porcine gene then acceptable? This remains a controversial issue.

Even if the GMO product is safe, if Muslim consumers feel that introducing pig genes into plants violates their religious responsibility, then such food is considered doubtful. The consumer has the right to accept or reject the reasoning behind the change. However, the industry, government agencies, and scholars have an obligation to educate the consumers about such issues (Man and Sazili, 2010).

The next point to be made concerns religious prohibition versus personal inhibition. People might not find a basis to call the food prohibited because that is the right of God alone. But they might still not want to eat something because they are not sure or it makes them feel uneasy. This will not make GM foods haram, but the concerns of the consumer must be respected.

Does the condition of necessity overrule prohibitions of GM foods? Hunger is still prevalent in the world. GM foods certainly offer tangible alternatives that may increase the food supply. Most Americans believe that biotechnology will benefit them or their families in the next five years, according to a survey conducted by the International Food Information Council. Consumers expect benefits from GMO plants such as improved health and nutrition; improved quality, taste, and variety of foods; reduced chemical and pesticide use on plants; reduced cost of food; and improved crops and crop yields (Langen, 2002). However, officials in some European countries do not share these views.

Islam teaches caution and moderation to Muslims when eating food. GM foods and GM ingredients may not be haram, but many Muslims may avoid them anyway because they do not feel comfortable consuming them. The introduction of animal genes into plants presents a considerable ethical challenges and difficulties for consumers, Muslims and non-Muslims alike.

Theoretically, a donor gene can be from any of biological source, such as plants, microorganisms, insects, fish, or other animals (Zailani et al., 2010). How does the source of donor gene affect the acceptability of the resultant GM products?

- A plant-to-plant gene transfer that makes a plant increase the amount of a conventional food ingredient it produces that is then purified and does not contain genetic material is acceptable.
- A plant-to-plant gene transfer where the resulting crop is consumed which, therefore, includes the genetic material also is acceptable.
- A transfer of an animal gene from a halal animal to bacteria that then is used to manufacture enzymes and other bioactive ingredients is also acceptable as long as the safety of such ingredients is established beyond doubt and the production process is halal. Many of the enzymes produced these days use this new technology.
- The plant or bacteria contain genes from a prohibited animal, such as the pig. These are probably not acceptable even when used on an animal and not directly used on humans. The use of porcine somatotropin for muscle mass buildup in beef cattle falls in this category (Chaudry and Regenstein, 1994).

GM rennet, called chymosin, is so widely used in cheese making that it has become the major coagulant. According to David Berrington of Chr. Hansen, Inc., about 80% of the cheese produced in the U.S. and UK, and 40% of the world's cheese, is now made with genetically engineered chymosin (Avery, 2001) although newer non-GMO microbial enzymes are also being widely used. This process has been accepted as halal as long as the production is halal.

The use of genes from haram animals put into halal animals or plants is going to be hard to gain acceptance. It will be difficult to convince Muslim consumers about the benefits of these GMO. It is better for the industry to avoid such products (MS 1500, 2009). Other biotechnology-related issues such as cloning animals for food use and designing new species of animals are going to be equally challenging for Muslims. There is no official information yet about how Muslim opinions on these issues will develop. It will be necessary for GM products with foreign genetic material from non-halal animals to be reviewed and evaluated by religious scholars.

A purely synthetic gene made through recombinant technology, which is similar to a porcine lipase, may be acceptable to produce halal GM products, because there is really no porcine material in the gene but the mere fact that it is a porcine gene may cause Muslims to reject it. Again, this will require extensive study by religious scholars.

REFERENCES

Al-Qaradawi, S. Y. (2007). *The Halal and Haram in the Private Life of a Muslim. The Lawful and Prohibited in Islam*. Lahore, Pakistan: Al-Falah Foundation.

Avery, D. T. (2001). Genetically modified organisms can help save the planet. In *Genetically Modified Organisms in Agriculture*. Ed. Nelson, G. London: Academic Press, 205–215.

Chaudry, M. M. and Regenstein, J. M. (1994). Implications of biotechnology and genetic engineering for kosher and halal foods. *Trends in Food Science and Technology*, 5(5), 165–168.

Kurien, D. (2002). Malaysia: studying GM foods' acceptability of Islam. Kuala Lumpur, Malaysia: August 9, Dow Jones Online News via News Edge Corporation.

Langen, S. (2002). IFIC conduct consumer food biotech survey, *Food Technology*, 56(1), 12.

Man, Y. B. C. and Sazili, A. Q. (2010). Food production from the halal perspective. *Handbook of Poultry Science and Technology*. Hoboken, NJ: Wiley, 183–215.

MS 1500. (2009). Halal food—production, preparation, handling and storage—general guideline. Cyberjaya: Department of Standards Malaysia, 1–13.

Nelson, G. C. (2001a). Introduction., In *Genetically Modified Organisms in Agriculture*. Ed. Nelson, G. London: Academic Press, 3–6.

Nelson, G. C. (2001b). Traits and techniques of GMOs. In *Genetically Modified Organisms in Agriculture*. Ed. Nelson, G. London: Academic Press, 7–13.

Nelson, G. C. and Bullock, D. (2001). The economics of technology adoption, In *Genetically Modified Organisms in Agriculture*. Ed. Nelson, G. London: Academic Press, 15–17.

Papazyan, T. T. and Surai, P. F. (2007). EU clearance of Sel-Plex®: Expanding the possibilities for new nutraceutical foods. In *Nutritional biotechnology in the feed and food industries: Proceedings of Alltech's 23rd Annual Symposium. The new energy crisis: food, feed or fuel?* Stamford, UK: Alltech UK, pp. 193–201.

Pickthall, M. M. (1994). *Arabic Text and English Rendering of the Glorious Quran*. Chicago, IL: Library of Islam, Kazi Publications.

Zailani, S. H. M., Ahmad, Z. A., Wahid, N. A., Othman, R., and Fernando, Y. (2010). Recommendations to strengthen halal food supply chain for food industry in Malaysia. *Journal of Agribusiness Marketing*, Special edition, October 2010, 91–105.

21 Animal Feed and Halal Food

Mian N. Riaz and Munir M. Chaudry

Muslims are supposed to make every effort to obtain wholesome halal foods. For non-Muslim consumers, halal foods often are perceived as carefully selected and processed to achieve the highest standards of quality. Halal foods must be free from any component that Muslims are prohibited from consuming (Wilson and Liu, 2010). Thus, there is a huge potential globally for halal meat because of the large growth in the Muslim population and the number of people of other religions that appreciate the benefits of halal meat. There is a lot of halal meat available in North America and Europe, although such meat most often does not include a concern for what the animals were fed, which is of concern to Muslims (Nakyinsige et al., 2012). Most animal feed contains animal by-products. These animal by-products should be of concern for those providing halal meat, and halal eggs and milk. Many Muslim consumers believe that the meat of animals that are fed feed with animal by-products and blood meal are haram. Satisfying the demand for halal feeds for food animals presents both a challenge and an opportunity for entrepreneurs in the animal feed industry.

An ideal halal feed will contain products from plants and microbial processes only. However, for economic reasons, animals are being fed with all kinds of animal-based ingredients that would not be fit for human consumption. For Muslims, this abhorrence is heightened if these animal by-products come from a pig. As a Muslim, one should therefore be concerned about what is being fed to food animals. Most consumers are not aware that simply doing the proper slaughter of animals according to Islamic law is not enough. The feed must also be halal, at least for some period of time prior to slaughter (Omar and Jaffar, 2011).

Some of the Gulf countries have already started to ask that some of their suppliers use only plant-based feeds for chickens. There have been several cases where imported livestock were rejected because cattle were fed pig-based feeds. From a marketing point of view, it will be beneficial for those involved in the meat trade to start to migrate to using halal feeds to gain a larger market share. Exporters of halal meat to Muslim countries should be ready to have their feed programs comply with halal principles as these countries require that the feed fed to the animals should be halal certified. In 2009, the Pakistan Supreme Court issued orders to destroy imported poultry feed that was contaminated with pig products and directed that action be taken against those responsible for importing the feed. In Malaysia, a committee has been formed to ensure that the chicken feed used by commercial farms in the country is free from any pig enzymes or other by-products from pig. Recently, Muslims living in Belgium and other parts of Europe were concerned about consuming fish that had been fed pork fat and protein. The European Union (EU) approved legislation allowing EU fish farmers to use fats and proteins from animals including pigs starting in June 2013. The decision has triggered alarm among Muslim communities living in Europe and prompted Islamic organizations in Brussels, the Belgian capital, to ask religious institutions to issue an official verdict.

Also of particular abhorrence to Muslims is the use of blood meal. The first step would be to eliminate the use of swine blood, but the use of other blood meals is also of serious concern. Currently animal proteins are often a major part of the diet of food animals, but now with the greater availability of vegetable feeds, those wishing to serve the growing Muslim market can migrate to slaughtered animal free diets. (Milk and egg by-products are not a concern.)

The processing of ruminant (which are all halal acceptable animals) by-products and extracts in animal feed has become more restricted in many countries due to mad cow disease, properly known as bovine spongiform encephalopathy (BSE) in the 1990s. But these may still be used in some cases for poultry and seafood feeding.

The following discussion of feed mill inspections is meant to point out deficiencies in the implementation of the new regulations, which can impact the halal status of farmed animals. To help prevent the establishment and amplification of BSE through feed in the U.S., the Food and Drug Administration (FDA) implemented a final rule that prohibits the use of most mammalian protein in feeds for ruminant animals. This rule, Title 21 Part 589.2000 of the Code of Federal Regulations, became effective on August 4, 1997. What does the above ruling mean by "most mammalian protein"? It seems that a certain amount of pork by-products and other non-ruminant mammalian by-products might still be used to formulate ruminant feeds. About a year after the initial compliance deadline, the FDA did a thorough inspection of the industry. The results are given in Table 21.1 (FDA, 2003).

Various segments of the feed industry had different levels of compliance with the feed ban regulation. About 27% of the firms handling feeds for ruminants were inspected. Of these, 2653 out of 9867 were handling materials prohibited for use in ruminant feed (FDA, 2003). Of those handling prohibited materials, 653 (25%)

TABLE 21.1
FDA Summary of Inspections

Category	Total Number	Reported	Failed[a]	Status
Renderers	264	241	25	USDA licensed
Feed mills	1240 est	1176	76	FDA licensed
Feed mills	6000-8000	4783	421	Not licensed
Other firms	Unknown	4094	110	Unknown

[a] Out of compliance

had a problem, which indicated that they might not be properly segregating materials or not labeling properly. Thus, the possible inclusion of materials of concern to Muslims might still be occurring in ruminants. Obviously, over time, the use of illegal material should become less of an issue for ruminants, but this does not cover the allowed mammalian products or other animal products, particularly poultry.

It seems that some cattle farmers and almost all poultry farmers still have the opportunity to feed the animals "protein supplements" made from rendered animal parts, including swine. Although data about the exact use of by-products is not available, it is probably safe to assume that the amount of by-products being fed to ruminants has decreased as a result of the FDA ban. Hence, it is likely that slaughterhouse by-products from cattle and pigs are more available, which may be exported to other countries, including those with a significant Muslim population. So the problem of by-products in feeds is a universal one, not just a U.S. issue.

Current halal standards in the U.S. and other countries require that the animal be slaughtered according to Islamic guidelines but they are not generally concerned with on-farm conditions (Hussain, 2002). It is important to note that Islamic scholars differ about what is "unclean" animal feed. Some scholars feel that haram animal parts fed to halal animals make them unclean and unfit for slaughter, whereas other scholars believe that an animal has to live in filth and eat filth regularly to meet the condition of unclean. The Arabic word used for this concept is jalalah (Malaysian Standard, 1500 2009). Jalalah refers to the situation where animals are living in and around heaps of filth and manure, and eating major portions of its feed out of such heaps for a good part of its life. The Prophet Muhammad (PBUH) forbade the consumption of meat and milk of such animals (Khan, 1991). It is important to keep in mind that all animals eat some dirt and filth, especially when they graze outdoors. Rendered feed containing animal organs after a high heat treatment does not seem to fit this description of filth. It would seem something of an exaggeration to apply the rule of jalalah to properly rendered animal feed in the U.S. and other countries.

Many Muslims however, feel very strongly that the feed for halal animals, whether raised for meat, milk, or eggs, must primarily be of plant origin, which the animals have been used to eating for centuries. In most cases, the majority of the feed is still plant-based. Other Muslim consumers object only to pork by-products but not to formulated feeds containing animal products. A few years back, Saudi Arabia banned products from Europe on the suspicion that the animals were given feed containing prohibited animal parts (Al-Zobaidy, 2002).

It seems that the FDA ruling that makes it unlawful to include some mammalian by-products in feeds for ruminants simply does not address the concerns of Muslim consumers. If the Muslim community is concerned about the feed of food animals, it will have to work with the halal certifiers to obtain animals that meet this standard. However, halal consumers should expect that such a standard will increase the cost of halal meat, milk, and eggs. It may also mean that milk and especially eggs, which are not currently under halal supervision, will require such supervision, which means such milk and eggs will not always be available everywhere.

REFERENCES

Al-Zobaidy, O. (2002). Imports of EU poultry, soft drinks banned. *Arab News*, July 27. Available online: www.arabnews.com/print.asp?id=.

FDA. (2003). Ruminant feed (BSE) enforcement activities. Rockville, MD: Center for Veterinary Medicine, Office of Management and Communications. Available online: www.fda.gov/cvm.

Hussain, M. (2002). Demand halal, consume halal. *HalalPak*, Summer Issue, 6–7.

Khan, G. M. (1991). *Slaying Animals for Food the Islamic Way: Al-Dhabh*. Jeddah, Saudi Arabia: Abdul-Qasim Bookstore.

MS 1500. (2009). Halal food—production, preparation, handling and storage—general guideline. Cyberjaya, Malyasia: Department of Standards Malaysia, 1–13.

Nakyinsige, K., Che Man, Y. B., Sazili, A. Q., Zulkifli, I., and Fatimah, A. B. (2012). Halal meat: A niche product in the food market. *Second International Conference on Economics, Trade and Development IPEDR*, Kuala Lampur, Malaysia, 36, 167–173.

Omar, E. N. and Jaafar, H. S. (2011). Halal supply chain in the food industry—A conceptual model. In *2011 IEEE Symposium on Business, Engineering and Industrial Applications (ISBEIA)*, Langkawi, Malaysia, 384–389. IEEE.

Wilson, J. A. and Liu, J. (2010). Shaping the halal into a brand? *Journal of Islamic Marketing*, 1(2), 107–123.

22 Halal Cosmetics

Mian N. Riaz and Munir M. Chaudry

CONTENTS

Guidelines for Halal Cosmetics 234
Partial List of Potentially Questionable Ingredients 235
    Allantoin 235
    Ambergris 235
    Animal Hair 235
    Arachidonic Acid 235
    Arachidyl Propionate 236
    Boar Bristles 236
    Carotene, Pro-vitamin A, Beta Carotene 236
    Castor—Castoreum 236
    Chitosan 236
    Cholesterol 236
    Civet Oil 236
    Collagen 236
    Colors and Dyes 237
    Cysteine, L-Form 237
    Cystine 237
    Emu Oil 237
    Fatty Acids and Fatty Acid Mixtures 237
    Feathers 237
    Gelatin 237
    Glycerin/Glycerol 237
    Hyaluronic Acid 237
    Hydrolyzed Animal Protein and Hydrolyzed Vegetable Protein 238
    Keratin 238
    Lanolin, Lanolin Acids, Wool Fat, and Wool Wax 238
    Lard 238
    Mink Oil 238
    Musk (Oil) 238
    Myristic Acid 238
    Oleic Acid 238
    Palmitic Acid 238
    Panthenol, Dexpanthenol. Vitamin B-Complex Factor and Pro-vitamin B-5 239
    Placenta 239
    Polypeptides 239
    Pristane 239
    Snail Slime 239

Stearic Acid 239
Stearyl Alcohol 239
Turtle Oil 239
References 239

About 25 to 30 years ago, the term halal was not used by the world's cosmetic industry. Today, it has become an important consideration of the industry. Just as Muslims are careful about what they eat, more and more Muslim women are now conscious of what they put on their skin, leading to the rise in halal-certified skin care and makeup products (Swidi et al., 2010).

There are various estimates of the current size of the halal cosmetic market ranging from USD $5 billion to $14 billion per year. In 2012, the estimated expenditure for halal cosmetics by Muslim consumers was $26 billion and by 2018 this is expected to reach $39 billion (Shikoh, 2013). These estimates probably vary because of different data collection methods in different countries and because all products have been considered halal in many predominately Muslim countries. About half of these sales are in the Middle East, with $2.1 billion in Saudi Arabia alone. Halal products are very quickly entering the mainstream markets in Europe and the U.S.

Halal-certified cosmetics and skin care products are expected to be pure and wholesome. Halal personal care products in the market today include hair shampoos, conditioners, bath and shower gels, cleansers, creams, lotions, talc and baby powders, toners, make up, perfumes, eau de colognes, and oral care products (Ahmad et al., 2015). These products do not contain alcohol or pork-derived ingredients, and are manufactured in a humane and cruelty-free manner. Muslims are becoming more discerning consumers, and along with rising income, has led to growth of this market, which then represents a new opportunity for cosmetic and personal care companies (Ireland and Abdollah Rajabzadeh, 2011). Non-Muslims also purchase halal cosmetics because of the perception that halal cosmetic products are safer to use than non-halal products (Hornby and Yucel, 2009; Shah Alam and Mohamed Sayuti, 2011). Muslim women love to be beautiful. But those who are devout can face obstacles. What if their cosmetics contain forbidden ingredients such as alcohol, beef- or pig-derived collagen, gelatin, or beef- or pig-derived fats and their many derivative products?

GUIDELINES FOR HALAL COSMETICS

The presence of ethanol in cosmetics is of concern to Muslim consumers; therefore, according to most authorities, only industrial alcohol is permitted (Hashim and Mat Hashim, 2013). See Chapter 16 for details. The other rules are similar to those for food, that is, all ingredients must be halal, and it must be produced in such a way that there is no contact with najis or haram materials. In addition, it must not harm the user (Mukhtar and Mohsin Butt, 2012).

Ingredients derived from the fur, hair, and related materials of land animals that are halal, soil, chemicals, synthetic materials, plants and microorganisms on land, air, or water, are all halal except those that are hazardous and/or mixed with najis (Hassali et al., 2015).

The cosmetic industry and the Muslim consumer needs to pay more attention to the chemicals and compounds listed as ingredients of cosmetics. Some of the ingredients that can be sourced from animals—oftentimes from pigs include: hyaluronic acid and collagen that may be derived from cow or pig placenta. Keratin is derived from animal hair, nails, and hooves. Stearic acid is a fatty substance that may come from a pig's stomach. In most countries, the sources of cosmetic ingredients do not have to be listed. One may be able to check the company's website or give them a call, but more reliable is for the product to be certified by a reliable halal certifying agency (Salleh and Hussin, 2013). Certified halal cosmetics are designed and formulated for Muslim woman but can be used by everyone. In fact, there are many consumers who are not Muslim that are drawn to these products for eco- and ethical reasons. In particular, the cruelty-free products.

The Food and Drug Administration (FDA), which oversees the cosmetics industry in the U.S., cannot force cosmetics companies to conduct safety assessments and largely allows the industry to make its own decisions about ingredients. Animal ingredients are widely used in cosmetics; some of the ingredients are unavoidable or are difficult to substitute. There is lack of legislative guidelines and some of these ingredients are masked by un-decipherable chemical names (Rajagopal et al., 2011).

PARTIAL LIST OF POTENTIALLY QUESTIONABLE INGREDIENTS

Allantoin

A botanical extract from the comfrey plant. It may also be derived from uric acid obtained from the urine of cows and other mammals. Requires halal supervision to be acceptable or halal certification required.

Ambergris

It is a solid, waxy, flammable, dull grey or blackish color product produced in the digestive system of sperm whales. Freshly produced ambergris has a marine, fecal odor. However, as it ages, it acquires a sweet, earthy scent commonly likened to the fragrance of rubbing alcohol without the vaporous chemical astringency.

Animal Hair

Animal hair (horse) can be used to make brushes, including cosmetic brushes.

Arachidonic Acid

A liquid unsaturated fatty acid that is found in the liver, brain, glands, and fat of animals and humans. Generally isolated from animal livers. Requires halal supervision to be acceptable or halal certification required.

Arachidyl Propionate

Arachidyl propionate is an amber-colored semisolid wax that can be obtained from animal fats. Arachidyl propionate melts upon contact with the human body, leaving a non-oily feeling to the skin (Khan and Abourashed, 2011). Requires halal supervision to be acceptable or halal certification required.

Boar Bristles

Hair from wild or captive hogs can be used to make "natural" toothbrushes and shaving brushes.

Carotene, Pro-vitamin A, Beta Carotene

A color pigment found in many animal tissues and in all plants. Requires halal supervision to be acceptable or halal certification required.

Castor—Castoreum

It is a creamy substance with a strong odor, originally from muskrat and beaver genitals but now typically made synthetically and as such can be halal (Khan and Abourashed, 2011).

Chitosan

A fiber derived from crustacean shells. This ingredient may be of concern for some Muslim consumers who do not accept crustaceans as halal (Jimtaisong and Saewan, 2014).

Cholesterol

A steroid alcohol in all animal fats and oils, nervous tissue, egg yolk, and blood. It can be derived from lanolin. Requires halal supervision to be acceptable or halal certification required.

Civet Oil

The unctuous secretion painfully scraped from a gland very near the genital organs of civet cats (Khan and Abourashed, 2011).

Collagen

It is the main structural protein of the various connective tissues in animals. It is a naturally occurring substance found in the skin, muscle, bones, and tendons of animals and generally obtained from pigs or cattle. In recent years, fish collagens have become available. Requires halal supervision to be acceptable or halal certification required.

Colors and Dyes

Some red pigments such as cochineal and carmine are obtained from crushed female cochineal insects and may not be acceptable to some Muslim consumers.

Cysteine, L-Form

An amino acid from hair that can come from duck feather and human hairs. Can also be synthesized. Requires halal supervision to be acceptable or halal certification required.

Cystine

An amino acid obtained from chemical reactions, urine, and horse-hair. Requires halal supervision to be acceptable or halal certification required.

Emu Oil

Obtained from a flightless ratite bird native to Australia that is now being farmed. It is a halal animal.

Fatty Acids and Fatty Acid Mixtures

Can be one or any mixtures of liquid and solid fatty acids derived from either plants or animals.

Feathers

These can be obtained from slaughtered birds and ground up for use as an ingredient. Requires halal supervision to be acceptable or halal certification required.

Gelatin

It is a translucent, colorless, brittle (when dry), flavorless solid substance. It is derived from collagen (see "collagen") (Schrieber and Gareis, 2007). Halal-certified gelatin will be acceptable for halal cosmetics.

Glycerin/Glycerol

Glycerin, also called glycerol, is an odorless chemical. It can be obtained from animal or plant sources. Halal certification is required for the glycerin.

Hyaluronic Acid

A protein found in umbilical cords and the fluids around the joints. May not be halal suitable.

Hydrolyzed Animal Protein and Hydrolyzed Vegetable Protein

Besides a concern for the source of the raw material, that is, the protein, the source of the enzymes used for the hydrolysis must also be considered. Requires halal supervision to be acceptable or halal certification required.

Keratin

Keratin is the key structural material making up the outer layer of human skin. A cow's skin, fur, nails, hooves, horns, feathers, and teeth all contain keratin. It can also be hydrolyzed. Requires halal supervision to be acceptable or halal certification required.

Lanolin, Lanolin Acids, Wool Fat, and Wool Wax

A product obtained from the oil glands of sheep and also extracted from their wool. If obtained from a live sheep or goat it is halal, but if from a slaughtered sheep, the sheep must be slaughtered halal.

Lard

The fat of pigs (Indrasti et al., 2010) and cannot be used in halal cosmetics.

Mink Oil

The mink is not a halal animal. Therefore, mink oil will not be suitable for halal cosmetics.

Musk (Oil)

Dried secretion obtained from musk deer, beaver, muskrat, civet cat, and otter genitals. All of these are haram animals.

Myristic Acid

This is a fatty acid typically derived from nut oils but occasionally can be obtained from animal sources. Requires halal supervision to be acceptable or halal certification required.

Oleic Acid

Obtained from various animal and vegetable fats and oils. Commercially it is often obtained from edible or inedible tallow, an animal fat. Requires halal supervision to be acceptable or halal certification required.

Palmitic Acid

It is the most common fatty acid (saturated) found in animals, plants, and microorganisms. To make sure its halal status, a certificate will be required for halal cosmetics.

Panthenol, Dexpanthenol. Vitamin B-Complex Factor and Pro-vitamin B-5

It can come from animal, plant, or synthetic sources. Therefore, a halal certificate will be required to use in halal cosmetics.

Placenta

Obtained from the uterus of animals and used in cosmetics. If animal is halal slaughtered then it can be used for halal cosmetics.

Polypeptides

See "Hydrolyzed Animal Protein and Hydrolyzed Plant Protein." To make sure of its halal status, a certificate will be required.

Pristane

Obtained from the liver oil of sharks and from whale ambergris.

Snail Slime

It is an external bodily secretion of snails.

Stearic Acid

It is more abundant in animal fat than vegetable fat. Mostly obtained from cows and sheep, as well as fatty substance taken from the stomachs of pigs. From vegetable sources it will be halal. From animal sources, as long as animal is slaughtered according to halal requirement, it can be used in halal cosmetics.

Stearyl Alcohol

Stearyl alcohol is a fatty alcohol generally derived from stearic acid (see "Stearic Acid"). Requires halal supervision to be acceptable or halal certification required.

Turtle Oil

The oil derived from the muscles and genital glands of the giant sea turtle. It melts at about 25°C. Turtles are haram and its oil should not be used for halal cosmetics.

REFERENCES

Ahmad, A. N., Rahman, A. A., and Ab Rahman, S. (2015). Assessing knowledge and religiosity on consumer behavior towards halal food and cosmetic products. *International Journal of Social Science and Humanity*, 5(1), 10.

Hashim, P. and Mat Hashim, D. (2013). A review of cosmetic and personal care products: Halal perspective and detection of ingredient. *Pertanika Journals of Science and Technology*, 21(2), 281–292.

Hassali, M. A., Al-Tamimi, S. K., Dawood, O. T., Verma, A. K., and Saleem, F. (2015). Malaysian cosmetic market: Current and future prospects. *Pharmaceutical Regulatory Affairs*, 4(155), 2.

Hornby, C. and Yucel, S. (2009). Halal food going mainstream in Europe: Nestle. The Hague: Reuters, November.

Indrasti, D., Man, Y. B. C., Mustafa, S., and Hashim, D. M. (2010). Lard detection based on fatty acids profile using comprehensive gas chromatography hyphenated with time-of-flight mass spectrometry. *Food Chemistry*, 122(4), 1273–1277.

Ireland, J. and Abdollah Rajabzadeh, S. (2011). UAE consumer concerns about halal products. *Journal of Islamic Marketing*, 2(3), 274–283.

Jimtaisong, A. and Saewan, N. (2014). Utilization of carboxymethyl chitosan in cosmetics. *International Journal of Cosmetic Science*, 36(1), 12–21.

Khan, I. A. and Abourashed, E. A. (2011). *Leung's Encyclopedia of Common Natural Ingredients: Used in Food, Drugs and Cosmetics.* Hoboken, NJ: Wiley.

Mukhtar, A. and Mohsin Butt, M. (2012). Intention to choose Halal products: The role of religiosity. *Journal of Islamic Marketing*, 3(2), 108–120.

Rajagopal, S., Ramanan, S., Visvanathan, R., and Satapathy, S. (2011). Halal certification: Implication for marketers in UAE. *Journal of Islamic Marketing*, 2(2), 138–153.

Salleh, M. F. M. and Hussin, R. (2013). Halal assurance system requirements and documentation in cosmetics industry. In *Proceedings of International Conference on Halal Issueas and Policies,* Makasar, pp. 1–6.

Schrieber, R. and Gareis, H. (2007). *Gelatine Handbook: Theory and Industrial Practice.* Hoboken, NJ: Wiley.

Shah Alam, S. and Mohamed Sayuti, N. (2011). Applying the Theory of Planned Behavior (TPB) in halal food purchasing. *International Journal of Commerce and Management*, 21(1), 8–20.

Shikoh, R. (2013). State of the global Islamic economy 2013 report. Toronto, Canada: Thomson Reuters.

Swidi, A., Cheng, W., Hassan, M. G., Al-Hosam, A., Kassim, M., and Wahid, A. (2010). *The Mainstream Cosmetics Industry in Malaysia and the Emergence, Growth, and Prospects of Halal Cosmetics.* Kedah: College of Law, Government and International Studies, Universiti Utara Malaysia, 1–20.

23 Labeling, Packaging, and Coatings for Halal Foods

Mian N. Riaz and Munir M. Chaudry

CONTENTS

Labeling Products for the Muslim Market 241
Labels and Printing Directly on Food Products 243
Packaging Food in a Halal Environment 244
Packaging Materials and Containers 244
Edible Coating and Edible Films 244
References 245

This chapter deals with some aspects of food production that wrap around the food in different ways. These varied aspects might seem unrelated and out of place, but these are issues that affect the halal status of foods and the labeling of products for Muslim markets.

- Halal labeling and printing issues
- Packaging food in a halal environment
- Packaging materials and containers
- Waxes, coatings, and edible films

LABELING PRODUCTS FOR THE MUSLIM MARKET

Label benefits are for consumers and, therefore, should be descriptive, clear, and meaningful. Halal packaged products should have the brand name of the product, minimum matrix content information of manufacture ingredient list, a code that represent the batch information, manufacturer and expiry date, and most importantly, the halal logo from a recognized certified agency (Soong, 2007).

Usually, the ingredient label does not list the origin of the ingredients. Other ingredients, such as processing aids, anticaking agents, carriers, and incidental ingredients from various sources, are also not labeled (Potter and Hotchkiss, 1995). These issues create problems for Muslim consumers. For example, in Europe, manufacturers can use up to 5% vegetable or animal fat in products labeled as pure chocolate. Thus, halal certification of the product and proper halal markings and logos on products are the only way that consumers can be certain a product is halal. Reading of a product label unfortunately is not usually sufficient to determine that a processed product is acceptable (Talib et al., 2010).

If alcohol is a part of the food formulation, then alcohol must be included on the label as an ingredient. If alcohol is found naturally or is a carrier of other ingredients, then it is an incidental ingredient and is not labeled. Some incidental additives may still be present in foods at insignificant levels. If they do not have any technical or functional effect in the food, they are exempt from food labeling requirements (Riaz, 1997). By looking for certified halal products, Muslims no longer need to learn a list of mysterious E-numbers in Europe or chemical jargon in the U.S. each time they go shopping. Even then, the halal status of the manufacturing equipment and of other ingredients that are not labeled make it questionable to ever determine if a product is halal without a halal marking or specific guidance from the appropriate halal authorities (Rezai et al., 2012).

The voluntary information provided by food companies includes the religious symbols used to indicate that the product has qualified for halal and kosher certification. Figure 23.1 shows a sample label with halal markings and Figure 23.2 shows one that is dual halal and kosher certified.

Some commercial label terminology might actually confuse Muslim consumers. "Red wine vinegar" might be a positive gourmet statement in Western countries, but it is not viewed favorably by Muslim consumers. Although there is no wine or significant amount of alcohol left in this product, some people might mistakenly think that all derivatives of alcohol/wine are haram and might not purchase a product containing wine vinegar (red or white). It is better to label the ingredients as flavored vinegar in these countries if permitted by the country's rules.

FIGURE 23.1 Typical meat product with halal logo.

Labeling, Packaging, and Coatings for Halal Foods

FIGURE 23.2 Food with dual certification: Halal and kosher.

If manufacturers want to attract additional consumers, such as Muslim, Jewish, vegetarian, and vegan, then labeling the source of ingredients that can be obtained from both animal and plants sources would increase the chances that a consumer might purchase these products. However, the best option available to the company is to have the product certified as halal, kosher, vegetarian, or vegan by a reputable organization. Additionally, if the target market is other than English-speaking customers, it might be beneficial to make the labels bilingual or even multilingual, including the marking of the halal-certifying agency. Note that if a label includes languages other than English, in the U.S., there are rules concerning what mandatory information required on a label must then also be presented in the additional language(s) (Mohammed et al., 2008). However, just putting the certification logo on the package in multiple languages does not require the use of additional languages elsewhere on the label.

LABELS AND PRINTING DIRECTLY ON FOOD PRODUCTS

Paper and plastic labels, glue used for pressure-sensitive labels, hot-melt glues, and edible printing dyes used directly on food are of concern because they might contain ingredients that would not be permissible in halal food. These may migrate into the food in minute quantities (Uddin, 1994). This may raise a question about its halal status that will require a decision by the halal certifier. The decision in these cases may differ with different halal-certifying agencies. These are just the type of issues that lead to lack of acceptability of products from some certifying agencies by other certifying agencies. Such problems can be avoided by avoiding using such materials in the first place.

PACKAGING FOOD IN A HALAL ENVIRONMENT

In North America, halal food products are generally made in facilities that also produce non-halal products. Most of the workers in the production areas are non-Muslim, who are not familiar with halal (Minkus-McKenna, 2007). Manufacturers can work with their halal certifying agency to establish standard operating procedures to accommodate halal production requirements:

- Keeping halal products in a separate room
- Scheduling production to avoid cross-contamination
- Not switching workers from non-halal to halal packing areas
- Properly marking areas to identify halal production
- Ensuring that workers do not bring personal food into production areas, wash their hands before entering the facility, and so on

PACKAGING MATERIALS AND CONTAINERS

Some packaging materials are questionable with regard to their halal status. In some cases, stearates, which can be of animal or vegetable origin are used in the production of plastic bags and containers. Waxes and coatings applied to plastic, paper, and Styrofoam cups and plates might contain animal fats that are used following any very hot production stage that would be hot enough to nullify any animal products used prior to that step. Each halal certifying agency would currently have to determine what that temperature is, although hopefully this is one of those issues that might be standardized in the future. Metal cans and drums can be contaminated with animal fats. Formation, rolling, and cutting of steel sheets to make containers requires the use of oils to aid in their manufacturing and protect the steel during transportation. Such oils can also be animal derived (Cannon, 1990). Steel drums, which are often reused, can be used to carry foods containing pork or pork fat, which, despite rigorous cleaning practices, might be retained in small amounts to contaminate halal products otherwise thought pure or where the cleaning process, even if rigorous, does not meet halal requirements.

EDIBLE COATING AND EDIBLE FILMS

Although the use of edible films in food products seems new, their use in the food industry actually started many years ago. In England, during the 16th century, larding, that is, coating food products with fat, was used to prevent moisture loss in foods (Labuza and Contrereas-Medellin, 1981). Currently, edible films and coatings are being used in a variety of applications, including casing for sausages, chocolate coating for nuts and fruits, and wax coating for fruits and vegetables.

As the food industry develops edible films and coatings, it will be important for it to understand how some of their choices will affect the halal status of food products. This is particularly true for products that traditionally do not have an ingredient label and cannot be easily certified, such as fruits and vegetables. The Nutritional Labeling and Education Act has a series of regulations dealing with potential coatings of fruits and vegetables, including a mandatory label on packing boxes and signage at the point of sale. The wording of the labels was designed to give consumers meaningful category information, for example, vegetable, mineral, petroleum, lac-resin (shellac), or animal-based ingredients. All these categories except animal-based ingredients are acceptable for halal coatings. In addition, many consumers do not expect to find "animal" products on fresh or frozen fruits and vegetables. Thus, the selection of materials appropriate for these products in particular is important (Regenstein and Chaudry, 2002). The supermarket industry in the U.S. has actively promoted keeping animal-based ingredients out of such products. However, in other countries such products might still contain animal-based products.

REFERENCES

Cannon, C. (1990). Islamic market spells opportunity for processor. *National Provision*, August 17, 1990.

Labuza, T. and Contreras-Medellin, R. (1981). Prediction of moisture protection requirements for foods. *Cereal Foods World (USA)*, 26: 335–343.

Minkus-McKenna, D. (2007). The pursuit of halal. *Progressive Grocer*, 86(17), 42.

Mohamed, Z., Rezai, G., Shamsudin, M. N., and Eddie Chiew, F. C. (2008). Halal logo and consumers' confidence: What are the important factors. *Economic and Technology Management Review*, 3(1), 37–45.

Potter, N. N. and Hotchkiss, J. H. (1995). *Governmental Regulation of Food and Nutrition Labeling*, 5th edn. New York, NY: Chapman & Hall, 567–569.

Regenstein, J. M. and Chaudry, M. M. (2002). Kosher and halal issues pertaining to edible films and coating, In *Protein Based Film and Coating*. Ed. Gennadios, A. Boca Raton, FL: CRC Press, 601–620.

Rezai, G., Mohamed, Z., and Shamsudin, M. N. (2012). Assessment of consumers' confidence on halal labelled manufactured food in Malaysia. *Pertanika Journal of Social Science and Humanity*, 20(1), 33–42.

Riaz, M. N. (1997). *Alcohol: The Myths and Realities, A Handbook of Halaal and Haraam Products*, Vol. 2, Uddin, Z., Ed. Richmond Hill, NY: Publication Center for American Muslim Research and Information, 16–30.

Soong, S. F. V. (2007). Managing halal quality in food service industry. UNLV Theses, Dissertations, Professional Papers, and Capstones. 701.

Talib, Z., Zailani, S., and Zainuddin, Y. (2010). Conceptualizations on the dimensions for halal orientation for food manufacturers: A study in the context of Malaysia. *Pakistan Journal of Social Sciences*, 7(2), 56–61.

Uddin, Z. (1994). *A Handbook of Halaal and Haraam Products*. Richmond Hill, NY: Center for American Muslim Research and Information.

24 How to Get Halal Certified

Mian N. Riaz and Munir M. Chaudry

CONTENTS

What is a Halal Certificate? 248
Types of Halal Certificates 248
Who is Authorized to Issue Halal Certificates? 249
Which Products Can Be Certified? 250
The Halal Certification Process 250
Review of Steps Involved In Halal Certification 252
Use of Halal Markings 252
References 259

As discussed in previous chapters, there are about 1.8 billion Muslims in the world (Hackett et al., 2015). Globally, the halal market is gaining popularity, and currently, the halal market constitutes a 16% share of the entire global market and will account for 20% of world trade for food products in the near future (Nestle, 2009). Europe, Africa, and Asia account for 10%, 24%, and 63% of the global market share (Hashim, 2010). Southeast Asia has 250 million Muslim halal consumers. Indonesia, Malaysia, Singapore, and many other countries have government mandates to only import halal-certified products. Recently, other countries in the region, such as Thailand and the Philippines, have initiated regulations to encourage both the export and import of halal products. In these countries, halal is considered as a symbol of quality and wholesomeness not only by Muslims but also by non-Muslims (Junaini and Abdullah, 2008). A halal program was started in Malaysia during the early 1980s with the passage of the Halal/Haram Act and formulation of a high government level halal/haram committee. Under Malaysian regulations, meat and non-meat food products have to be certified halal by a Malaysian government recognized halal authority. The Malaysian Department of Islamic Affairs has created two lists of approved entities for dealing with meat and poultry products exported to Malaysian from each country that wishes to export to Malaysia: (1) approved Islamic supervision organizations and (2) approved meat and poultry slaughterhouses or abattoirs.

The halal certification program started in Indonesia during the early 1990s and has developed into one of the strictest programs in the world. The program is run by an organization known as LP-POM or the Assessment Institute for Food, Drugs and Cosmetics (Van Spiegel et al., 2012). This institution operates under the authority of the Indonesian Council of Ulama (ICU), generally known as Majelis Ulama Indonesia (MUI). The institute enjoys the full backing of various government departments and agencies. Malaysia, Indonesia, Singapore, and Thailand have designed their own official government supported halal logo for the products offered for retail or food service (Bonne and Verbeke, 2008). A company wishing to apply each country's logo must make an application to that country, even when the product is manufactured in another country. However, certificates and logos of other recognized agencies are acceptable as well.

North America has a Muslim population of approximately 8 million: about 7 million in the U.S. and 1 million in Canada (Cornell University Survey, 2002). Thus, there is a tremendous economic opportunity for companies to meet the needs of Muslims for halal food products. Perception of the word halal varies among different groups of Muslims. People from the Middle East associate halal with meat and poultry only, whereas people of South Asian and Southeast Asian origin contend that all food and consumable products have to be halal. The latter group calls the meat and poultry slaughtered by Muslims according to the traditional method as zabiha (or dhabiha). Over the past 30 years, many halal markets, ethnic stores, and restaurants have sprung up in non-Muslim-majority countries, mainly in major metropolitan areas in North America and Europe. The main group of products sold as halal or zabiha halal at these stores are fresh meat and poultry as well as imported ethnic products (Van der Spiegel et al., 2012). The U.S. food industry, for the most part, has ignored this population group and has concentrated its efforts on exports for the Muslim-majority countries. In many communities, Muslims working with local retailers have been slaughtering their own animals, and the concept of halal certification was foreign to them. In the late 1990s, however, small- to mid-size companies recognized the vacuum and the need to capture this niche market. For retail and multilevel marketed products, halal products certified by just one of the certifiers (i.e., the Islamic Food and Nutrition Council of America) has increased from less than 2000 five years ago to now over 5000 (Riaz, 2017). Halal certification is now becoming as popular for domestic products as it has been for exported products (Riaz, 2002). According to Ab Talib, et al. (2016) proper application of resources, in case of halal certification, could positively influence logistics performance of the company.

WHAT IS A HALAL CERTIFICATE?

A halal certificate is a document issued by an Islamic organization certifying that the products listed on it meet Islamic dietary guidelines, as defined by that certifying agency (Ariff, 2009).

TYPES OF HALAL CERTIFICATES

- *Registration of a site certificate*: This type of certificate signifies that a plant, production facility, food establishment, slaughterhouse, abattoir, or any establishment handling food has been inspected and approved to produce, distribute, or market halal food. This does not mean that all food products made or handled at such a facility are halal certified. A site certificate should not be used as a halal product certificate. The site certificate

should be dated, ideally with both a start and end date (ideally one year or less, subject to renewal) (Hanzaee and Ramezani, 2011).
- *Halal certificate for a specific product for a specific duration*: This type of certificate signifies that the listed product or products meet the halal guidelines formulated by the certifying organization. Such a certificate may be issued for a certain time period (ideally one year or less, subject to renewal) or for a specified quantity of the product destined for a particular distributor or importer. If the certificate is for a specific quantity, it may be called a batch certificate or a shipment certificate. Meat and poultry products, for which each batch or consignment has to be certified, generally receive a batch certificate. A batch certificate issued for each consignment is valid for as long as that specific batch or lot of the product is in the market, generally up to product expiration date or "Use By" date (Hanzaee and Ramezani, 2011).
- *Yearly certification*: May be automatically renewed contingent on passing the annual inspection through halal compliance and payment of the certification fee. Often a system of occasional unannounced plant visits is used to confirm the plant's status.

WHO IS AUTHORIZED TO ISSUE HALAL CERTIFICATES?

Any individual Muslim, Islamic organization, or agency can issue a halal certificate, but the acceptability of the certificate depends on the country of import or the Muslim community served through such certification. For example, to issue a halal certificate for products exported to Malaysia and Indonesia, the body issuing the halal certificate must be listed on each country's approved list (Noordin et al., 2014). More than 100 organizations issue halal certificates in the U.S., but only 5 to 10 have been approved by MUI and 15 to 20 by the Jabatan Kemajuan Islam Malaysia (JAKIM). Fifty percent of the certificate organizations approved by JAKIM over the years are not even active in issuing halal certificates according to JAKIM sources (Personal communication, 2016) (Alter, 2006).

The food industry not only needs to understand halal requirements for different countries and the principles of halal but also needs an understanding of the organizations that would best meet their needs—organizations that can service their global needs and are acceptable to the countries of import (Panisello and Quantick, 2001).

Malaysia and Indonesia are the only countries that have a formal program to approve a halal-certifying organization. Other countries such as Saudi Arabia, Singapore, Kuwait, United Arab Emirates (UAE), Egypt, and Bahrain also do approvals of organizations for specific products or purposes. Currently, the UAE has become more active in this arena and is encouraging countries to develop accreditation, that is, to have a body that can accredit the various certification organizations.

Some of the major voices in halal recognition worldwide include:

- Jabatan Kemajuan Islam Malaysia (JAKIM), Malaysia
- Majelis Ulama Indonesia (MUI), Indonesia
- Majlis Ugama Islam Singapura (MUIS), Singapore
- Muslim World League (MWL), Saudi Arabia

WHICH PRODUCTS CAN BE CERTIFIED?

With the complexity of manufacturing systems and the utilization of all animal by-products, any product consumed by Muslims may be certified, whether the product is consumed internally or applied to the body externally. Medicines and pharmaceutical products that are used for health reasons need not be certified; however, knowledgeable consumers look for products that are halal-certified or at least meet halal guidelines, even for medicines and pharmaceuticals. In recent years, packaging has become more of an issue. The products that may be certified but not limited include:

- Meat and poultry fresh, frozen, and processed products
- Meat and poultry ingredients
- Dairy products and ingredients
- Prepared foods and meals
- Enzymes
- Flavor
- Beverages and Juices
- Fruits and vegetables
- Cereal-based products
- Confectionary, gums, chocolate
- Prepared meal
- All kinds of snacks
- Fish and seafood products
- Cooking oil, butter, and margarine
- All other packaged food products
- Cosmetics and personal care products
- Pharmaceuticals
- Nutritional and dietary supplements
- Packaging materials
- Lubricants and greases

THE HALAL CERTIFICATION PROCESS

The halal certification process starts with choosing an organization that meets the company's needs for the markets to be served. If the target is a specific country, it is best to use an organization that is approved, recognized, or acceptable in that country. If the market areas are broader or even global, then an organization with an international scope would better meet the company's needs (Lam and Alhashmi, 2008).

After the preliminary search for an agency and possibly some conversation with one or more certifying agencies, the formal process generally starts with filling out an application explaining the production process, the products to be certified, and regions the products will be sold or marketed in, along with specific information about the component ingredients and manufacturing process and information about other products manufactured in the same facility. Most certifying organizations review the information and if all is in order, they set up an audit of the

How to Get Halal Certified

facility. This preliminary audit may have a charge associated with it. If the audit is successful, then it is advisable to negotiate the fees and clearly understand the costs involved.

During the review of the ingredient information or the facility audit, the certifying organization might ask for replacement of any ingredients that do not meet its guidelines. They will often be able to assist in finding the acceptable replacement ingredients. Generally, the company and the halal-certifying agency sign a multiyear supervision agreement. Then a halal certificate can be issued for a specific shipment of a product or for a given period of a few months to several years, although even if the contract is for multiple years, the working certificate should probably be only for a maximum of one year. Overall, the process for halal certification of almost all food products is not complicated, as explained in Figure 24.1.

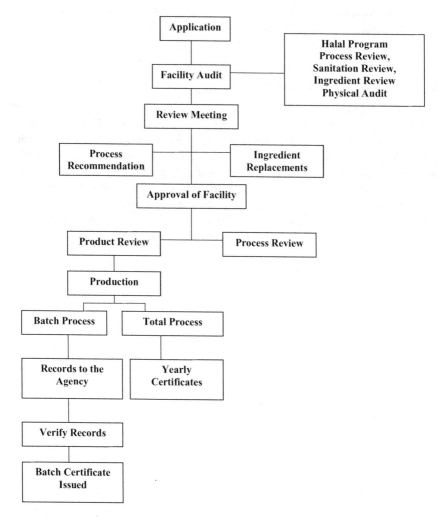

FIGURE 24.1 Halal certification process flow-chart.

REVIEW OF STEPS INVOLVED IN HALAL CERTIFICATION

- Filling out an application to the certification organization on paper or on the Internet. Figure 24.2 shows a typical application.
- Review of the information by the organization, especially the type of product and its components.
- Inspection and approval of the facility. This includes a review of the production equipment, inspection of ingredients, cleaning procedures, sanitation, and potential for cross-contamination.
- For slaughterhouses, inspection involves a review of the holding areas, method of stunning, actual slaying, pre- and post-slaughter handling, etc.
- Determining the cost and fees involved and signing of the contract. (Note: The contract normally includes the responsibility of both parties to maintain confidentiality.)
- Payment of fees and expenses.
- Issuance of the halal certificate.

USE OF HALAL MARKINGS

When a product is certified halal, a symbol is normally printed on the package. For example, the Islamic Food and Nutrition Council of America (IFANCA) uses the Crescent M symbol, which signifies "good for Muslims." Generic symbols, such as the word halal in Arabic, often inside a circle, are used by some companies with or without the endorsement of a certifying organization. However, a product is better accepted by Muslim consumers if the logo signifies a reputable halal certification organization who can be held accountable. Figure 24.3 shows a typical food with a halal symbol. Figure 24.4 gives some of the halal markings and logos used by different countries and organizations (Hanzaee and Ramezani, 2011).

How to Get Halal Certified

ifanca Islamic Food and Nutrition Council of America	Doc. Number: IFANCA-Frm-31	Revision: 3.0
	Document Name: Application Form for Halal Certification	

Type of application: ☐ New Company ☐ New plant(s) ☐ Add new product(s)

Date of Application		IFANCA Application No.	

(for office use only)

Company Information:

Company Name	
Address	
City	
State/Province	
Country	
Postal/Zip Code	
Web Address	

Primary Contact	
Position/Title	
E-mail Address	
Telephone No.	
Fax No.	

*Point of contact in the USA, if different than above:

Application Authorized by: _____ Date Authorized: _____
(please print)

Position/Title of Individual: _____
(please print)

Manufacturing Facility Information: (if different than above)
*Please provide all production locations. Include the full address, contact number and contact person for each additional location.

Company/Plant Name	
Address	
City	
State/Province	
Country	
Postal/Zip Code	
Gov't Plant Code	

Plant Contact	
Position/Title	
E-mail Address	
Telephone No.	
Fax No.	

Access and Travel Information:

Name of the nearest major city and airport to the location	
Distance between the airport and location to be certified	Kilometers — Miles

| IFANCA Confidential | Effective Date: 01-17-16 | Revision Date/By: 10-31-2017/MB | Approved Date/By: 10-31-2017/MMC | Page 1 of 8 |

FIGURE 24.2 A typical application for halal certification.

ifanca Islamic Food and Nutrition Council of America	Doc. Number: IFANCA-Frm-31	Revision: 3.0
	Document Name: Application Form for Halal Certification	

Product Information:

(1) Has the company ever applied for Halal certification previously?	☐ Yes ☐ No
If yes, please state the Halal agency that was previously applied to	
(2) Has the factory ever been supervised before, either on a yearly basis or for a specific batch production for another buyer?	☐ Yes ☐ No
If yes, please state the Halal agency that was certifying	
(3) Please state all food safety programs implemented at the factory (Please include a copy of each food safety program certificate with this application)	☐ HACCP ☐ ISO-22000 ☐ GMP ☐ Organic Food ☐ Other
(4) Marketing type	☐ Food Service (Bulk) ☐ Retail ☐ Direct Marketing ☐ Industry ☐ Other: _____
(5) Is the Brand Name	☐ Owned ☐ Private Label ☐ Other: _____
(6) Do you produce product using pork or pork derivative at this facility? yes, IFANCA will require more details from your company)	☐ Yes ☐ No
(7) Do you produce product using animal meat or animal derivatives such as beef, chicken, deer or mutton? (If yes, site inspection must be done and cost of inspection is borne on the applicant)	☐ Yes ☐ No
(8) Do you use gelatin or capsule in your product? (If yes, please provide a copy of all Halal certificates for each with this application)	☐ Yes ☐ No
(9) If this application is for food product, does the product contain alcohol exceeding 0.1%? (If yes, discussion is required for further modification of the formula)	☐ Yes ☐ No
(10) If this application is for flavor as a final product, does the product contain alcohol exceeding 0.5%? (If yes, discussion is required for further modification of the formula)	☐ Yes ☐ No
(11) Do you produce product using glycerine or its derivatives? (If yes, please provide a copy of all Halal certificates for each with this application)	☐ Yes ☐ No
(12) Please check scheme(s) under which you like to get Halal certificate:	☐ SMIIC (OIC) ☐ MUI (INDONESIA) ☐ JAKIM (MALAYSIA) ☐ ESMA (UAE) ☐ GSO (GCC) ☐ MUIS (SINGAPORE) ☐ OTHERS

| IFANCA Confidential | Effective Date: 01-17-16 | Revision Date/By: 10-31-2017/MB | Approved Date/By: 10-31-2017/MMC | Page **2** of **8** |

FIGURE 24.2 (Continued)

How to Get Halal Certified

FIGURE 24.2 (Continued)

ifanca — Islamic Food and Nutrition Council of America

Doc. Number: IFANCA-Frm-31
Revision: 3.0
Document Name: Application Form for Halal Certification

(16) Please provide information on all processing aids.

- Please include the full address of the manufacturer or supplier.
- Please provide a copy of all HALAL certificates covering the material.
- Please make additional copies of this page if more ingredients need to be listed.

No.	PROCESSING AIDS (full name of ingredient)	COMPOSITION (provide a complete description)	MANUFACTURER/SUPPLIER (full name and address)	PACKAGING METHOD (Poly-bag, truck tanker, drums, etc.)	HALAL Certification
1					☐ Yes ☐ No
2					☐ Yes ☐ No
3					☐ Yes ☐ No
4					☐ Yes ☐ No
5					☐ Yes ☐ No
6					☐ Yes ☐ No
7					☐ Yes ☐ No
8					☐ Yes ☐ No
9					☐ Yes ☐ No
10					☐ Yes ☐ No

COMMENTS
Please provide any additional details relevant to this certification process

Please Note: IFANCA agrees that the information submitted in this application will be dealt with in strict confidentiality and will not be used for anything other than evaluating this product for certification.

PLEASE E-MAIL THIS APPLICATION AND SUPPORTING DOCUMENTATION TO: Halal@ifanca.org OR BY FAX TO: (847) 993-0038

| IFANCA Confidential | Effective Date: 01-17-16 | Revision Date/By: 10-31-2017/MB | Approved Date/By: 10-31-2017/MMC | Page 5 of 8 |

---

ifanca — Islamic Food and Nutrition Council of America

Doc. Number: IFANCA-Frm-31
Revision: 3.0
Document Name: Application Form for Halal Certification

(17) Please provide information on all sanitation, cleaning chemicals and materials.

- Please include the full address of the manufacturer or supplier.
- Please provide a copy of all HALAL certificates covering the material.
- Please make additional copies of this page if more ingredients need to be listed.

No.	SANITATION, CLEANING CHEMICALS (full name of ingredient)	COMPOSITION (provide a complete description)	MANUFACTURER/SUPPLIER (full name and address)	PACKAGING METHOD (Poly-bag, truck tanker, drums, etc.)	HALAL Certification
1					☐ Yes ☐ No
2					☐ Yes ☐ No
3					☐ Yes ☐ No
4					☐ Yes ☐ No
5					☐ Yes ☐ No
6					☐ Yes ☐ No
7					☐ Yes ☐ No
8					☐ Yes ☐ No
9					☐ Yes ☐ No
10					☐ Yes ☐ No

COMMENTS
Please provide any additional details relevant to this certification process

Please Note: IFANCA agrees that the information submitted in this application will be dealt with in strict confidentiality and will not be used for anything other than evaluating this product for certification.

PLEASE E-MAIL THIS APPLICATION AND SUPPORTING DOCUMENTATION TO: Halal@ifanca.org OR BY FAX TO: (847) 993-0038

| IFANCA Confidential | Effective Date: 01-17-16 | Revision Date/By: 10-31-2017/MB | Approved Date/By: 10-31-2017/MMC | Page 6 of 8 |

FIGURE 24.2 (Continued)

How to Get Halal Certified

ifanca — Islamic Food and Nutrition Council of America

Doc. Number:	Revision:
IFANCA-Frm-31	3.0

Document Name: Application Form for Halal Certification

(18) Please provide information on all lubricants.

- Please include the full address of the manufacturer or supplier.
- Please provide a copy of all HALAL certificates covering the material.
- Please make additional copies of this page if more ingredients need to be listed.

No.	LUBRICANTS (full name of ingredient)	COMPOSITION (provide a complete description)	MANUFACTURER/SUPPLIER (full name and address)	PACKAGING METHOD (Poly-bag, truck tanker, drums, etc.)	HALAL Certification
1					☐ Yes ☐ No
2					☐ Yes ☐ No
3					☐ Yes ☐ No
4					☐ Yes ☐ No
5					☐ Yes ☐ No
6					☐ Yes ☐ No
7					☐ Yes ☐ No
8					☐ Yes ☐ No
9					☐ Yes ☐ No
10					☐ Yes ☐ No

COMMENTS — Please provide any additional details relevant to this certification process

Please Note: IFANCA agrees that the information submitted in this application will be dealt with in strict confidentiality and will not be used for anything other than evaluating this product for certification.

PLEASE E-MAIL THIS APPLICATION AND SUPPORTING DOCUMENTATION TO: Halal@ifanca.org **OR BY FAX TO:** (847) 993-0038

IFANCA Confidential	Effective Date: 01-17-16	Revision Date/By: 10-31-2017/MB	Approved Date/By: 10-31-2017/MMC	Page 7 of 8

---

ifanca — Islamic Food and Nutrition Council of America

Doc. Number:	Revision:
IFANCA-Frm-31	3.0

Document Name: Application Form for Halal Certification

(19) Please provide information on all packaging materials.

- Please include the full address of the manufacturer or supplier.
- Please provide a copy of all HALAL certificates covering the material.
- Please make additional copies of this page if more ingredients need to be listed.

No.	PACKAGING MATERIALS (full name of ingredient)	COMPOSITION (provide a complete description)	MANUFACTURER/SUPPLIER (full name and address)	PACKAGING METHOD (Poly-bag, truck tanker, drums, etc.)	HALAL Certification
1					☐ Yes ☐ No
2					☐ Yes ☐ No
3					☐ Yes ☐ No
4					☐ Yes ☐ No
5					☐ Yes ☐ No
6					☐ Yes ☐ No
7					☐ Yes ☐ No
8					☐ Yes ☐ No
9					☐ Yes ☐ No
10					☐ Yes ☐ No

COMMENTS — Please provide any additional details relevant to this certification process

Please Note: IFANCA agrees that the information submitted in this application will be dealt with in strict confidentiality and will not be used for anything other than evaluating this product for certification.

PLEASE E-MAIL THIS APPLICATION AND SUPPORTING DOCUMENTATION TO: Halal@ifanca.org **OR BY FAX TO:** (847) 993-0038

IFANCA Confidential	Effective Date: 01-17-16	Revision Date/By: 10-31-2017/MB	Approved Date/By: 10-31-2017/MMC	Page 8 of 8

FIGURE 24.2 (Continued)

FIGURE 24.3 Typical ready to eat food with halal logo.

FIGURE 24.4 Halal logos of different halal certifying agencies.

REFERENCES

Ab Talib, M. S., Abdul Hamid, A. B., and Chin, T. A. (2016). Can halal certification influence logistics performance? *Journal of Islamic Marketing*, 7(4), 461–475.

Ariff. (2009). Importance of halal certification. Available online: www.halaljournal.com/article/4262/importance-of-halal-certification.

Bonne, K. and Verbeke, W. (2008). Religious values informing halal meat production and the control and delivery of halal credence quality. *Agriculture and Human Values*, 25(1), 35–47.

Cornell University Survey. (2002). Study on American Muslim. Survey sponsored by Bridges TV, Orchard Park, NY.

Hackett, C., Connor, P., Stonawski, M., Skirbekk, V., Potancoková, M., and Abel, G. (2015). *The Future of World Religions: Population Growth Projections, 2010–2015*. Washington, DC: Pew Research Center's Forum on Religion and Public Life.

Hashim, D. D. (2010). The quest for a global halal standard. In *Meat Industry Association of New Zealand Annual Conference*, Christchurch, 19–20.

Hanzaee, K. H. and Ramezani, M. R. (2011). Intention to halal products in the world markets. *Interdisciplinary Journal of Research in Business*, 1(5), 1–7.

Junaini, S. N. and Abdullah, J. (2008). MyMobiHalal 2.0: Malaysian mobile halal product verification using camera phone barcode scanning and MMS. In *International Conference on Computer and Communication Engineering*, Kuala, Lumpar, Malaysia, 2008, 528–532. IEEE.

Lam, Y. and Alhashmi, S. M. (2008). Simulation of halal food supply chain with certification system: A multi-agent system approach. In *Pacific Rim International Conference on Multi-Agents*. Eds. Bui, T. D, Ho, T. V and Ha, Q. T. Berlin, Heidelberg: Springer, 259–266.

Majelis ulama indonesia. (2001). List of Islamic organizations approved by AIFDC-ICU, Mayid Istiglal, Jakarta, Indonesia, 1.

Nestle. (2009). Nestle and halal. Retrieved 16.12.10, Available online: www.nestle-tasteofhome.com/NR/rdonlyres/B519F379-6669-428E-872B-6DCA9DFA2B6B/46458/NestleandHalal_Nov09.pdf.

Noordin, N., Noor, N. L. M., and Samicho, Z. (2014). Strategic approach to halal certification system: An ecosystem perspective. *Procedia-Social and Behavioral Sciences*, 121, 79–95.

Panisello, P. J. and Quantick, P. C. (2001). Technical barriers to hazard analysis critical control point (HACCP). *Food Control*, 12(3), 165–173.

Riaz, M. N. (2002). Halal production standards and plant inspections requirements. Paper presented at the 4th International Halal Food Conference on Current and Future Issues in Halal. Toronto, Canada, April 21–23.

Riaz, M. N. (2017). Personal Communication, rothman@ifanca.org and www.ifanca.org.

Van der Spiegel, M., Van der Fels-Klerx, H. J., Sterrenburg, P., Van Ruth, S. M., Scholtens-Toma, I. M. J., and Kok, E. J. (2012). Halal assurance in food supply chains: Verification of halal certificates using audits and laboratory analysis. *Trends in Food Science & Technology*, 27, 109–119.

25 Comparison of Kosher, Halal, and Vegetarianism

Mian N. Riaz and Munir M. Chaudry

CONTENTS

Meat of Animals Killed by the Ahlul-Kitab	262
Kosher Laws	263
Halal Laws	263
Kosher Dietary Laws	264
Allowed Animals for Kosher	264
Prohibition of Blood	265
Prohibition of Mixing of Milk and Meat	266
Kosher: Special Foods	267
Passover Requirements	268
Equipment Kosherization	268
Halal Dietary Laws	268
Prohibited and Permitted Animals	271
Prohibition of Blood	271
Prohibition of Alcohol and Intoxicants	272
Halal Cooking, Food Processing, and Sanitation	273
Kosher and Halal	273
Gelatin	273
Biotechnology	274
Vegetarianism	275
Food Standards for Vegetarians	276
Animal Products	276
Kitchen and Hygiene Standards	276
Comparison of Kosher, Halal, and Vegetarian	276
References	279

Why are we discussing kosher and vegetarian issues in a halal book? There are two reasons:

- Permitted food of the Jews is called kosher. There are both similarities and differences between kosher and halal. Similarly, many halal consumers might think that because vegetarian products are from vegetable sources, they are halal. But again there are important differences (Regenstein et al., 2003).
- Many people in the U.S. food industry are familiar with the word kosher and what is required to make the products kosher. By comparing halal with

kosher, food industry professionals can understand each concept better and provide foods that better serve both communities.
- Before undertaking a more comprehensive comparative discussion, the permissibility for Muslims to eat of the meat of animals killed by the Ahlul-Kitab (Jews and Christians) will be discussed.

MEAT OF ANIMALS KILLED BY THE AHLUL-KITAB

There has been much discussion among Muslim consumers as well as Islamic scholars about the permissibility of consuming meat of animals killed by the Ahlul-Kitab (people of the book), meaning mainly Jews and Christians. (There are few small sects in the Middle East that also would qualify.) This generally implies that the animal was killed by an Ahlul-Kitab, but the Islamic method of slaughter while invoking the name of God as required under the Islamic guidelines was not followed.

In the holy Quran [Arabic text and English rendering by Pickthall (1994)] this issue is presented only once, in the following words:

This day are (all) good things made lawful for you. The food of those who have received the Scripture is lawful for you, and your food is lawful for them.

> This verse addresses the Muslims and seems to have been set in a social context where Muslims, Jews, and Christians had to interact with each other. It points to two sides of the issue, first, "the food of the people of the book is lawful for you," and, second, "your food is lawful for them."

Chapter V, Verse 5

As regards the first part of the ruling, Muslims are allowed to eat the food of the Jews and Christians as long as it does not violate the opening statement of the verse, "This day all good and wholesome things have been made lawful for you."

The majority of Islamic scholars are of the opinion that the food of the Ahlul-Kitab must meet the criteria established for halal and wholesome food, including proper slaughtering of animals. They believe that the following verse from the Quran establishes a strict requirement for Muslims.

> And eat not of that whereupon Allah's name hath not been mentioned, for lo! It is abomination.

Chapter VI, Verse 121

However, some Islamic scholars such as Al-Qaradawi (1984) are of the opinion that this verse does not apply to the food of Ahlul-Kitab. They opine that meat of halal animals sold in Western countries is acceptable for Muslims. They contend that God's name may be pronounced at the time of eating rather than at the time of slaughtering of an animal. Note that additional issues need to be considered. The method of kosher slaughter of animals is functionally the same, while the secular slaughter is not. Even in the U.S., one cannot assume that the secular slaughter person is someone who is a Christian. On the other hand, the Jewish slaughter person does invoke the name of God before slaughtering a group of animals. In addition, some major American rabbis have given permission for the Jewish slaughter person to say the Takbir in Arabic at the time of slaughter over each animal. Regulatory agencies in halal-food-importing countries, halal certifiers, or individual Muslim consumers can accept or reject products based on this reasoning.

For Muslims who want to strictly follow the requirements of Chapter VI, Verse 121, no meat containing products of the Ahlul-Kitab meets the Islamic standard. According to Jackson (2000), most kosher food processors think that Muslims accept kosher as meeting halal standards and requirement other than meats and alcohol containing products. Religiously, Muslims prefer not to accept kosher certification as a substitute for halal certification although they will note a kosher certification, which means that many of the issues being discussed with respect to halal have been accounted for.

KOSHER LAWS

Kosher dietary laws determine which foods are fit or proper for consumption by Jewish consumers who observe these laws. The laws are Biblical in origin, coming mainly from the original five books of the Holy Scriptures (the Torah). At the same time Moses received the Ten Commandments on Mount Sinai, Jewish tradition teaches that he also received the oral law, which was eventually written down many years later in the Talmud. This oral law is as much a part of Biblical law as the written text. Over the years, the meaning of the Biblical kosher laws have been interpreted and extended by rabbis to protect Jews from violating any of the fundamental laws and to address new issues and technologies. The system of Jewish law is referred to as Halacha (Regenstein and Chaudry, 2001; Regenstein et al., 2003).

HALAL LAWS

Halal dietary laws determine which foods are lawful or permitted for Muslims. These laws are found in the Quran and the books of hadith (the traditions). These guidelines are referred to as Islamic laws and has been interpreted by Muslim scholars over the years. The basic principles of Islamic laws remain definite and unaltered. However, their interpretation and application might change according to time, place, and circumstances. Some of the issues Muslim scholars are dealing with include biotechnology, unconventional sources of ingredients, synthetic materials, and innovations in animal slaughter and meat processing (Department of Standards Malaysia, 2009; Qureshi et al., 2012).

Although many Muslims purchase kosher food in the U.S., these foods, as shown later, do not always meet the needs of Muslim consumers. The most common areas of concern for Muslim consumers when considering purchasing kosher products are the use of various questionable gelatins in products produced by more lenient kosher supervisions and the use of alcohol in cooking food and as a carrier for flavors.

KOSHER DIETARY LAWS

Kosher dietary laws predominantly deal with three issues, all focused on the animal kingdom:

- Allowed animals
- Prohibition of blood
- Prohibition of mixing of milk and meat

Additionally, for the week of Passover (in late March or April) restrictions on chometz, the prohibited grains (wheat, rye, oats, barley, and spelt), and rabbinical extensions of this prohibition lead to a whole new set of regulations, focused in this case on the plant kingdom. In addition, separate laws deal with grape juice, wine, and alcohol derived from grape products; Jewish supervision of milk; Jewish cooking, cheese making, and baking; equipment kosherization; purchasing new equipment from non-Jews; and old and new flour (Regenstein and Chaudry, 2001; Regenstein et al., 2003).

ALLOWED ANIMALS FOR KOSHER

Ruminants with split hoofs that chew their cud, traditional domestic birds, and fish with fins and removable scales are generally permitted. Pigs, wild birds, sharks, dogfish, catfish, monkfish, and similar species are prohibited, as are all crustacean and molluscan shellfish. Almost all insects are prohibited such that carmine and cochineal (natural red pigments) are not used in kosher products by most rabbinical supervisors. With respect to poultry, traditional domestic birds such as chicken, turkey, squab, duck, and goose are kosher. Birds in the ratite category (ostrich, emu, and rhea) are not kosher as the ostrich is specifically mentioned in the Bible. However, it is not clear as to whether the animal of the Bible is the same animal that is known today as an ostrich. A set of criteria is sometimes referred to in trying to determine whether a bird is kosher. The kosher bird has a stomach (gizzard) lining that can be removed from the rest of the gizzard. It cannot be a bird of prey. Another issue deals with tradition, for example, newly discovered or developed birds might not be acceptable. Some rabbis do not accept wild turkey, whereas some do not accept the featherless chicken.

The only animals from the sea that are permitted are those with fins and scales. All fish with scales have fins, so the focus is on scales. These must be visible to the human eye and must be removable from the fish skin without tearing the skin. A few fish remain controversial, probably swordfish being the most discussed (Regenstein and Regenstein, 2000; Regenstein et al., 2003).

Most insects are not kosher. The exception includes a few types of grasshoppers, which are acceptable in the parts of the world where the tradition of eating them has not been lost. Edible insects are all in the grasshopper family identified as permitted in the Torah due to their unique movement mechanism. Again, only visible insects are of concern; an insect that spends its entire life cycle inside the food is not of concern. The recent development of exhaustive cleaning methods to prepare prepackaged salad vegetables eliminates a lot of the insects that are sometimes visible, rendering the product kosher and, therefore, usable in kosher foodservice establishments and in the kosher home, without requiring extensive special inspection procedures. Although companies in this arena go through a great deal of effort to produce an insect-free product, some kosher supervision agencies remain unconvinced and only certify those products (or particular lots) that meet their more stringent requirements (Regenstein and Regenstein, 1988; Regenstein et al., 2003). The prohibition of insects focuses on the whole animal. If one's intent is to make a dish where the food will be chopped up in a food processor, then one may skip the elaborate inspection of fruits and vegetables for insects and assume that only insect parts are present rather than whole insects, which does not render the food non-kosher. There are guidebooks describing which fruits and vegetables in particular countries need inspection; recommended methods for doing this inspection are included. Kosher consumers have appreciated the use of pesticides to keep products insect-free as well as the use of prepackaged vegetables that have been properly inspected. Modern IPM (integrated pest management) programs that increase the level of insect infestation in fruits and vegetables can cause problems for the kosher consumer. Examples of problems with insects that one might not think about include insects under the triangles on the stalks of asparagus, under the greens of strawberries, and thrips on cabbage leaves. Because of the difficulty of properly inspecting them, many Orthodox consumers do not use Brussel sprouts (Regenstein and Regenstein, 1988; Regenstein et al., 2003).

PROHIBITION OF BLOOD

Ruminants and fowl must be slaughtered according to Jewish law by a specially trained religious slaughter person (shochet), using a special knife designed for the purpose (chalef). The knife must be extremely sharp, without any nicks, and have a very straight blade that is at least twice the diameter of the neck of the animal to be slaughtered. The animal is not stunned prior to slaughter. If the slaughter is done in accordance with Jewish law and with good animal-handling practices, the animal will die without showing any signs of stress. With respect to kosher supervision, slaughtering is the only time a blessing is said, and it is said before commencing slaughter. The slaughter person says a traditional blessing. The blessing is not said over each animal. The rules for slaughter are very strict and the shochet checks the chalef before and after the slaughter of each animal. If any problem occurs with the knife, such as even one small nick, the animal becomes treif (not kosher). The shochet also checks the cut on the animal's neck after each slaughter to make sure it was done correctly. Slaughtered animals are subsequently inspected, particularly the lungs, for defects by rabbinically trained inspectors. If an animal is found to have a defect, the animal is deemed unacceptable and becomes treif. There is no trimming of defective portions as generally permitted under secular law. The general rule is that the defect would not lead to a situation where the animal could be expected to die within a year (Regenstein et al., 2003). Consumer desire for more stringent kosher meat inspection requirements in the U.S. has led to the development of a standard for kosher meat that meets a stricter inspection requirement, mainly with respect to the condition of the animal's lungs. As the major site of halachic defects, the lungs must always be inspected. Other organs are spot-checked or examined when a potential problem is observed. Meat that meets this stricter standard is referred to as glatt (smooth) kosher, referring to the fact that the animal's lungs do not have any adhesions (sirkas). The bodek (inspector of internal organs) is trained to look for lung adhesions in the animal both before and after its lungs are removed. To test a lung, the bodek first removes all sirkas and then blows up the lung by using normal human air pressure. The lung is then put into a water tank and the bodek looks for air bubbles. If the lung is still intact, it is kosher. In the U.S., a glatt kosher animal's lungs generally have fewer than two adhesions, which permit the task of lung inspection to be done carefully in the limited time available in large plants (Regenstein and Chaudry, 2001; Regenstein and Regenstein, 1979, 1988; Regenstein et al., 2003). Any meat that is declared treif, if it passes secular inspection, is moved into the secular meat supply.

Meat and poultry must be further prepared by properly removing certain veins, arteries, prohibited fats, blood, and the sciatic nerve. In practical terms, this means that only the front quarter cuts of kosher red meat are used in the U.S. and most Western countries. Again, the hindquarters are moved to the secular meat supply. However, some of this meat is also moved to the halal market, with or without a blessing over each animal. Although it is very difficult and time consuming to remove an animal's sciatic nerve, necessity demands that this deveining be done in parts of the world where the hindquarter is needed in the kosher food supply. In some animals such as deer, it is relatively easy to devein the hindquarter. However, if there is no tradition of eating any hindquarter meat within a community, some rabbis have rejected the deer hindquarters for their community. To further remove the prohibited blood, red meat and poultry must then be soaked and salted within 72 hours of slaughter. The salted meat is then rinsed three times (Regenstein and Chaudry, 2001; Regenstein and Regenstein, 1988; Regenstein et al., 2003). Any ingredients or materials that might be derived from animal sources are generally prohibited because of the difficulty of obtaining them from kosher animals. This includes many products that might be used in foods and dietary supplement, such as emulsifiers, stabilizers, and surfactants, particularly those materials that are derived from fat. Very careful rabbinical supervision is necessary to ensure that no animal-derived ingredients are included. Almost all such materials are available in a kosher form derived from plant oils (Regenstein and Chaudry, 2001; Regenstein et al., 2003). This is where kosher inspections and the extensive infrastructure globally to determine the status of products starting in many cases at the field level, can serve the interests of the Muslim consumer.

PROHIBITION OF MIXING OF MILK AND MEAT

"Thou shalt not seeth the kid in its mother's milk." This passage appears three times in the Torah and is therefore considered a very serious admonition. The meat side of the equation has been rabbinically extended to include poultry. The dairy side includes all milk derivatives.

To keep meat and milk separate in accordance with kosher law requires that processing and handling of all materials and products fall into one of three categories:

- Meat product
- Dairy product
- Pareve (parve, parev), or neutral product

The pareve category includes all products that are not classified as meat or dairy. All plant products are pareve along with eggs, fish, honey, and lac resin (shellac). These pareve foods can be used with either meat products or dairy products. However, if they are mixed with meat or dairy, they take on the identity of the product they are mixed with; for example, an egg in a cheese soufflé becomes dairy.

To ensure the complete separation of milk and meat, all equipment, utensils, pipes, steam, and so on, must be of the properly designated category. If plant materials (e.g., fruit juices) are run through a dairy plant, they will be considered a dairy product religiously. Some kosher supervision agencies permit such a product to be listed as dairy equipment (DE) rather than dairy. The DE tells the consumer that it does not contain any intentionally added dairy ingredients, but that it was made on dairy equipment (see discussion on allergy). If a product with no meat ingredients is made in a meat plant (e.g., a vegetarian vegetable soup), it may be marked meat equipment (ME). Although one may need to wash the dishes before and after use, the DE food can be eaten on meat dishes and the ME food on dairy dishes. Thus, the equipment used in kosher productions are carefully monitored and fairly comprehensive procedures are required to change the status of equipment. Again, this means that the equipment used for pareve or dairy productions, along with non-labeled food ingredients, are carefully checked.

A significant wait is normally required to use a product with dairy ingredients after one has eaten meat [i.e., from one to six hours depending on the customs (minhag) of the area the husband came from]. With the DE listing, the consumer can use the DE product immediately before or after a meat meal but not with a meat meal. Following dairy, the wait before eating meat is much less, usually from a rinse of the mouth with water to one hour. Certain dairy foods require the full wait of one to six hours; for example, when a hard cheese (defined as a cheese that has been aged for over six months or one that is particularly dry and hard, such as many of the Italian grating cheeses) is eaten, the wait is the same as that for meat to dairy. Thus, most companies producing cheese for the kosher market usually age their cheese for less than six months, although with proper package marking this is not a religious requirement. If one wants to make an ingredient or product truly pareve, the plant equipment must undergo a process of equipment kosherization (Regenstein and Chaudry, 2001; Regenstein et al., 2003).

KOSHER: SPECIAL FOODS

Rules governing grape products, yashon (old) and chodesh (new) flour, Jewish milk, and other foods are beyond the scope of this chapter.

Passover Requirements

The Passover holiday comes in spring and requires observant Jews to avoid eating the usual products made from five prohibited grains: wheat, rye, oats, barley, and spelt (Hebrew: chometz). Those observing kosher laws can eat only the specially supervised unleavened bread from wheat (Hebrew: matzos) that is prepared especially for the holiday. Once again, some matzos (schmura matzos) are made to a stricter standard with rabbinical inspection of the wheat beginning in the field. For other Passover matzo, supervision does not start until the wheat is about to be milled into flour. Matzo made from oats and spelt is now available for consumers with allergies including "schmura" matzos.

Special care is taken to ensure that the matzo does not have any time or opportunity to rise. In some cases, this literally means that products are made in cycles of less than 18 minutes. This is likely to be the case for handmade schmura matzo. In continuous large-scale operations, the equipment is constantly vibrating so that there is no opportunity for the dough to rise (Regenstein and Chaudry, 2001; Regenstein and Regenstein, 1988).

Equipment Kosherization

There are three ways to make equipment kosher or to change its status back to pareve from dairy or meat. (Rabbis generally frown on going from meat to dairy or vice-versa. Most conversions are from dairy to pareve or from treif to one of the categories of kosher.) There are a range of process procedures to be considered, depending on the equipment's prior production history. These are at least equal to modern sanitation requirements but also include covering the entire equipment with boiling water or using a blow torch to heat all relevant surfaces.

HALAL DIETARY LAWS

Halal dietary laws deal with the following five issues; all except one are in the animal kingdom:

- Prohibited animals
- Prohibition of blood
- Method of slaughtering and blessing
- Prohibition of carrion
- Prohibition of intoxicants

Islamic dietary laws are derived from the Quran, a revealed book; the hadith, the traditions of Prophet Muhammad (PBUH); and through extrapolation of and deduction from the Quran and the hadith by Muslim jurists.

The Quran states:

> Forbidden Unto you (for food) are carrion and blood and swine-flesh, and that which hath been dedicated unto any other than Allah, and the strangled, and the dead through

beating, and the dead through falling from a height, and that which hath been killed by (the goring of) horns, and the devoured of wild beasts, saving that which ye make lawful (by the death-stroke), and that which hath been immolated unto idols. And (forbidden is it) that ye swear by the divining arrows. This is an abomination. This day are those who disbelieve in despair of (ever harming) your religion; so fear them not, fear Me! This day have I perfected your religion for you and completed my favor unto you, and have chosen for you as religion AL-ISLAM. Whoso is forced by hunger, not by will, to sin: (for him) Lo! Allah is Forgiving, Merciful.

Chapter V, Verse 3

The Quran also states:

Ye who believe! Eat of the good things wherewith We have provided you, and render thanks to Allah, if it is (indeed) He whom you worship.

Chapter II, Verse 172

Eleven generally accepted principles pertaining to halal (permitted) and haram (prohibited) in Islam provide guidance to Muslims in their customary practices:

- The basic principle is that all things created by Allah are permitted, with a few exceptions that are prohibited. Those exceptions include pork, blood, meat of animals that died of causes other than proper slaughtering, food that has been dedicated or immolated to someone other than God, alcohol, and intoxicants.
- To make lawful and unlawful is the right of Allah alone. No human being, no matter how pious or powerful, may take it into his or her own hands to change things.
- Prohibiting what is permitted and permitting what is prohibited is similar to ascribing human partners to Allah. This is a sin of the highest degree that makes one fall out of the sphere of Islam.
- The basic reasons for the prohibition of things are impurity and harmfulness. A Muslim is not supposed to question exactly why or how something is unclean or harmful in what Allah has prohibited. There might be obvious reasons and there might be obscure reasons. Some of the reasons that have been proposed are indicated below although many of them may not be consistent with modern scientific thinking.
- Carrion and dead animals are unfit for human consumption because the decay process leads to the formation of chemicals harmful to humans (Awan, 1988).
- Blood that is drained from an animal contains harmful bacteria, potentially harmful products of metabolism, and toxins (Hussaini and Sakr, 1983).
- Swine serves as a vector for pathogenic worms to enter the human body. Infections by *Trichinella spiralis* and *Traeniasolium* are not uncommon (Awan, 1988). Intoxicants are considered harmful for the nervous system, affecting the senses and human judgment, leading to social and family problems, and might even lead to death (Al-Qaradawi, 1984; Awan, 1988).

- These reasons and other similar explanations may sound reasonable to a layperson but become more questionable under scientific scrutiny (Sakr, 1991). If the meat of dead animals were prohibited due to harmful chemicals in decaying meat, then dead fish should have been prohibited. If pork contains *Trichinae*, beef might contain *E. coli*. If pork fat is bad, so are trans fatty acids. The underlying principle, it seems, behind the prohibitions is not scientific reasons but the divine order "forbidden unto you are."
- What is permitted is sufficient and what is prohibited is then superfluous. Allah prohibited only things that are unnecessary or dispensable, while providing better alternatives. People can survive and live better without consuming unhealthful carrion, unhealthful pork, unhealthful blood, and the root of many vices—alcohol.
- Whatever is conducive to the prohibited is in itself prohibited. If something is prohibited, anything leading to it is also prohibited.
- Falsely representing unlawful as lawful is prohibited. It is unlawful to make flimsy excuses, to consume something that is prohibited, such as drinking alcohol for supposedly medical reasons.
- Good intentions do not make the unlawful acceptable. Whenever any permissible action of the believer is accompanied by a good intention, his action becomes an act of worship. In the case of haram, it remains haram, no matter how good the intention or how honorable the purpose. Islam does not endorse employing a haram means to achieve a praiseworthy end. Islam indeed insists that not only the goal be honorable, but also the means chosen to achieve it be lawful and proper. Islamic laws demand that right should be secured through just means only.
- Doubtful things should be avoided. There is a gray area between clearly lawful and clearly unlawful. This is the area of "what is doubtful." Islam considers it an act of piety for the Muslims to avoid doubtful things, for them to stay clear of unlawful. Prophet Muhammad (PBUH) said: "The halal is clear and the haram is clear. Between the two there are doubtful matters concerning which people do not know whether they are halal or haram. One who avoids them in order to safeguard his religion and his honor is safe, while if someone engages in a part of them, he may be doing something haram."
- Unlawful things are prohibited to everyone alike. Islamic laws are universally applicable to all races, creeds, and sexes. There is no favored treatment of a privileged class. Actually, in Islam, there are no privileged classes; hence, the question of preferential treatment does not arise. This principle applies not only among Muslims, but between Muslims and non-Muslims as well.
- Necessity dictates exceptions. The range of prohibited things in Islam is quite limited, but emphasis on observing these prohibitions is very strong. At the same time, Islam is not oblivious to the exigencies of life, to their magnitude, or to human weakness and capacity to face them. A Muslim is permitted, under the compulsion of necessity, to eat a prohibited food in quantities sufficient to remove the necessity and thereby survive (Chaudry, 1992; Regenstein and Chaudry, 2001; Riaz, 1999a).

PROHIBITED AND PERMITTED ANIMALS

Meats of pigs, boars, and swine are strictly prohibited, and so are meats of carnivorous animals such as lions, tigers, cheetahs, dogs, and cats, and birds of prey such as eagles, falcons, ospreys, kites, and vultures.

Meat of domesticated animals such as ruminants with split hoof, such as cattle, sheep, goat, and lamb, is allowed for food, and so are meats of camels and buffaloes. Also permitted are meats of birds that do not use their claws to hold down food, such as chickens, turkeys, ducks, geese, pigeons, doves, partridges, quails, sparrows, emus, and ostriches. Some of the animals and birds are permitted only under special circumstances or with certain conditions. The animals fed unclean or filthy feed such as feeds formulated with sewage (bio-solids) or protein from dead animals must be kept separate and placed on clean feed for 3 to 40 days (Awan, 1992).

Foods from the sea, namely fish and seafood, are the most controversial among various denominations of Muslims. Certain groups accept only fish with scales as halal, while others consider everything that lives in water all the time as halal. Consequently, prawns, lobsters, and clams are halal for most Muslims but may be detested (makrooh) by some and hence not consumed.

There is no clear status for insects established in Islam except that the locust is specifically mentioned as halal. Among the by-products from insects, use of honey was very highly recommended by Prophet Muhammad (PBUH). Other products such as royal jelly (from bees), bee's wax, shellac, and carmine are acceptable to be used without restrictions by most Muslims; however, some might consider shellac and carmine makrooh or offensive to their psyche.

Eggs and milk from permitted animals are also permitted for Muslim consumption. Milk from cows, goats, sheep, camels, and buffaloes is halal. Unlike kosher, there is no restriction on mixing meat and milk (Chaudry, 1992; Regenstein and Chaudry, 2001).

PROHIBITION OF BLOOD

According to Quranic verses, blood that pours forth from an animal when it is slaughtered is prohibited for consumption. It includes blood of permitted and non-permitted animals alike. There is general agreement among Muslim scholars that anything made from the blood of any animal, including fish, is unacceptable. Products such as blood sausage and ingredients such as blood albumin are either haram or questionable at best, and should be avoided in product formulations (Riaz, 1996).

Proper Slaughtering of Permitted Animals
There are special requirements for slaughtering animals:

- Animal must be of a halal species.
- Animal must be slaughtered by an adult and sane Muslim.
- The name of Allah must be pronounced at the time of slaughter.
- Slaughter must be done by cutting the throat in a manner that induces rapid and complete bleeding, resulting in the quickest death. The generally accepted method is to cut at least three of the four passages in the neck, that is, carotids, jugulars, trachea, and esophagus.

- The meat of animals thus slaughtered is called zabiha or dhabiha meat. Many Muslims do not want halal meat slaughtered in a facility that also slaughters pigs, even if that is done on a different day and the facility has been properly cleaned.

Islam places great emphasis on gentle and humane treatment of animals, especially before and during slaughter. Some of the conditions include giving the animal proper rest and water, avoiding conditions that create stress, not sharpening the knife in front of the animals, and using a very sharp knife to cut the neck. After the blood is allowed to drain completely from the animal and the animal has become lifeless, only then may dismemberment begin. Unlike kosher, post-slaughter religious inspection, deveining, and soaking and salting of the carcass is not required for halal. Hence, halal meat is treated no different than commercial meat. Animal-derived food ingredients such as emulsifiers, tallow, and enzymes must be made from animals slaughtered by a Muslim to be halal.

Hunting of permitted wild animals, such as deer and elk, and birds, such as doves, pheasants, and quails, is permitted for the purpose of eating, but not merely for deriving pleasure out of killing an animal. Hunting by any means by tools such as guns, arrows, spears, or trapping is permitted. Trained dogs or birds of prey may also be used for catching or retrieving the hunt as long as the hunting animal does not eat any of the prey. The name of Allah may be pronounced at the time of ejecting the hunting tool rather than the actual catching of the hunt. The hunted animal has to be bled by slitting the throat as soon as it is caught. If the blessing is made at the time of pulling the trigger or the shooting of an arrow and the hunted animal dies before the hunter reaches it, it would still be halal as long as slaughter is performed and some blood comes out. Fish and seafood may be hunted or caught by any reasonable means available as long as it is done humanely.

The requirements of proper slaughtering and bleeding are applicable to land animals and birds. Fish and other creatures that live in water need not be ritually slaughtered. Similarly, there is no special method of killing locusts.

The meat of a permitted animal that dies of natural causes or diseases, from being gored by other animals, by being strangled, by falling from a height, through beating, or by being killed by wild beasts is unlawful to be eaten unless one saves such animals by halal slaughtering before they actually become lifeless. Fish that dies of itself, if floating on water, or if lying out of water, is still halal as long as it does not show any signs of decay or deterioration.

An animal must not be slaughtered after dedication to someone other than Allah and immolated to anybody other than Allah under any circumstances. This is a major sin (Chaudry, 1992; Regenstein and Chaudry, 2001; Chaudry and Regenstein, 2000).

Prohibition of Alcohol and Intoxicants

Consumption of alcoholic drinks and other intoxicants is prohibited according to the Quran as follows:

> Ye who believe! Strong drink and games of chance and idols and divining arrows are only an infamy of Satan's handiwork. Leave it aside in order that ye may succeed.
>
> **Chapter V, Verse 90**

> Satan seeketh only to cast among you enmity and hatred by means of strong drink and games of chance, and to turn you from the remembrance of Allah and from (His) worship. Will ye then have done?
>
> **Chapter V, Verse 91**

The Arabic term used for alcohol in the Quran is khamr, which means that which has been fermented, and implies not only to alcoholic beverages such as wine, beer, whiskey, or brandy but also to all things that intoxicate or affect one's thought process. Although there is no allowance for added alcohol in any beverage such as soft drinks, the small amount of alcohol contributed from food ingredients might be considered an impurity and hence ignored. Synthetic or grain alcohol can be used in food processing for extraction, precipitation, dissolving, and other reasons as long as the amount of alcohol remaining in the final product is very low, generally below 0.1%. However, each importing country has its own guidelines that must be understood by exporters and strictly adhered to (Chaudry, 1992; Regenstein and Chaudry, 2001; Riaz, 1997). More details can be found in chapter 16 on alcohol.

Halal Cooking, Food Processing, and Sanitation

There are no restrictions about cooking in Islam, as long as the kitchen is free from haram foods and ingredients. There is no need to keep two sets of utensils, one for meat and the other for dairy, as in kosher.

In food companies, haram materials should be kept segregated from halal materials. The equipment used for non-halal products has to be thoroughly cleansed by using proper techniques of acids, bases, detergents, and hot water. As a general rule, kosher clean-up procedures are adequate for halal too. If the equipment is used for haram products, especially pork, it must be properly cleaned, sometimes by using an abrasive material, and then be blessed by a Muslim inspector by rinsing it with hot water seven times (Regenstein and Chaudry, 2001).

KOSHER AND HALAL

Gelatin

Important in many food products, gelatin is probably the most controversial of all modern kosher and halal ingredients. Gelatin can be derived from pork skin, beef bones, or beef skin (Cheng et al., 2012; Hermanto and Fatimah, 2013; Jaswir et al., 2009). In recent years, some gelatins from fish skins have also entered the market. As a food ingredient, fish gelatin has many similarities to beef and pork gelatin, such as a similar range of bloom strengths and viscosities (Schrieber and Gareis, 2007). However, depending on the species from which fish skins are obtained, its melting point can vary over a much wider range of melting points than beef or pork gelatin (Gómez-Estaca et al., 2009; Karim and Bhat, 2009). This offers some unique opportunities to the food industry, especially for ice cream, yogurt, dessert gels, confections, and imitation margarine. Fish gelatins can be produced kosher and halal with proper supervision, and be acceptable to almost all the mainstream religious supervision organizations.

Most of the currently available gelatin—even if called kosher—is not acceptable to the mainstream U.S. kosher supervision organizations and to the Muslim community. Many gelatins are, in fact, totally unacceptable to halal consumers because they might be pork based.

A recent development has been the manufacture of kosher gelatin from the hides of kosher-slaughtered cattle. It has been available in limited supply at great expense, and this gelatin has been accepted by the mainstream kosher supervisions and even some of the stricter kosher supervision agencies. The acceptability of this gelatin would presumably follow the acceptability of al-kitabe meat products.

The gelatin companies produce gelatins of different bloom strengths, and both soft and hard capsules of various sizes. This is an important new development that should be of interest to nutraceutical and drug markets. Similarly, at least two major manufacturers are currently producing certified halal gelatin from cattle bones of animals that have been slaughtered by Muslims. Halal-certified hard and soft gelatin capsules are available at competitive prices. Hard, two-piece and soft, one-piece capsules made with different vegetable materials are also available, most of which are certified as halal, kosher, and vegetarian (Al-Mazeedi et al., 2013; Blech 2008; Regenstein 2012; Regenstein et al., 2003).

One finds a wide range of attitudes toward gelatin among the lenient kosher supervision agencies. The most liberal view holds that gelatin, being made from bones and skin, is not being made from a food (flesh). Further, the process used to make the product goes through a stage where the product is so unfit that it is not edible by humans or dogs, and as such becomes a new entity. Rabbis holding this view might accept pork gelatin. Most water gelatin desserts with a generic K follow this ruling.

Other rabbis permit gelatin only from beef bones and hides, and not pork. Still other rabbis only accept "Indian dry bones" as a source of beef gelatin. These bones, found astray in India, are aged and become degreased over time and are considered "dry as wood" by rabbis. Kosher religious laws exist for permitting these materials. Again, none of these products is accepted by the mainstream kosher or halal supervisions, and therefore, they are not accepted by a significant part of the kosher and halal community (Regenstein and Chaudry, 2001; Riaz, 1999b).

BIOTECHNOLOGY

Rabbis and Islamic scholars currently accept products made by simple genetic engineering; for example, chymosin (rennin) was accepted by rabbis about a half a year before the Food and Drug Administration (FDA) accepted it. Production conditions in fermenters must still be kosher or halal; that is, the ingredients, the fermenter, and any subsequent processing must use kosher or halal equipment and ingredients of the appropriate status (Malaysian Standard 1500, 2009). A product produced in a dairy medium, for example, extracted from cow's milk, would be kosher dairy. Mainstream rabbis may approve porcine lipase made through biotechnology when it becomes available, if all the other conditions are kosher. The Muslim community is still considering the issue of products with a porcine gene; although a final ruling has not been established, the leaning seems to be toward rejecting such materials. Religious leaders of both communities have not yet determined the status of more complex genetic manipulations (Chaudry and Regenstein, 1994; Kurien, 2002; Regenstein and Chaudry, 2001; Riaz, 1999a). An interesting question that the Muslim community may need to address is the status of a porcine (pig) gene that was created in the laboratory, that is, knowing the sequence of a gene, it can be produced totally synthetically.

VEGETARIANISM

Vegetarianism encompasses a variety of options and choices, based on lifestyles, philosophies, and religions. The preferences vary from eating nothing but the parts of plants that be picked without destroying the plant to eating everything except flesh (red meat and poultry)

According to a nationwide survey in 2006, approximately 2.3% (4.9 million) of adults are following a vegetarian diet, and never eat fish, meat, or chicken; 1.4% of the U.S. adult population was vegan. According to a report of nationwide survey in 2005, 3% of children between the ages of 8 and 18 and adolescents were vegetarian, and close to 1% were vegan (Stahler, 2006, 2009). Types of vegetarians vary from lenient to the most strict include (The Vegetarian Society, 2002):

- *Pesco vegetarians*: Eat fish, eggs, and dairy products, but avoid poultry and meat products, i.e., products that are generally categorized as being slaughtered.
- *Lacto-ovo vegetarians*: Consume all types of vegetable products, eggs, and milk products, but avoid all forms of flesh, including meat, poultry, and fish. People who do not eat eggs but eat dairy products are called lacto-vegetarians, whereas ovo vegetarians consume eggs but not dairy products.
- *Vegans*: Do not eat anything of animal origin. A vegan therefore avoids all meats, poultry, and any other animal products and their derivatives, such as gelatin, eggs, milk, cheese, yogurt, and other dairy products; and fish, shellfish, crustaceans, and other marine animal products. Vegans also try to avoid honey, royal jelly, and cochineal and other insect-derived products. In addition, vegans do not knowingly consume hidden animal ingredients (Craig and Mangels, 2009; The Vegan Society, 2002).

Fruitarians follow a type of vegan diet of fruits, vegetables, seeds, and nuts that is minimally processed or cooked. Fruitarians believe that only plant foods that can be harvested without killing the plant should be eaten (Craig and Mangels, 2009; The Vegan Society, 2002).

Vegetarianism is not a particular religion. Believers of many religious denominations including Muslims, Jews, Seventh Day Adventist, Christians, Latter Day Saints (Mormons), Hindus, Buddhists, and Jains might practice vegetarianism to some extent.

Mainstream vegetarianism is usually defined as lacto-ovo; however, veganism is becoming more popular in the West. The comparison to be made here is limited to vegan and lacto-ovo vegetarianism rather than other types of vegetarianism.

FOOD STANDARDS FOR VEGETARIANS

ANIMAL PRODUCTS

For both vegans and lacto-ovo vegetarians, all products of animal flesh including food ingredients from meat, poultry, and seafood must be avoided. Moreover, for vegans, products and by-products from live animals, such as milk and eggs, and products from insects, such as honey and royal jelly, must also be avoided. Lacto-ovo vegetarians who consume egg and dairy products avoid incidental ingredients of animal origin, including enzymes such as rennet. However, most of the enzymes for cheese manufacture at least in the U.S. are from microbial or genetically modified microbes, which has found acceptance among this group (Lea et al., 2006).

Ingredients not acceptable as vegetarian include vitamin D from sheep wool, gelatin used in juice processing for clarification, and anchovies in Worcestershire sauce (The Vegan Society, 2002). However, the source of such ingredients may not be carefully traced in products that claim to be vegan unless they are subject to certification by a vegan certifying body that indicates that it checks such issues (Lea et al., 2006).

KITCHEN AND HYGIENE STANDARDS

Dishes and utensils used for preparing and serving vegetarian products must be separate from non-vegetarian dishes or at a minimum must be thoroughly washed. It is recommended that cross-contamination from non-vegetarian food be avoided. But how well this is enforced in practice must be ascertained by any Muslim considering vegetarian or vegan foods (Wilson et al., 2011).

COMPARISON OF KOSHER, HALAL, AND VEGETARIAN

The view of kosher present here are those of the Orthodox Jewish community and not those of the Conservative or Reform movements. Table 25.1 gives a fairly complete list of kosher, halal, and vegetarianism. There are both similarities and differences, especially between kosher and halal. In both religions, pork and pork products derived from pigs and swine are prohibited. Also, carnivorous animals and birds are not allowed in either religion. Because it is understood that animal products are against the philosophy of vegetarianism, there is no need to mention vegetarianism specifically in this comparison.

Of the permitted animals, ruminants and poultry have to be killed by a Jew to make them kosher and killed by a Muslim to make them halal. In kosher, all animals have to be hand-slaughtered without stunning. However, an animal may be stunned after slaying to facilitate bleeding in some cases, but it is not acceptable as glatt kosher by the normative mainstream, that is, only the liberal Orthodox and Conservative communities accept post-slaughter stunning. The Reform Movement generally does not consider kosher issues. For halal, interventions using the non-penetrating captive bold or by reversible electrical stunning is accepted by some Muslims as long as the animal is alive at the time of slaughter. For halal, many halal certifiers accept mechanical slaughter of poultry if supervised by a Muslim with the

TABLE 25.1
Comparative Summary of Kosher, Halal, and Vegetarian Guidelines

	Kosher	Halal	Vegetarian
Pork/pig/swine and carnivorous animals	Prohibited	Prohibited	Not applicable
Ruminants and poultry	Slaughtered by a trained Jew	Slaughtered by an adult Muslim	Not applicable
Blessing/invocation	Blessing before entering slaughtering area. Not on each animal.	Blessing on each animal while slaughtering	Not applicable
Slaughtering by hand	Mandatory	Preferred	Not applicable
Mechanical slaughtering	Not allowed	May be done for poultry under supervision	Not applicable
Stunning before slaying	Sometimes permitted	Permitted	Not applicable
Stunning after slaying	Permitted		
Other restrictions about meat	Only front quarters used. Soaking and salting required	Whole carcass used. No salting required	Not applicable
Blood of any animal	Prohibited	Prohibited	Not applicable
Fish	With scales only	Most accept all fish, some only with fish scales	Not applicable
Seafood	Not permitted	Varying degree of acceptance	Not accepted
Microbial enzymes	Accepted	Accepted	Accepted
Biotech-derived enzymes	Accepted	Accepted	Accepted
Animal enzymes	Kosher slaughtered only	Accepted sometimes	Sometimes accepted
Porcine enzymes	Maybe accepted	Not accepted	Sometimes accepted
Cattle gelatin	From Kosher-slaughtered animals	From halal-slaughtered animals	Not applicable
Fish gelatin	Kosher fish only	Any fish	Not applicable
Pork gelatin	Allowed by liberal orthodox rabbis	Not permitted	Not applicable
Dairy products Whey	Made with Kosher enzymes	Made with halal enzymes	No restrictions
Addition of cheese culture	Must be added by a Jew	No restriction	No restrictions
Alcohol	Permitted	Not permitted	Accepted
Combining meat and dairy	Not permitted	Not applicable	Not applicable
Insects and by-products	Grasshoppers accepted. By-products not accepted	Locusts and by-products accepted	By-products acceptable

(Continued)

TABLE 25.1 (CONTINUED)
Comparative Summary of Kosher, Halal, and Vegetarian Guidelines

	Kosher*	Halal**	Vegetarian***
Plant materials	All permitted	Intoxicants and alcohol not permitted	All acceptable
Sanitation of equipment	Cleaning Idle period required Kosherization/ritual cleaning	Thorough cleaning No idle period required	Thorough cleaning
Special occasion	Additional restrictions during Passover	Same rules apply year-round	Same rules apply year-round

back-up person, namely, the person slaughtering any birds that were missed by the mechanical equipment, being a Muslim.

In kosher, a blessing is made before the start of the slaughter operation. However, for halal, an invocation of the name of God must be made at the time of slaughtering each animal. Only front quarters of ruminants are used for kosher meat. Meat is also soaked in water, and covered with salt for an hour to remove the blood. For halal, the entire animal is permitted to be consumed. There is no requirement for soaking or salting of meat for halal. Blood and blood by-products are not permitted under kosher as well as halal rules.

Among fish and seafood, only fish with fins and removable scales are kosher, whereas all fish and seafood that spend their entire life in the water are halal. However, some Muslim denominations do not accept fish without scales or seafood, or both, as halal.

Enzymes derived from microbial or biotech (GMO) sources are acceptable as kosher, halal, and vegetarian. Enzymes extracted from kosher-killed animals are accepted as kosher and enzymes extracted from halal-killed animals are accepted as halal. Some liberal Orthodox rabbis might accept enzymes from non-kosher-killed animals as well as porcine enzymes. Enzymes from non-halal-killed animals might be accepted by some groups and countries, but not all. Porcine enzymes are generally not accepted by Muslims (Riaz, 1999a).

Gelatin prepared from kosher-killed animals is accepted by Orthodox rabbis as kosher and gelatin from halal-slaughtered animals is acceptable as halal. Some countries also accept regular bovine gelatin under certain conditions. Gelatin from fish with removable scales are kosher, but gelatin from any fish is halal.

For dairy ingredients and cheese, the reader is referred to the discussion on enzymes (Chapter 11). Additionally, for cheese to be kosher, a Jew must add the cultures to the milk for mainstream kosher. There is no such restriction for halal, in which any person may add the cultures to the milk.

Alcohol, especially alcoholic drinks, is not allowed for Muslims, but Jews consider alcohol as kosher; however, some restrictions exist about the source of alcoholic drinks in kosher. All alcoholic products are vegetarian. It is also important for kosher not to combine meat and dairy products, but no such restriction exists for halal.

The equipment for halal and vegetarian production must be thoroughly cleaned and can be used immediately after cleaning. For kosher, after thorough cleaning, the equipment might have to be left idle for a period of up to 24 hours as explained earlier in this chapter.

Insects are not kosher (except grasshoppers and locusts), or not vegetarian. Insect by-products such as carmine and cochineal are not considered kosher by the majority of Orthodox rabbis, but some rabbis permit them. Some insect by-products are considered halal.

All plant materials are kosher and vegetarian; however, some restrictions exist in kosher regarding insect infestations and sources of certain alcoholic products, as explained earlier in this chapter. Plant materials are halal as long as they do not contain significant amounts of alcohol or intoxicants.

The rules for halal and vegetarianism apply year-round; however, for kosher, during Passover, additional rules might apply beyond everyday kosher. Table 25.1 show a comparative summary of kosher, halal, and vegetarian guidelines.

REFERENCES

Al-Mazeedi, H. M., Regenstein, J. M., and Riaz, M. N. (2013). The issue of undeclared ingredients in halal and kosher food production: A focus on processing aids. *Comprehensive Reviews in Food Science and Food Safety*, 12(2), 228–233.

Al-Qaradawi, Y. (1984). *The Lawful and Prohibited in Islam*. Beirut, Lebanon, The Holy Quran Publishing House.

Awan, J. A. (1988). Islamic food laws-I: Philosophy of the prohibition of unlawful foods. *Science and Technology in the Islamic World*, 6(3), 151–165.

Awan, J. A. (1992). Islamic codex Alimentarius. *The World of Islam: Its Science and Technology*, 10(1), 7–18.

Blech Z. (2008). The story of gelatin. In: *Kosher Food Production*, 2nd edn. Hoboken, NJ: Wiley Blackwell, 317–322.

Chaudry, M. M. (1992). Islamic food laws: Philosophical basis and practical implications. *Food Technology (USA)*, 46(10), 92–93, 104.

Chaudry, M. M. and Regenstein, J. M. (1994). Implications of biotechnology and genetic engineering for kosher and halal foods. *Trends in Food Science and Technology*, 5(5), 165–168.

Chaudry, M. M. and Regenstein, J. M. (2000). Muslim dietary laws: Food Processing marketing. Encyclopedia *of Food Sciences*, Ed. Francis, F. J. 2nd edn. New York, NY: Wiley. 1682–1684.

Cheng, X. L., Wei, F., Xiao, X. Y., Zhao, Y. Y., Shi, Y., Liu, W., Zhang, P., Ma, S. C., Tian, S. S., and Lin, R. C. (2012). Identification of five gelatins by ultra performance liquid chromatography/time-of-flight mass spectrometry (UPLC/Q-TOF-MS) using principal component analysis. *Journal of Pharmaceutical and Biomedical Analysis*, 62, 191–195.

Craig, W. J. and Mangels, A. R. (2009). Position of the American Dietetic Association: vegetarian diets. *Journal of the American Dietetic Association*, 109(7), 1266–1282.

Gómez-Estaca, J., Montero, P., Fernández-Martín, F., and Gómez-Guillén, M. C. (2009). Physico-chemical and film-forming properties of bovine-hide and tuna-skin gelatin: A comparative study. *Journal of Food Engineering*, 90(4), 480–486.

Hermanto, S. and Fatimah, W. (2013). Differentiation of bovine and porcine gelatin based on spectroscopic and electrophoretic analysis. *Journal of Food and Pharmaceutical Sciences*, 1(3), 68–73.

Hussaini, M. M. and Sakr, A. H. (1983). *Islamic Dietary Laws and Practices*. Bedford Park, IL: Islamic Food and Nutrition Council of America.

Jackson, M. A. (2000). Getting religion: For your products, that is. *Food Technology*, 54(7), 60–66.

Jaswir, I., Mirghani, M. E. S., Hassan, T., and Yaakob, C. M. (2009). Extraction and characterization of gelatin from different marine fish species in Malaysia. *International Food Research Journal*, 16, 381–389.

Karim, A. A. and Bhat, R. (2009). Fish gelatin: Properties, challenges, and prospects as an alternative to mammalian gelatins. *Food Hydrocolloids*, 23(3), 563–576.

Kurien, D. (2002). Malaysia: Studying GM foods' acceptability of Islam. Kuala Lumpur, Malaysia: August 9, Dow Jones Online News via News Edge Corporation. http://www.merid.org/en/Content/News_Services/Food_Security_and_AgBiotech_News/Articles/2002/08/14/Malaysia_Studying_GM_Foods_Acceptability_Under_Islam.aspx. Accessed on June 26, 2018.

Lea, E. J., Crawford, D., and Worsley, A. (2006). Public views of the benefits and barriers to the consumption of a plant-based diet. *European Journal of Clinical Nutrition*, 60(7), 828.

MS1500, M.S. (2009). *Halal Food Production, Preparation, Handling and Storage, General Guideline*. Cyberjaya, Malaysia: Department of Standards Malaysia, 1–13.

Pickthall, M. M. (1994). *Arabic Text and English Rendering of The Glorious Quran*. Chicago, IL: Kazi Publications, Library of Islam.

Qureshi, S. S., Jamal, M., Qureshi, M. S., Rauf, M., Syed, B. H., Zulfiqar, M., and Chand, N. (2012). A review of Halal food with special reference to meat and its trade potential. *Journal of Animal and Plant Sciences*, 22(2 Suppl), 79–83.

Regenstein, J. M. (2012). The politics of religious slaughter-how science can be misused. In *65th Annual Reciprocal Meat Conference at North Dakota State University in Fargo, ND*.

Regenstein, J. M. and Chaudry, M. (2001). A brief introduction to some of the practical aspects of kosher and halal laws for the poultry industry. In *Poultry Meat Processing*. Ed. Owens, C. M. 2nd edn. Boca Raton, FL: Taylor and Francis Group, 409–426.

Regenstein, J. M., Chaudry, M. M., and Regenstein, C. E. (2003). The kosher and halal food laws. *Comprehensive Reviews in Food Science and Food Safety*, 2(3), 111–127.

Regenstein, J. M. and Regenstein, C. E. (1979). An introduction to the kosher dietary laws for food scientists and food processors. *Food Technology (USA)*, 33(1), 89–99.

Regenstein, J. M. and Regenstein, C. E. (1988). The kosher dietary laws and their implementation in the food-industry. *Food Technology*, 42(6), 86.

Regenstein, J. M. and Regenstein, C. E. (2000). Kosher foods and food processing. *Encyclopedia of Food Sciences*, Ed. Francis, F. J. 2nd Edn. New York, NY: Wiley, 1449–1453.

Riaz, M. N. (1996). Hailing halal. *Prepared Foods*, 165(12), 53–54.

Riaz, M. N. (1997). Alcohol: The myths and realities, *A Handbook of Halaal and Haraam Products*, Vol. 2, Uddin, Z., Ed. Richmond Hill, NY: Publication Center for American Muslim Research and Information, 16–30.

Riaz, M. N. (1999a). A halal food processing and marketing, In *10th World Congress of Food Science and Technology*, Book of Abstracts. Sydney: Australian Institute of Food Science and Technology, 44.

Riaz, M. (1999b). Examining the halal market. *Prepared Foods*, 168(10), 81–83.

Sakr, A. H. (1991). *Pork: Possible Reasons for its Prohibition*. Lombard, IL: Foundation for Islamic Knowledge.

Schrieber, R. and Gareis, H. (2007). *Gelatine Handbook: Theory and Industrial Practice*. Hoboken, NJ: Wiley.

Stahler, C. (2006). How many adults are vegetarian? The Vegetarian Resource Group asked in a 2006 national poll. *Vegetarian Journal*, 25(4), 42–4.

Stahler, C. (2009). How many adults are vegetarian? The Vegetarian Resource Group Website. https://www.vrg.org/nutshell/Polls/2016_adults_veg.htm. Accessed June 26, 2018.
The Vegan Society. (2002). Donald Watson House, UK. Available online: www.vegansociety.com/.
The Vegetarian Society. (2002). Park Dale, UK. Available online: www.vegsoc.org/info/definitions.html.
Wilson, S. (2011). The dietitian's guide to vegetarian diets: Issues and applications. *Journal of Nutrition Education and Behavior*, 43(3), 207–e3.

26 Globalization of Halal Certification

From an Industrial Perspective

Jes Knudsen, Mian N. Riaz and Munir M. Chaudry

CONTENTS

The Early Days of Global Halal Certification 283
Halal is Not Just Halal 284
Multiple Certifications 285
Reasons for Lack of Mutual Recognition 285
Forums for Discussion 286
A Global Standard 286
Multiple Standards 287
Accreditation of Certification Bodies 287
Conclusion 288
References 288

THE EARLY DAYS OF GLOBAL HALAL CERTIFICATION

Halal certification was not very common outside of Southeast Asia until the turn of the millennium. Manufacturers of food ingredients could often handle requests about the halal compliance of a product by issuing statements regarding the absence of ethanol and no ingredients of animal origin in the product. Furthermore, products from these companies were often kosher certified and it was a common misunderstanding that kosher certification would also ensure compliance with halal requirements. Another common misunderstanding was that halal acceptability was only a question about which animal species were used and that the slaughtering methods were appropriate (Omar and Jafaar, 2011). Thus, the need to determine if a product was halal was only relevant for actual meat products and not for other products and ingredients without apparent (or actual) contents of animal origin. For these reasons, halal certification outside Southeast Asia was then mainly used by companies that exported poultry and meat products to Muslim countries.

However, the need for halal certification of other types of food products and food ingredients grew dramatically around 2001. This was to a large extent triggered by an incident where it was discovered that a multinational company's plant in Indonesia had introduced a porcine-derived ingredient in the production of sodium glutamate, a widely used condiment, especially in oriental cooking. This did lead to an increased awareness of food products' halal status among many Muslim consumers—and did cause major problems for this company's reputation and sales in Muslim countries (Mohammed et al., 2008; Rezai et al., 2012). Consequently, many manufacturers of food products and food ingredients became concerned. They began to investigate obtaining a credible halal certification. As part of this certification process, many of their suppliers were asked to provide halal certification for their ingredients (Abdul Latiff et al., 2013; Mohamed et al., 2013).

Most large food manufacturing companies had for many years been used to working with different certification schemes such as ISO 9000, ISO 14000, and kosher. Even though these standards are open to some interpretation, they were generally perceived as being relatively well defined. Even though the standards might change over time, such changes would generally only be introduced with prior notice that would enable certified companies to adjust in a reasonable amount of time. For each of these certifications, it was often possible for multinational companies to select one certification body that could audit all subsidiaries and issue certificates that were generally acceptable throughout the world (Latif et al., 2014).

Based on the experience with these other types of certification, it was thought that halal certification could be handled similarly. Companies that only had plants within a limited geographical area would generally choose a local halal certification body, whereas multinational companies might choose a certification body that could offer global auditing and certification (Fischer, 2015).

HALAL IS NOT JUST HALAL

In the past, halal certifications could not always source ingredients that had halal certification. Therefore, they often had to accept such ingredients based on other documentation of halal acceptability in the form of statements, ingredient lists, flow charts, and so on. This was inefficient for the ingredient supplier and it was clearly easier to have a reliable halal certification. So the number of halal certified products and halal certification bodies grew significantly with the hope that this would make the halal certification process easier for all parties (Anir et al., 2008; Bonne and Verbeke, 2008; Ziegler, 2007).

Unfortunately, this was not to be the case. Manufacturers of halal certified products often experienced the following problems, some of which companies had experienced with kosher also (Halim and Salleh, 2012; Shafie and Othman, 2006; Van Waarden and Van Dalen, 2011).

- Ingredients could not be approved because the halal certificate was not accepted by the company's own halal certification body. The certifier might then still require statements, ingredient lists, and so on. This required significant resources from both the supplier and the customer—and often suppliers were not willing to provide this detailed information.
- In some cases, even the documentation would not be sufficient and the certification body would require an actual audit of the supplier. This also required that the supplier was willing to be audited by yet another halal certification body.

- Ultimately, a company might be required to find an alternative supplier with acceptable certification—provided this existed. However, changing suppliers is not always possible due to contractual obligations—and may have a negative impact on, for example, quality, sustainability, pricing, or security of supply.
  - Products could not be exported to certain countries (mainly Indonesia, Malaysia, and Singapore) because the halal certificate was not accepted by the national halal councils (MUI, JAKIM, and MUIS, respectively). Even if the products were certified by an organization that was recognized by MUI, JAKIM, or MUIS, this recognition could be discontinued overnight. When this happens, there is apparently no formal notification system in place, so a company will most likely discover this when their products fail to clear customs.

MULTIPLE CERTIFICATIONS

Many companies tried to reduce the problem by obtaining additional halal certifications (Fauzi and Mas'ud, 2009). Unfortunately, it is usually not easy to determine which certificates are accepted by other halal certification bodies and national councils, although MUI and JAKIM do have a "positive list" of recognized certifiers. Even when one certifier is accepted by another organization, this may change overnight—usually without any notification.

Furthermore, it is not a simple task to obtain an additional certification. When a company needs certification from several certifiers, there will most likely be situations where an ingredient is not accepted by all their certifiers. This makes it harder to educate staff as these distinctions confuse non-Muslim employees and become much more difficult to learn (Lam and Alhashmi, 2008; Van der Spiegel et al., 2012).

REASONS FOR LACK OF MUTUAL RECOGNITION

There are several reasons for this lack of mutual recognition among halal organizations:

- *Different definitions*: Organizations may have different definitions about the details of what is halal and what is not halal (Zailani, 2010).
- *Different system requirements*: Organizations may have different requirements for managing the system for certification, for example, each agencies Halal Assurance System (HAS) (Noordin et al., 2009).
- *Requirements for proximity*: Some organizations will only accept certificates from organizations from the same country or the same region as the manufacturer, and will, for example, not accept an American certification of products from a European plant (Shafie and Othman, 2006).
- *Desire for local control*: Some organizations want to set the agenda for halal certification in their own region and are therefore unwilling to let other organizations get too influential (Noordin et al., 2009).
- *Competition*: Halal certification is also a business with competition about market shares. Even though—or exactly because—this is a growing market,

some organizations may therefore reject others' certificates in the hope that this can keep others out of the market and generate additional business for themselves (Shafie and Othman, 2006).
- *Mistrust*: Some organizations are unwilling to accept others' certificates because they are concerned about, for example, technical competencies, or the thoroughness or integrity of other organizations (Noordin et al., 2009; Shafie and Othman, 2006).
- *Personal disputes*: Some of the disagreements and conflicts above can in some cases lead to personal disputes to an extent where it is not even possible for two organizations to meet and discuss how to remedy the situation (Shafie and Othman, 2006). [Note that this is NOT unique to halal, it exists in kosher, organic, and other general certification schemes.]

In the specific situation where one organization will not accept another's certification, it is usually due to a combination of several of the problems listed previously. However, it is often not possible for the halal certified food manufacturing companies to evaluate what the real reasons are—and close to impossible for the companies to do anything about it (Prabowo et al., 2015).

FORUMS FOR DISCUSSION

The halal organizations themselves are very well aware of the problems of lack of mutual recognition and several attempts to improve the situation have been made. Two structured forums for discussing the problems exist: the WHFC (World Halal Food Council) and the WHC (World Halal Council). These councils both have common standards and mutual acceptance as part of their vision, but even though they have existed for many years, it is difficult for the industry to see any progress and positive outcome of the discussions. WHC and WHFC have not been able to develop an official reciprocity either within their organizations or between them. Some halal organizations have arranged halal conferences with participation from certification bodies, national councils, and industry. At these conferences, halal certification bodies and other Muslim organizations will often focus on the huge market for halal certified products and the benefits of halal certification—whereas representatives from the industry will with steadily increasing intensity highlight the desperate need for mutual acceptance. So far with little noticeable success (Prabowo et al., 2015; Wan Hassan, 2007).

A GLOBAL STANDARD

As the problems with lack of mutual recognition have worsened over the last decade, most manufacturers would like to see a single global standard, but this is probably impossible because Muslims do not have a single standard. Many of the "political/business" issues discussed previously may be resolved, but some degree of lack of acceptability will always remain. Although some look to an ISO standard, that is, a standard that uses the same framework as other global standards that the industry is used to complying with, some Muslims do not want halal standards to be controlled by a non-Muslim organization (Shafie and Othman, 2006). Whether a Muslim organization such as WHC or WHFC can agree to a single standard, with clear internal differences that would have to be highlighted remains to be seen. A major example of such a fundamental disagreement is that of whether a pre-slaughter intervention is permitted. This will not be resolved but may possibly be dealt with by establishing a label scheme to distinguish between the two options (Nakyinsige et al., 2012).

Most halal organizations may agree about 95% of the requirements, but it will probably be extremely difficult—if not impossible—to reach agreement about the last 5%. There is a risk that such standardization efforts would simply require the strictest interpretation, which could cause compliance problems. On the other hand, some companies may prefer that reliability of standards but that is no guarantee of mutual recognition for the reasons mentioned earlier. It might be noted that the kosher certification industry continues to thrive despite having both a lack of standardization and some lack of mutual recognition (Prabowo et al., 2015).

MULTIPLE STANDARDS

One possible compromise solution could be to accept that several standards do exist and then establish an accreditation setup where halal certification bodies could be accredited to audit and certify based on the appropriate standards of the targeted user. This does mean that a certifying agency might be in a position where it needs to audit to a standard it does not itself accept. They will have to then decide if such a system might compromise their certification.

Another compromise might see some organizations agree on consolidating standards so that only a limited number of standards exist (Adams, 2011; Iberahim et al., 2012).

ACCREDITATION OF CERTIFICATION BODIES

Another aspect of certification that might be usefully addressed is to use an accreditation system, presumably at the national level, to authenticate that a certifying agency is able and is actually doing what it claims, that is, to focus on procedural issues (Zailani, 2010). Such a process would also permit a certifying body to audit for standards elsewhere by establishing that it is properly following the other certifying body's protocol. Some of the issues that could be subject to accreditation include: audit frequency; audit techniques; qualification of auditors; number of auditors relative to the number of companies being supervised; and audit reporting either to their own standard or to the outside standard (Iberahim et al., 2012).

Such an accreditation system should be impartial and only evaluate whether a given body meets the defined requirements. The accreditation body would therefore also have to be respected nationally (and internationally) and be free of any conflict-of-interests that would impact its credibility. Ideally, this would be a nongovernmental agency considered representative of all Muslims within the country. The accreditation system would gain additional credibility if it were affiliated with the appropriate secular agency within the country that is involved in accreditation across many other platforms. Such a system might provide companies with both greater certainty and a flexibility in choosing an appropriate certifying body that they currently might not have (Adams, 2011). The halal status has increasingly been considered a certification standard for quality and is under greater scrutiny than ever before (SGS, 2017).

CONCLUSION

The present unpredictable, chaotic situation where halal certificates are often not accepted by other certifying organizations is unacceptable to industry. Although the ideal solution would be one global halal standard, this is not realistic. However, the standards that exist should be clearly defined and a system for accreditation of certification bodies to audit and certify based on these standards should be established.

REFERENCES

Abdul Latiff, Z. A., Mohamed, Z. A., Rezai, G., and Kamaruzzaman, N. H. (2013). The impact of food labeling on purchasing behavior among non-Muslim consumers in Klang Valley. *Australian Journal of Basic and Applied Sciences*, 7(1), 124–128.

Adams, I. A. (2011). Globalization: Explaining the dynamics and challenges of the halal food surge. *Intellectual Discourse*, 19(1), 123.

Anir, N. A., Nizam, M. N. M. H., and Masliyana, A. (2008). The users' perceptions and opportunities in Malaysia in introducing RFID system for Halal food tracking. *WSEAS Transactions on Information Science and Applications*, 5(5), 843–852.

Bonne, K. and Verbeke, W. (2008). Religious values informing halal meat production and the control and delivery of halal credence quality. *Agriculture and Human Values*, 25(1), 35–47.

Fauzi, A. M. and Mas'ud Z. A. (2009). "Instrumentation techniques for potential application in halal products authentication." In *3rd IMT-GT International Symposium on Halal Science and Management*, Kuala Lumpar, Malaysia, December, 21–22.

Fischer, J. (2015). Islam, standards, and technoscience: In global Halal zones, Ed. Fischer, J. New York, NY: Routledge, Taylor & Francis Group.

Halim, M. A. A. and Salleh, M. M. M. (2012). The possibility of uniformity on halal standards in organization of Islamic countries (OIC) country. *World Applied Sciences Journal*, 17(17), 6–10.

Hassan, W. W. (2007). Globalising halal standards: Issues and challenges. *The Halal Journal*, 2007, 38–40.

Iberahim, H., Kamaruddin, R., and Shabudin, A. (2012). Halal development system: The institutional framework, issues and challenges for halal logistics. In *2012 IEEE Symposium on Business, Engineering and Industrial Applications (ISBEIA)*, Bandung, Indonesia, 760–765.

Lam, Y. and Alhashmi, S. M. (2008). Simulation of halal food supply chain with certification system: A multi-agent system approach. In *Pacific Rim International Conference on Multi-Agents,* Berlin, Heidelberg: Springer, 259–266.

LLatif, I. A., Mohamed, Z., Sharifuddin, J., Abdullah, A. M., and Ismail, M. M. (2014). A comparative analysis of global halal certification requirements. *Journal of Food Products Marketing*, 20(sup1), 85–101.

Mohamed, Z., Rezai, G., Shamsudin, M. N., and Eddie Chiew, F. C. (2008). Halal logo and consumers' confidence: What are the important factors. *Economic and Technology Management Review*, 3(1), 37–45.

Mohamed, Z., Shamsudin, M. N., and Rezai, G. (2013). The effect of possessing information about halal logo on consumer confidence in Malaysia. *Journal of International Food and Agribusiness Marketing*, 25(sup1), 73–86.

Nakyinsige, K., Man, Y. B. C., and Sazili, A. Q. (2012). Halal authenticity issues in meat and meat products. *Meat Science*, 91(3), 207–214.

Noordin, N., Noor, N. L. M., Hashim, M., and Samicho, Z. (2009). Value chain of Halal certification system: A case of the Malaysia Halal industry. In *European and Mediterranean Conference on Information Systems*, Izmir, Turkey, No. 2008, 1–14.

Omar, E. N. and Jaafar, H. S. (2011). Halal supply chain in the food industry-A conceptual model. In *2011 IEEE Symposium on Business, Engineering and Industrial Applications (ISBEIA)*, Langkawi, Malaysia, 384–389.

Prabowo, S., Abd Rahman, A., Ab Rahman, S., and Samah, A. A. (2015). Revealing factors hindering halal certification in East Kalimantan Indonesia. *Journal of Islamic Marketing*, 6(2), 268–291.

Rezai, G., Mohamed, Z., and Shamsudin, M. N. (2012). Assessment of consumers' confidence on halal labelled manufactured food in Malaysia. *Pertanika Journal of Social Science and Humanity*, 20(1), 33–42.

SGS. (2017). Global halal certification: Key Trends, challenges and opportunities. Available online: www.sgs.com/en/events/2017/04/global-halal-certification-key-trends-challenges-and-opportunities.

Shafie, S. and Othman, M. N. (2006). Halal Certification: An international marketing issues and challenges. In *Proceeding at the International IFSAM VIIIth World Congress*, Penang, Malaysia, 28–30.

Van der Spiegel, M., Van Der Fels-Klerx, H. J., Sterrenburg, P., Van Ruth, S. M., Scholtens-Toma, I. M. J., and Kok, E. J. (2012). Halal assurance in food supply chains: Verification of halal certificates using audits and laboratory analysis. *Trends in Food Science and Technology*, 27(2), 109–119.

Van Waarden, F. and Van Dalen, R. (2011). Hallmarking halal. The market for halal certificates: Competitive private regulation. *Jerusalem Papers in Regulation and Governance Working Paper*, (33), 4–14.

ZAILANI, S. H. B. D. M. (2010). Halal traceability and halal tracking systems in strengthening halal food supply chain for food industry in Malaysia. *Journal of Food Technology*, 8(3), 74–81.

Ziegler, P. (2007). Germany still a developing country? The halal label booms in the world market. *Fleischwirtschaft*, 87(9), 29–32.

27 Testing Non-Halal Materials

Winai Dahlan, Mian N. Riaz, and Munir M. Chaudry

CONTENTS

The Halal Forensic Science Laboratory 291
Testing of Gelatin 292
    Introduction 292
Presence of Alcohol 292
    Introduction 292
Fatty Acid Composition 293
    Introduction 293
DNA for Species Identification 294
    Detection of Porcine DNA in Food Products 294
References 296
Gelatin 297
Alcohol 297
Fatty Acids 297
    DNA 297

THE HALAL FORENSIC SCIENCE LABORATORY

Nowadays, raw materials used for producing halal food could be contaminated with najis (filth), mashbooh, or haram substances at any point in the food chain, either intentionally or accidentally. When small amounts of haram substances are in contact with food products, these cannot be detected with the naked eyes, or by smell or taste (Syahariza et al., 2005). This is where laboratory testing can strengthen halal compliance. Laboratory analysis will also support the preparation of a more powerful raw materials database. Of the 348 E-numbers, 92% are raw materials that are widely used by the halal food industry. Only 59%, however, are considered halal by all authorities. The remaining 33% of E-numbers (or 106 compounds) either are mashbooh or can be obtained from both halal and haram sources.

    The testing procedures will depend on what are the concerns, for example, porcine DNA, source of proteins, source of the lipids, ethyl alcohol, low molecular weight compounds, or hormones (Cordella et al., 2002). Four non-halal materials, that is, animal gelatin, fatty acids, ethyl alcohol, and porcine DNA will be described in some detail.

TESTING OF GELATIN

INTRODUCTION

Gelatin, the partially hydrolyzed collagen tissue of various animal parts, is a multifunctional ingredient widely used in many food products. The crucial point for halal scientific laboratory analysis is that gelatin can be derived from halal or haram sources. There are two main types of gelatin. Type A gelatin is made exclusively from pork skins and is haram. Type B gelatin is made either from cattle and calf skins or demineralized cattle bones and is halal unless the animal slaughtering process did not comply with Islamic regulation. Fish skins are an alternative source of gelatin, which is halal for all Muslims if the fish are in category 1 (Chapter 10). Gelatin is deficient in tryptophan and has a small amount of methionine. It also has the unique amino acid hydroxyproline found almost exclusively in gelatin/collagen.

There are two general approaches used to quantitate gelatin in food products (Shabani et al., 2015). First, the picric acid test is used to detect gelatin in milk and can also be used with sour, fermented, cultured, or very old samples of milk, cream, or buttermilk. Precipitates produced using picric acid in the absence of gelatin flocculate, separate readily and do not adhere to the walls of a container and are easily removed by rinsing with water. When gelatin is present, the precipitate will remain in suspension long after the flocculent has settled, but on standing overnight, the characteristic sticky deposits will be found adhering to the bottom and sides of the test vessel (Hermanto and Fatimah, 2013).

Second, the Woessner hydroxylproline assay can be used both qualitatively and quantitatively. The hydrolysis of products with acid, oxidizing agent, and colorimetric reagents gives a red purple color in the solution when gelatin is the present. However, neither of these tests can identify which animal the gelatin was obtained from. Thus, future work will need to be done to attempt to do species identification (AOAC, 1996a). Given that gelatin is initially a hydrolysate and that many laboratories are studying the use of materials that are further hydrolyzed, this will remain a difficult problem. However, the presence of gelatin, particularly if not indicated on the ingredient statement, that is, used as a processing aid such as in clarifying juices, may alert a halal certifying agency that it needs to do more inspection of the ingredients and the process (Figure 27.1).

PRESENCE OF ALCOHOL

INTRODUCTION

Alcohol is haram. The FAO/WHO Codex Alimentarius regulations for food labeling on the use of the term "halal" mentions that producers should avoid using liquor in their products. Alcohol, that is, ethyl alcohol or ethanol, has the chemical formula of CH_3CH_2OH. Alcohols are a class of chemicals but only ethanol is of concern with respect to halal. The concentration of ethyl alcohol in various liquors can vary generally from 3.5% to 40% ethanol. The "proof" of alcoholic beverages is double the percentage alcohol. Ethanol is produced by yeast fermentation of sugar. Often the starch must first be broken down to sugar. This can be done by various microbes. However, in some foods containing sugar and other carbohydrates such as fruit juice, natural fermentations such as soy sauce, or other sauces, ethanol in small amounts may be unavoidable (AOAC, 1996b). Vinegar,

Testing Non-Halal Materials

FIGURE 27.1 A typical gelatin analysis set up.

although normally prepared from ethanol is halal as almost all of the ethanol has been converted almost completely to acetic acid. Low amounts, that is., less than 0.5% may remain. This level of natural ethanol is not considered harmful to the body so that less than that amount is allowed in most countries. In addition to its presence in food, ethanol is used as a disinfectant, both to sanitize the workers' hands, and utensil and equipment surfaces. Ethanol (70%; v/v) used for this purpose is allowed.

If the alcohol is obtained from a liquor, even one drop will not be accepted in halal food products. But synthesized or extracted ethanol (industrial alcohol) used as a solvent for dissolving flavor or color before adding into the food products can be acceptable. IFANCA only permits 0.1% in final product with 0.5% in any single ingredient. Natural alcohol between 0.1% and 0.5% is acceptable. When testing for alcohol content, it is important to know something about the product being tested and also to understand the standards of the different importing countries. At this time, there is no single standard (FAO/WHO, 2001; TACFS, 2007; Yarita et al., 2002).

The current preferred test for ethyl alcohol uses gas chromatography with a flame ionization detector. The area of the peak in the proper place for alcohol would be quantitated using a calibration curve that is grounded with a carefully measured sample of n-propanol that serves as the internal standard. Because one is looking at low levels of alcohol, sample handling and good analytical technique are needed to be comfortable before "accusing" a sample of being above the required level of the halal certifying agency (ASEAN, 2005; Marikkar et al., 2005) (Figure 27.2).

FATTY ACID COMPOSITION

INTRODUCTION

Triglycerides (TG) are the main constituents of vegetable oils and animal fats. They are also referred to as triacylglycerol (TAG), that is, a chemical compound formed by one molecule of glycerol and three molecules of fatty acids.

FIGURE 27.2 A typical alcohol analysis set up.

The fatty acids esterified to the glycerol backbone usually have carbon chains of 4 to more than 20 carbon atoms. In addition, the fatty acid can also contain one or more double bonds at specific positions leading to unsaturated and polyunsaturated fatty acids.

Animal fats tend to have a larger proportion of long chain saturated fatty acids, so they are solid at room temperature. In contrast, fats from plant sources contain a higher proportion of unsaturated fatty acids and are usually liquid at room temperature (Ensminger et al., 1994).

The fatty acids are commonly analyzed using gas chromatography. The separation of fatty acids depends on the difference in their boiling point. The TG are chemically hydrolyzed to free fatty acids, which are separated from the glycerol, and then normally converted to fatty acid methyl esters (FAME). The mixture of FAME is extracted using the organic solvent hexane prior to separation in the gas chromatograph. Methyl pentadecanoic acid (C15:0) can be added as an internal standard (Kowalski, 1989).

The proportion of the different fatty acids are characteristic of their source. However, they do change within a single species depending on diet, overall health, and age of the animal or plant. However, the proportion of fatty acid in animals and plants are quite difference, so the test can generally identify whether the fat is from animals or plants. Sometimes, one can also determine the actual species, but this becomes much harder if the sample contains a mixture of sources in a large variety of ratios. So at this time, it can be used to determine if there are potential problems but again cannot replace on-site inspection and chain-of-custody records (Figure 27.3).

DNA FOR SPECIES IDENTIFICATION

DETECTION OF PORCINE DNA IN FOOD PRODUCTS

Every living organism has a unique DNA and it can in practice be used for species identification. The most important use of this technique for halal foods is to

Testing Non-Halal Materials

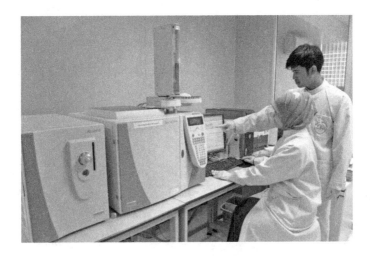

FIGURE 27.3 A typical fatty acid analysis set up.

determine the presence of DNA from non-halal species. It cannot be used to determine if the products of a halal animal came from an animal that was or was not slaughtered according to halal requirements. The most important species of concern is the pig (Man and Mirghani, 2001; Man et al., 2005).

The polymerase chain reaction (PCR) is a relatively efficient method with high specificity and high sensitivity. The initial DNA sample will be amplified (i.e., copied to make the concentration greater so that it is detectable and analyzable). The key to this method for halal is to have a piece of DNA that is absolutely unique to the pig as the DNA template that must be amplified. However, this is a requirement that can never actually be reached with 100% certainty. In addition, other materials are needed so that the amplification reaction is possible, that is, more strands of the DNA target are produced (Aida et al., 2005).

The DNA of interest will be amplified continuously for a total of 30 to 50 cycles leading to millions of copies. These are then normally observed using gel electrophoresis that can be difficult to quantitate but which clearly shows the presence of the amplified gene in the appropriate position on the gel.

The PCR technique has been used to detect pork in the presence of other meats in both raw and cooked products. Obviously, any purified material, for example, lard, which does not have any DNA, cannot be tested using this method. In conclusion, the PCR technique is very useful for detecting contamination with pig products, although there have been some cases of false positives, that is, the test is positive, but the product is unlikely to have been contaminated. Careful laboratory technique and confirmatory testing should be used before a company's reputation is harmed.

A further development has been the development of "real-time" PCR, which allows the initial amount of DNA (or cDNA or RNA depending on the primer) to be quantitatively measured. It is based on the measurement of the fluorescence produced by a "reporter" molecule that increases as the reaction proceeds so that the initial amount of DNA can be obtained by extrapolation back to the sample. The

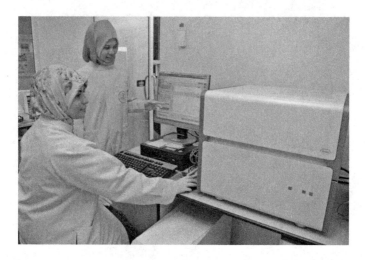

FIGURE 27.4 A typical DNA analysis set up.

fluorescent compounds include material such as SYBR® Green, which is non-specific and binds to any DNA or sequence specific probes such as Molecular Beacons, TaqMan®, FRET Hybridization Probes, and Scorpion® Primers.

Real-time PCR is now replacing classical PCR and is being automated. Using reverse transcription, the less stable RNA molecules can be converted to the more stable DNA analog. Further developments in DNA/RNA detection for the benefit of halal compliance testing can be expected (Figure 27.4).

REFERENCES

Cordella, C., Moussa, I., Martel, A.C., Sbirrazzuoli, N. and Lizzani-Cuvelier, L. (2002). Recent developments in food characterization and adulteration detection: Technique-oriented perspectives. *Journal of Agricultural and Food Chemistry*, 50(7), pp. 1751–1764.

Hermanto, S. and Fatimah, W. (2013). Differentiation of bovine and porcine gelatin based on spectroscopic and electrophoretic analysis. *Journal of Food and Pharmaceutical Sciences*, 1(3), 68–73.

Kowalski, B. (1989). Sub-ambient differential scanning calorimetry of lard and lard contaminated by tallow. *International Journal of Food Science & Technology*, 24(4), pp. 415–420.

Man, Y.C., Gan, H.L., NorAini, I., Nazimah, S.A.H. and Tan, C.P. (2005). Detection of lard adulteration in RBD palm olein using an electronic nose. *Food Chemistry*, 90(4), pp. 829–835.

Man, Y.C. and Mirghani, M.E.S. (2001). Detection of lard mixed with body fats of chicken, lamb, and cow by fourier transform infrared spectroscopy. *Journal of the American Oil Chemists' Society*, 78(7), pp. 753–761.

Marikkar, J.M.N., Ghazali, H.M., Man, Y.C., Peiris, T.S.G. and Lai, O.M. (2005). Distinguishing lard from other animal fats in admixtures of some vegetable oils using liquid chromatographic data coupled with multivariate data analysis. *Food Chemistry*, 91(1), pp. 5–14.

Shabani, H., Mehdizadeh, M., Mousavi, S.M., Dezfouli, E.A., Solgi, T., Khodaverdi, M., Rabiei, M., Rastegar, H. and Alebouyeh, M. (2015). Halal authenticity of gelatin using species-specific PCR. *Food Chemistry*, 184, pp. 203–206.

Syahariza, Z.A., Man, Y.C., Selamat, J. and Bakar, J. (2005). Detection of lard adulteration in cake formulation by Fourier transform infrared (FTIR) spectroscopy. *Food Chemistry*, 92(2), pp. 365–371.

GELATIN

AOAC. (1996a). *Official Methods of Analysis*, 990.26, 16th edn. Washington, DC: Association of Official Analytical Chemists.

ALCOHOL

AOAC. (1996b). *Official Methods of Analysis*, 973.23, 16th edn. Washington, DC: Association of Official Analytical Chemists.

Committee of National Bureau of Agricultural Commodity and Food Standards and Central Islamic Committee of Thailand. (2007). Thailand Halal Standard, TACFS 8400-2007 (Translation).

FAO/WHO. (2001). General guidelines for use of the term "HALAL," CAC-GL 24/1997. Codex alimentarius commission: Food labeling complete text. Revised. Joint FAO/WHO Food Standards Programme. Rome: FAO, 43–46.

ASEAN Cooperation in Food (Agriculture and Forestry). (2005). ASEAN General guidelines preparation and the handling of halal food. Available from: www.halalrc.org/images/Research%20Material/Report/Processing%20of%20Halal%20Food.pdf.

FATTY ACIDS

Ensminger, A. H., Ensminger, M. E., Konlande, J. E., and Robson, J. R. K. (1994). *Food and Nutrition Encyclopedia*, 2nd edn. London: CRC Press.

Yarita, T., Nakajima, R., Otsuka, S., Ihara, T., Takatsu, A., and Shibukawa, M. (2002). Determination of ethanol in alcoholic beverages by high-performance liquid chromatography–flame ionization detection using pure water as mobile phase. *Journal of Chromatography A*, 976(1–2), 387–391.

DNA

Aida, A. A., Man, Y. C., Wong, C. M. V. L., Raha, A. R., and Son, R. (2005). Analysis of raw meats and fats of pigs using polymerase chain reaction for Halal authentication. *Meat Science*, 69(1), 47–52.

28 Potential Hazards and Sanitation of Halal Facilities

Mian N. Riaz and Munir M. Chaudry

Halal means permitted and wholesome means good, healthy, safe, clean, and nutritious. The concept of wholesome means morally good: promoting the health and well-being of the mind or spirit, and of the body. Islam teaches about the importance of cleanliness in every part of life, including during food preparation and processing. According to one hadith, cleanness is half of the Muslim faith. In fact, today's halal consumers are more concerned about halal food quality and safety than ever before (Ismail and Phoon, 2007). Although the Muslim food supervision organizations have always monitored compliance with the halal food system, adulteration, contamination, and other hazards continue to be detected with laboratory tests (Wahab, 2004).

In general, there are four potential types of hazards in the halal food manufacturing process (as is true for all food production): physical, chemical, biological, and allergenic. A physical hazard is any extraneous object or foreign matter in halal foods that may cause illness or injury to the consumer. These foreign objects may include, but are not limited to, bone or bone chips, metal flakes or fragments, injection needles, pieces of product packaging, stones, glass or wood fragments, insects or other filth, personal items, or any other foreign material not normally found in halal food. Such contaminants may come into the plant with raw materials, may fall into foods from badly maintained facilities and equipment, or may reflect poor employee practices.

Chemical hazards can occur at any point in the food chain. When toxic chemicals are used for pest control or for cleaning and sanitizing food contact surfaces and these surfaces come into contact with food, the food may be contaminated with those chemicals. Toxic metals used to make the production equipment can be a source of chemical contamination. Gray enamelware containers may be plated with anatomy or cadmium that can make acidic foods such as orange juice, tomato sauce, or pickles poisonous. Pottery dishes with lead glazes should not be used to prepare or serve food.

Microbiological hazard occurs when food becomes contaminated with microorganisms found in the air, food, water, soil, animals, and the human body. Many microorganisms are helpful and necessary for life. However, given the right conditions, some microorganisms may cause a foodborne illness or spoilage. Microorganisms commonly associated with foodborne illnesses include bacteria, viruses, and parasites.

Food allergen hazards can be a significant health risk to those consumers who have allergic reactions to specific foodstuffs. The challenge is to prevent any allergens not being used in a particular product from becoming part of that product.

All of these hazards may be monitored and controlled by applying different food standards and guidelines such as Good Manufacturing Practices (GMP), Hazard Analysis Critical Control Points (HACCP), the British Retail Consortium (BRC) standards, the Global Food Safety Initiative (GFSI) standards, International Standards Organization (ISO) standards, and Codex Alimentarious (Codex) international food standards (FAO/WHO) (Talib et al., 2008). For example, Codex currently has 49 different guidelines for different food products each including hygienic practices. Standard and Industrial Research Institute of Malaysia (SIRIM), have devised a comprehensive halal food standard named MS1500:2004 to help the food manufacturers to monitor, control, and provide guidance for processing of products to satisfy HACCP, Islamic law, and GMP. According to Janis (2004), the MS1500:2004 lays out comprehensive requirements according to Islamic law and also the requirements for food manufacturing and foodservice supply chains from processing to handling, distribution, storage, display, servings, packaging, and labeling. The aesthetic aspects, hygiene, sanitation, and food safety are also included as part of the requirement (Janis, 2004). Most of the developed countries and international food companies follow very strict rules for producing clean and wholesome foods. For example, Cargill, DuPont, and Nestle have their own guidelines for wholesome and clean foods. They are committed to never violating any rule that jeopardizes the safety or cleanliness of food. Whereas it may be a major issue for food companies in the emerging economies to produce wholesome foods. Sometimes, food companies in the emerging economies do not have enough resources to follow the wholesome rules for foods. Regulatory agencies may not strictly enforce the guidelines for producing clean and wholesome foods in some countries. Some companies may not understand the importance of cleanliness during the production of halal foods. These companies need to learn more about the hazards in food beyond the issue of contamination of halal foods with non-halal ingredients. According to a FAO (2015) report, 70% of deaths among children under 5 years of age were linked to contaminated foods in less developed countries (FAO, 2015).

With globalization, the internet, trade, and other advancements, our food supply is now global. That means anyone from anywhere can order any food at any time to be consumed anywhere in the world. That creates a challenge to produce wholesome and clean foods all over the world. The halal food industry needs to consider some additional hazards to assure that foods are wholesome. These can be identified as religious food hazards. One of the most serious hazards is cross-contamination with non-halal food ingredients. The focus should always be on avoiding non-halal ingredients (Bonne and Verbeke, 2008). During processing of halal and wholesome food, the focus should always be on avoiding non-halal ingredients (Khawanjah, 2001). This contamination can occur through hidden ingredients like lubricants, releasing agents, antifoaming agents, filtering agents, clarifying agents, and bleaching compounds; all being examples of processing aids that are not disclosed on the consumer label. Many of these ingredients may contain non-halal ingredients often from pork and its by-products. According to Meindertsma (2009), there are approximately 185 different products and ingredients that are made from pork and its by-products such as the blood, bones, and hides. One of the ingredients is the enzyme that is extracted from pork stomachs (porcine pepsin). This one enzyme can be used in making and processing of hundreds of food products (Omojola et al., 2009). According to Sustainable Swine Resources (2017), no other animal provides society with a wider range of products than the pig. It provides a vast array of high quality pig co-products that are used as sustainable resources in many human and animal industries. These industries include, but are not limited to, medical, human health, industrial and consumer, companion animal, and animal health companies. Products from the pig such as the pancreas, spleen, thyroid, trachea, and kidney are further processed into supplements for both human and animal consumption (Nuruddin, 2007). There are over 40 pharmaceuticals and medicines that are derived from pig co-products (Sustainable Swine Resources, 2017). Food becomes religiously unacceptable when contaminated with pork or its ingredients. Contamination of a halal product can occur by directly touching pork meat to the product, by not washing hands and then touching a halal meat product, by using the same surfaces, such as cutting boards or cleaning cloths that have not been cleaned after non-halal food handling, and by non-halal foods drippings on or touching halal foods.

An additional concern for managers of halal-certified plants is the prevention of cross-contamination of halal products with non-halal products. Examples of non-halal products are pork, the meat of carnivores, and alcohol. Although cross-contamination with non-halal product is not necessarily a public health hazard, this kind of contamination nullifies the halal status of the product and may reduce its value as a consumer product. Fortunately, there is a vast array of information available to food producers to help them produce food products that are safe to eat and properly certified as halal. Therefore, effective cleaning and sanitation procedures are essential for every halal food processing establishment. In halal-certified food production areas, additional precautions should be considered to help prevent contamination with non-halal substances. It is recommended that rooms or compartments in which halal products are processed, handled, or stored are separate from rooms or compartments in which non-halal products are processed, handled, or stored.

If a business processes both halal and non-halal foods, halal products should be processed before non-halal products when using the same facilities and equipment. The rational for this order of operation is clear: such practices greatly reduce the likelihood of halal products being exposed to non-halal products. If comingling does occur, a product can no longer be classified as halal. If halal products are processed with equipment that has previously been used to produce non-halal products (such as meat from livestock not slaughtered under halal conditions), complete pre-operation sanitation procedures must be done before halal products are processed. As stated earlier, halal products that have been cross-contaminated with non-halal products may no longer be certified as halal.

Regarding the safety of food—rather than halal status—the U.S. Department of Agriculture (USDA) Food Safety and Inspection Service (FSIS) regulates meat, poultry, and processed egg products for safety and wholesomeness. Regulatory requirements under the Federal Meat Inspection Act, the Poultry Products Inspection Act, and the Egg Products Inspection Act can be found on the USDA website (www.fsis.usda.gov). Similarly, the Food and Drug Administration (FDA) regulates foods other than meat, poultry, and processed egg products, such as fruits, vegetables, seafood, and grains. Their website is www.fda.gov. FDA's jurisdiction extends even further to include nutritional supplements, energy drinks, flavorings, vaccines and pharmaceuticals, and cosmetics. Many of these products are discussed in other chapters of this book.

REFERENCES

Bonne, K. and Verbeke, W. (2008). Muslim consumer trust in halal meat status and control in Belgium. *Meat Science*, 79(1), 113–123.

FAO. (2015). WHO's first ever global estimates of foodborne diseases find children under 5 account for almost one third of deaths. Dec 3. News release Geneva. Available online: http://www.who.int/news-room/detail/03-12-2015-who-s-first-ever-global-estimates-of-foodborne-diseases-find-children-under-5-account-for-almost-one-third-of-deaths.

FDA. (2003). Ruminant feed (BSE) enforcement activities. Rockville, MD: Center for Veterinary Medicine, Office of Management and Communications. Available online: www.fda.gov/cvm.

Ismail, K. and Phoon, O. F. (2007). *Consumer Intention in Purchasing Halal Food Products*. Available online: http://econ3.upm.edu.my/epms/view.php?cat=journal&jt=all.

Janis, Z. M. (2004). *Malaysian standard MS 1500 (2009): Halal Food – Production, Preparation, Handling And Storage – General Guidelines* (Second Revision). Malaysia: Dept. of Standards Malaysia, Ministry of Science, Technology and Innovation, 3–4.

Khawanjah, M. (2001). *Part of transcript of the talk given*, Killing methods (Question No. EFSA-Q-2003-093). *Scientific report of the Scientific Panel for Animal Health and Welfare*. Available online: www.halalfoodauthority.co.uk.

Meindertsma C. (2009). PIG 05049. Available online: https://www.dezeen.com/2009/08/29/pig-05049-by-christien-meindertsma/

Nuruddin, N. A. (2007). Islamic Manufacturing Process (IMP). Available online from Institute of Islamic Understanding Malaysia: www.ikim.gov.my.

Omojola, A. B., Fagbuaro, S. S., and Ayeni, A. A. (2009). Cholesterol content, physical and sensory properties of pork from pigs fed varying levels of dietary garlic (Allium sativum). *World Applied Sciences Journal*, 6(7), 971–975.

Sustainable Swine Resources. (2017). Available online: www.sustainableswineresources.com/.

Talib, H. H. A., Ali, K. M., Jamaludin, K. R., and Rijal, K. (2008). Quality assurance in halal food manufacturing in Malaysia: A preliminary study. In *Proceedings of International Conference on Mechanical and Manufacturing Engineering (ICME2008)*, Johar, Malaysia, 21–23.

Wahab, A. R. (2004). Halal guidelines for manufacturers, guidelines for the preparation of halal food and goods for the muslim consumers. *Amal Merge: Halal and Food Safety Institute*. Available online: www.halalrc.org/images/Research%20Material/Literature/halal%20Guidelines.pdf.

29 Halal Awareness and Education Schemes

Mian N. Riaz and Munir M. Chaudry

CONTENTS

Methodology for Raising Awareness of Halal 304
Methodology for Raising Religious Knowledge 304
Awareness Methodology for the Halal Food Industry 304
Awareness Methodology for Food Scientists and Technologists 305
Awareness Methodology for the Halal Consumers 306
Awareness Methodology Requirements for the Halal Food Supply Chain 306
References 307

Several countries are working hard to become the global halal hub. Halal has traditionally been associated with food products offered to consumers, but now the discussions are about developing a complete halal value chain: "from farm to fork." The goal is to assure that products are halal by starting at the source of all raw materials. Muslim consumers and the food industry need to be educated about the importance of the halal value chain. The education of those in the food industry is particularly important as these are the people who will make creating a halal value chain possible. At this time, very few universities in Muslim or non-Muslim countries are teaching about halal foods. We need to start raising consumer awareness of the need for greater integrity in halal food production by educating food science and nutrition students about halal food processing and production. There is a need for more halal food research and development (R&D), such as to how to create and use a halal food verification program, and how to use food analysis to solve halal problems (Dali et al., 2007). To achieve these goals, we also need to provide training to halal certification bodies and their auditors, to start formal halal degree programs, particularly at the masters' and PhD levels, to train people in the food industry about halal food production; to conduct more halal conferences and seminars that reach more people; to publish more peer-reviewed papers on halal issues; and to write more articles and books on the halal food supply chain. Government agencies need to promote halal trade practices and to provide an incentive for engaging in halal trading consistent with a respect for separation of church and state. The community needs halal consumer organizations to represent them with industry and the government, and the industry needs halal trade associations and certifying agencies that assist with global halal commerce.

METHODOLOGY FOR RAISING AWARENESS OF HALAL

Research and training is a vital part of the development of a smoothly functioning, transparent, and effective halal food supply chain. The current R&D framework of most progressive food industries is supported by governmental and private independent research organizations that could be enlisted to help in developing the halal food industry. Specialized, independent halal research and training institutes and/or university-based institutes to support the halal food industries are also required. This will open new opportunities for qualified food technologists, scientists, and researchers to serve the Muslim community.

METHODOLOGY FOR RAISING RELIGIOUS KNOWLEDGE

Halal food production guidelines should only be developed by qualified persons. The qualifications of such persons should include the proper knowledge of Islamic principles and the ability to handle the complex and complicated technical issues. They must also have wide general knowledge about what is happening in the food processing industry. Research on Islamic dietary laws requires that a team of people have a comprehensive understanding of all aspects of food and the many sources and complex production methods using to produce different foods. The research should first be directed toward providing answers to the existing issues affecting the halal supply chain and the development of international halal production guidelines. These guidelines when applied to modern food production and processing and distribution practices will provide a basic framework to ensure the integrity of the halal food industries. The present trend of dealing with each specific matter that arises in a specific situation by having one isolated Muslim religious leader delivering a religious ruling tends to confuse the issue and create further problems for the industry and the consumers. For example, declaring a certain food chain or product to be halal or haram without having a sound detailed knowledge of the entire set of circumstances and a proper understanding of the modern food supply chain is counterproductive to the further development of the halal food industries and also seems to be inconsistent with key principles of Islam.

AWARENESS METHODOLOGY FOR THE HALAL FOOD INDUSTRY

The infrastructure of the typical food production facility in the modern food industry involves using technology, a work force, various raw materials, and a process that meets appropriate quality standards. Industrialization of food production, processing, and distribution was initiated mostly in Western countries and has spread across the world with the development of food technology and the globalization of the food supply chain. The initial industrial standards for food production focused primarily on business needs. Consumer opinions, including ethical norms, were not given equal importance. Modernization of the food and agriculture industries in many Muslim countries was also based on the same industrial standards devoid of ethical and religious values. The transfer of technology without being adapted to local standards and values created complex problems that are now being faced by the halal food supply chain internationally. Solutions for many of these problems are not straightforward in part because of the absence of political will and the lack of financial resources in many Muslim countries. Relatively stable Muslim economies such as those of Malaysia, Indonesia, and the Gulf States led by Saudi Arabia are progressively dealing with issues related to the halal food supply chain. Demand for halal foods by significant Muslim minorities in developed Western countries such as the U.S. and the EU have created significant halal markets. Although some regulatory authorities require permits for religious slaughter, most of the halal food supply chain remains unregulated in terms of religious compliance. Private Islamic organizations have made valuable contributions to raising the visibility and integrity of the halal food and supply chains by establishing halal standards supported by the community and by providing credible and relevant certification and monitoring procedures. These worthy efforts may have serious limitations due in part to the lack of proper resources such as professionals with an understanding of both religious and technical issues, financial constraints, and a lack of support by the relevant governmental regulatory authorities. Furthermore, the absence of a unified discussion of halal food industry issues has contributed to variations in the standards for the production, certification, and monitoring of halal foods that may or may not reflect actual Muslim consumer needs. A lack of transparency has meant that a Muslim consumer cannot always determine if a product with a halal claim meets their standard. These problems are a major holdup in the development of an integrated global halal food supply chain, where products from various certifying bodies around the world need to be widely acceptable in various countries and between certifying bodies.

The modern food industry lacks the understanding of the requirements for halal food production. Understanding halal food production is becoming essential for the food industry, especially those exporting to Muslim countries and/or catering to their local Muslim communities. Research and training is vital in the halal food industry and many aspects of halal food production requires innovation, including the development of new halal products by sensitively incorporating the religious and customary habits of Muslim consumers. The availability of training for key workers and managers in all aspects of halal food production would help in developing halal food industries. Qualified training in halal food production would also develop the professional capabilities and increase the employability of these food professional in other parts of the world including Muslim countries.

AWARENESS METHODOLOGY FOR FOOD SCIENTISTS AND TECHNOLOGISTS

Halal food is becoming a global reality and food technologists are generally not being taught in universities about the fundamentals of halal food production. Education in halal foods needs to be introduced in the early stages of a food technologist's academic program. This will lead to a sustainable development of the halal food supply chain by providing adequately educated professionals who can carry out research and solve technical issues in the halal food industry. Higher education and research in halal food science and technology would provide the skills needed to obtain technical solutions to the problems that arise and to create innovative products and technologies. The training of food technologists in halal food production and developing qualified scientists acquainted with halal food standards would provide a valuable service for the rapid development of halal food industries. As kosher has also been expanding, there is a need for similar educational efforts in that area at the university level. By working together, the acceptability of such university training might be more acceptable in many countries.

AWARENESS METHODOLOGY FOR THE HALAL CONSUMERS

Research about consumer trends is vital for any successful food product. New halal food products not only require adhering to halal food standards but also to the social and cultural norms of consumers. Taking these considerations into account provides greater assurance for the success of new products in the halal market. Consumers also need to be aware of the nature and quality of food. A lack of proper information and training at the community level creates distrust and may lead consumers and religious leaders to discredit the halal food supply chain. The mass media and the internet could play a positive role in educating the consumer regarding halal food issues (Mohamed et al., 2008). Consumers should also be aware of their right to demand proper halal food from the industry with fraud enforcement from the regulatory authorities, that is, the legitimate role of government—enforcing "truth-in-labeling" and "consumers' right-to-know." Consumer training should also be focused on Islamic etiquette with respect to food consumption. Health-related Islamic guidelines and education of Muslim consumers will improve their nutritional and health status.

AWARENESS METHODOLOGY REQUIREMENTS FOR THE HALAL FOOD SUPPLY CHAIN

The future integrity and robustness of the halal food supply chain will depend on research and training. Proper coordination is required between halal food organizations in both the private, and where appropriate, in the governmental sector. This would help to improve technical and management practices and would also provide a platform for the exchange of ideas and experiences (Clifford and Stank, 2005). Research and training in the following areas is required to develop the halal food industry and strengthen the global halal food supply chain:

- Research to identify new halal markets, develop new halal products, and improve the existing halal-certified products.
- Develop laboratory-based research to help and support Islamic scholars. Provide evidence-based information on such controversial issues as slaughter of animals, genetically modified materials (GMO), and other complex technologies used in modern food production.
- Research into natural alternatives that are halal to replace any questionable or haram ingredients used in food processing.
- Develop techniques, tools, and processes to increase through-put in slaughterhouses without compromising the Islamic rules of slaughtering.

- Develop internet portals and dedicated software to improve the visibility and facilitate the growth of the halal food supply chain.
- Professional training of all key persons involved in the food supply chain, including regulatory authorities, to understand and implement halal guidelines and deal with halal issues.
- Research and training to improve the functional capabilities of the food supply chain.

Food scientists and technologists are generally not qualified to make a religious ruling, although they may often be the appropriate people to frame the question and provide important background information. Qualified Islamic scholars require proper education and training about the modern food supply chains to establish fact-based religious rulings.

The modern food industry requires research and basic training in halal food production to be able to successfully and sustainably prepare halal foods for both domestic and export purposes.

Consumer training about the halal status of foods in modern supply chains at the local level would create greater trust and establish better relations with the food industry and the halal certifying authorities.

REFERENCES

Clifford Defee, C. and Stank, T. P. (2005). Applying the strategy-structure-performance paradigm to the supply chain environment. *The International Journal of Logistics Management*, 16(1), 28–50.

Dali, N. R. S. M., Sulaiman, S., Samad, A. A., Ismail, N., and Alwi, S. H. (2007). Halal Products from the Consumers Perception. An Online Survey. In *Proceeding of the Islamic Entrepreneurship Conference*, Nilai, Negeri Sembilan, Malaysia, Kolej Universiti Islam Malaysia.

Mohamed, Z., Rezai, G., Shamsudin, M. N., and Eddie Chiew, F. C. (2008). Halal logo and consumers' confidence: What are the important factors? *Economic and Technology Management Review*, 3(1), 37–45.

30 Halal Food Model

Mian N. Riaz and Munir M. Chaudry

CONTENTS

Need for Development of a Widely Accepted Halal Food Model 310
How to Strengthen the Halal Food Model 312
    At the Farm 312
    Plants 313
    Animals 313
    At the Processing Stage 313
    At Slaughtering Facilities 313
    At Production/Processing Facilities 314
    Post-Production Mechanisms 314
Education 315
    Business 315
    Consumers 315
    Governments 315
References 316

The halal food trade is growing at a spectacular rate and so are concerns about its authenticity and safety. The differing halal standards and certification models used by various Muslim organizations and individuals lead to further consumer concerns. Ideally, the font certification process should have taken care of these concerns, but it has not in light of recent scandals that have impacted the halal food industry. As a direct result of these problems, halal authentication through accreditation, that is, having a respected Muslim national body that sets standards for the halal certifying bodies, has become an extremely important issue for both halal food producing companies and halal consumers. Adding to the concern are the situations where companies are self-certifying or obtaining their certification from third-party certifiers who do not follow the normative widely accepted certification standards of the Muslim community. A parallel concern about authenticity can be seen with changes occurring in the larger food industry, which faced somewhat similar credibility problems (Krishnan et al., 2017; Shafie and Othman, 2006). The Global Food Safety Initiative (GFSI) was launched in 2000 after the food industry went through a series of food safety crises. GFSI's procedures ensure that food safety auditing schemes and accreditation systems adhere to common "benchmarks" for safety, transparency and accountability. It has now become the gold standard for third-party food safety audits and now affects all aspects of the food supply chain.

    The halal certification industry can learn from the GFSI model in its attempt to strengthen and maintain consumer trust for halal. As a first step, it needs to harmonize halal standards by developing a widely acceptable set of halal food standards and a model of how a certifying agency should carry out its task. In addition, an effort needs to be made to reduce audit duplication throughout the supply chain—from farm to fork. Because at present there is no widely recognized halal standard or a model for how a certifying agency should act, there is an immediate need to develop such documents. Such a model needs to establish the standards for determining equivalency between existing halal auditing schemes, while retaining flexibility and choice in the market place. After such documents are developed, a mechanism should be developed to monitor, disseminate, and update this information by using modern communication technologies. This will probably involve:

1. The creation of a Digital Halal Database, which will list all halal certified ingredients with information about their certifier, halal certificate, and certifying agencies updates, a comprehensive list of halal certified retail and industrial products, and a comprehensive listing of recalls and other problems with specific products.
2. Regular, but unscheduled, on-site inspections and audits where the representatives of the halal supervision agency compare the database information with the products and ingredients actually found at the plants.
3. Development of experts on the production of the products and an ability to check that the plant's operations do not compromise halal such as tracing of pipes to assure no cross-contamination.
4. Development of tamper-proof trademarked halal certification logos on products to provide traceability and accountability. The latest technologies such as scannable logos (e.g., using QSR code) on smart phones, etc. should be used (Siti Hasnah et al., 2009; Norhabibah, 2011).

NEED FOR DEVELOPMENT OF A WIDELY ACCEPTED HALAL FOOD MODEL

The market for halal products is widely distributed throughout the world. This has led to an increased demand for halal products not only by Islamic countries but also by almost all non-Islamic countries. The modern food supply chain is a global enterprise and components of the food may be grown and processed in many different countries before reaching the end consumer (Mathew, 2014; Wibowo and Ahmad, 2016). Thus, developing a food product is no longer a simple process. For example, a raw material could be extracted in India; processed into powdered form (or into an ingredient) in China; shipped to the U.S. for production; and then sold in grocery stores anywhere in the world. This increases the probability of a halal problem with foods and cosmetic and pharmaceutical products unless subject to proper supervision throughout the supply chain.

Drinkable yogurt can be used as an example of the complexity of modern foods that on the surface would seem to be "obviously" halal. Milk is obtained from cow's using milking machines or by hand, processed and pasteurized, and shipped to a yogurt processing facility. The milk may go through many plants and transport vehicles. In addition, various ingredients and processing aids may be used.

(The requirements to be a "processing aid" will vary depending on the country of production and/or final sale.) Some of the ingredients that may be added along the way are sugar, milk powder, emulsifiers, flavors, and preservatives. Each ingredient has to be determined to be halal, even if used in small amounts (Al-Mazeedi et al., 2013). Among the more obvious concerns with processing aids are the lubricants and emulsifying agents that often are produced from materials derived from slaughtered animal by-products, an obvious halal concern. Finally, the yogurt is packaged. As innocuous as it may sound, packaging itself can give rise to halal contamination. It may be made from an undesirable source; sometimes it may be coated with a haram product, like a wax from an animal source.

Some progress has been made with the development of the Organization of Islamic Countries (OIC) General Guidelines on Halal Food in 2009. This document needs to be discussed further so as to obtain a version that provides both flexibility and choices with regards to differences in the Fiqh (deep understanding) of the different Muslim school of thoughts while also conforming to national laws and regulations of both Muslim majority and non-Muslim countries. A careful balance is needed to develop a document that strictly adheres to the guidelines obtained from the Qur'an and the Sunnah while allowing for differences on secondary issues (Department of Standards Malaysia, 2009). Halal food standards for a modern food supply chain are being developed in various countries according to the needs and requirements of industry and commerce in those countries. The import and export regulations of certain Muslim countries in the early 1970s focused largely on the requirement for halal slaughter. These halal food standards were mainly developed, regulated, and enforced by various secular governments. Muslim organizations, especially in North America and Europe, developed halal food standards in the early 1980s. These individual standards vary in scope and quality. Each variation in halal standards should be clearly identified so that other certifying agencies and consumers can make an informed choice. These standards should be widely disseminated to the general population as well as to the industry so that there is no ambiguities or confusion (Krishnan et al., 2017).

Currently there are several different halal food standards that are being used by more than 400 certifying agencies currently identified worldwide. Indonesian and Malaysian halal food standards emphasize the certification process and a halal assurance program rather than an emphasis on final food analysis. Their focus is on third-party auditing and verification so there is no doubt about any contamination or haram ingredients (Shafie and Othman, 2006). On the other hand, the United Arab Emirates (UAE) depends on standards that sends products to a testing laboratory to detect any haram materials. The different standards are based on the rulings of the different school of thoughts, some of which follow very strict rules whereas other schools of thought are much more flexible. A decision needs to be made as to whether a more broadly accepted standard should follow the strictest standards, which would allow everyone to use the products but would possibly eliminate products that others consider halal, or whether the standard should provide for multiple standards with guidance on how to clearly identify such products. The structuring of such standards can be guided by one of the generally international standards for good practices such as the family of ISO (International Standards Organization) practices.

HOW TO STRENGTHEN THE HALAL FOOD MODEL

The need to establish a widely accepted halal standard is currently challenging the halal food industry. The definition of halal, while generally agreed upon by Muslims, display significant differences when it comes to their application in the industry. Due to differing halal standards not only between countries but also within each country's certifying system, confusion, misunderstanding, and even abuse in the halal audit and certification process have occurred (Omar and Jafaar, 2011). For example, some of the certifying agencies in the U.S. use meat coming from kosher slaughtering and certify the meat products as halal. Some of the halal standards in the U.S. only allow hand slaughtering whereas some accept mechanical slaughtering (e.g., for poultry). At the same time, there is major debate going on in Europe and North America regarding the use of pre-slaughter or post-slaughter interventions in addition to the traditional cut for the halal slaughter (Nakyinsige et al., 2012a). Another example of a controversy in the processed food industry is gelatin. Some of the halal-certifying agencies are okay with pork gelatin based on their beliefs that the chemical structure changes during processing of bones into gelatin and the finished product sufficiently changed its chemical and physical nature. Whereas some of the agencies allows beef gelatin in food products based on their beliefs that it was slaughtered by Ahle-Kitab (People of the Book, the ones who have received the Scripture). Whereas some reputable and very strict certifying agencies will only allow gelatin, which is made from hand slaughtered animal bones. Therefore, different halal certifying bodies follow different Islamic rulings regarding issues such as gelatin, food flavorings, animal enzymes, phosphates, mechanical slaughter, stunning of animals, and the usage of a thoracic stick after the dhabiha slaughter. (The thoracic stick is a second cut that increases the speed of blood loss and decreases the time to unconsciousness.) This sometimes creates confusion for producers who may not know which authority to use to get their product certified for their target market. While the industry believes that a global halal standard would be ideal, depending on those standards, current products might no longer be possible or the cost of halal production may be significantly increased, which may not be desirable by the industry or the end user. To develop the most comprehensive set of standards will require a closer look at all the steps in getting halal products from farm to fork.

AT THE FARM

The halal supply chain is vital in ensuring that the final product is halal. This starts at the farm. Therefore, halal auditing should begin on the farm. At present, not much attention is paid by the halal food certifiers to the treatment of animals before the pre-slaughter stage, including what the animal is being fed (Nakyinsige et al., 2012b). The halal auditing should begin with the determination that the animals for halal use have been raised in such a way that they are fit for consumption, in terms of animal welfare and halal food regulations, for example, that the feed was pure (Hambrecht et al., 2004). If the animals are a result of any DNA manipulation or other artificial manipulations, this information should be noted and the legitimacy of the process determined by the Muslim scholars. The addition of the ethical dimensions of starting at the farm will add value to the end product (Kurien, 2002; Papazyan and Surai, 2007).

Similarly, other raw materials should also be audited and their growing practices be analyzed to ensure their compatibility with the general Islamic guidelines on their consumption. Some questions to be considered during the audit are:

PLANTS

1. Are the raw materials sourced from plants grown using organic farming practices or any other special agricultural method?
2. Details should be obtained if the plants have been genetically modified. Have any foreign genes been introduced into these plants? (For example, has a fish gene been added into tomatoes to fight frost?) Do these foreign genes come from halal sources?

ANIMALS

1. Do the animals come from a category fit for Muslim consumption?
2. Have any foreign genes been introduced into these animals?
3. Have the animals been treated humanely? (This needs to be defined from an Islamic perspective consistent with modern animal welfare science.)

AT THE PROCESSING STAGE

At the processing stage, the materials, especially those from animals or birds need to be monitored carefully as there is always a risk of contamination if they are processed in facilities where non-halal products or animals are processed. The multi-use of processing lines for halal and non-halal food processing must follow the standards for proper sanitization and monitoring with scientifically validated procedures, ensuring a zero level of cross-contamination between halal and non-halal food products (Ab Talib and Johan, 2012).

AT SLAUGHTERING FACILITIES

Most of the regulations for halal foods are related to foods derived from animals and include the species, method of slaughtering, and post-slaughter procedures. The modern methods of mechanical slaughter that have gradually been substituted for the traditional manual methods have raised new issues regarding their conformity to halal regulations. The application of new tools and technologies need to be thoroughly understood and innovative procedures need to be scientifically validated to show that they conform to Islamic dietary laws (Al-Qaradawi, 1994). Pre-slaughter interventions (stunning) are one of the most controversial issues in the modern halal food supply chain, and there are arguments in favor and against this procedure (Apple et al., 2005; Hambrecht et al., 2004; Kannan et al., 2003; Ljungberg et al., 2007; Schaefer et al., 2001) . Some of the other key points that need to be kept in mind are:

1. Whether only halal animals are slaughtered at the facility. If not what are the provisions/procedures for sanitizing the plant and avoiding cross-contamination?
2. Whether the animals are transported humanely/safely from the farm to the slaughtering facility.
3. What are the provisions for removing dead animals?
4. Whether a facility has governmental inspectors on-site monitoring the health and safety of the animals and the meat. The halal audit report should incorporate the findings of the governmental inspectors.
5. Whether the slaughtering practices adhere to the slaughtering requirements of the halal certifying body. What mechanisms are in place to assure compliance? How is noncompliance dealt with? How is any noncompliance conveyed to the consumers?
6. How are the slaughtering requirements of the halal certifying body conveyed to users further along in the food supply chain?

At Production/Processing Facilities

1. Are the halal compliance certificates from the farm and the slaughtering facility being properly monitored, including checking the validity of the certificate/markings at the time of receiving?
2. What mechanisms are in place to ensure that there is no contamination during transportation, such as the halal sanitization of transport vehicles prior to use?
3. What are the pre-processing preparatory cleaning systems in place at the processing facility? How is the plant being sanitized for halal product processing?
4. If the product is processed at more than one facility, what halal compliance mechanisms are in place to assure that all halal marked products are coming only from halal certified facilities?

Post-Production Mechanisms

Once the products are processed, the important thing is that they be labeled properly. The label should reflect not only its halal authenticity but also the halal standard used by the certifier. Food labeling and packaging are crucial stages in the halal food supply chain. Islamic norms require honest trading of food products and proper labeling helps to implement these norms.

The halal logo should be tamper-proof and embedded with the latest technologies in labeling (Regenstein and Chaudry, 2002; Soong, 2007). One developing area is the use of scannable logos (using QSR code) on smartphones. The smartphone should be able to read the logo and verify its authenticity immediately. At the processing facilities, the same smart code technology could be used for tracing the journey of the materials and products to assure compliance with the required halal certification standards throughout the supply chain.

This information should also be stored in the Digital Halal Database. While this list can, and should, be as expansive as possible, it is important to note that there are several limitations (Yahaya et al., 2011).

Due to trade secrets and privacy issues, it can only include information that can be disclosed to the public. While the industry can share this information with the halal certifier, it may be reluctant to disclose some of it publicly. The question as to how much information must be disclosed and how much can be disclosed needs to be carefully addressed by those developing the Digital Halal Database.

Most certifiers are still using handwritten audit forms, which take a lot of the auditor's time to fill out properly. They can and should be digitized so that they can easily be used with an ipad or other tablet product and the information transferred electronically to the appropriate long-term databases, namely, those within the certifying agency.

EDUCATION

A key ingredient in the effort to strengthen halal food compliance is education of the businesses, consumers, and governments involved in the halal food supply chain. A well-developed halal model, with room for variations, will benefit the stakeholders as follows (Hasnah, 2011):

BUSINESS

1. Provide financial incentives including the opening of new markets.
2. Provide acceptance and satisfaction for consumers.
3. Add value to the products while strengthening food safety.
4. Increase consumer confidence in the industry. At present, there is a lot of suspicion among consumers. Implementing the above will reverse the trend.

CONSUMERS

1. Confidence in their purchases.
2. Satisfaction and trust of products, certifying agencies, and businesses.

GOVERNMENTS

1. Governments, especially in the non-Muslim countries, will have a much clearer idea of the halal standards and will be more willing to use traditional consumer protection regulations to protect Muslim and non-Muslim consumers purchasing halal products.
2. At present, they are hearing different voices, which confuse them and makes them reluctant to undertake the development of regulations that would protect the halal consumer.

One of the main reasons why the halal industry cannot grow faster, despite rising demands for halal products worldwide, is because there is no consensus on standards. Hopefully, a more widely accepted and publicly available standard will increase confidence in halal and lead to further expansion of the halal market.

REFERENCES

Ab Talib, M. S. and Johan, M. R. M. (2012). Issues in halal packaging: A conceptual paper. *International Business and Management*, 5(2), 94–98.

Al-Mazeedi, H. M., Regenstein, J. M., and Riaz, M. N. (2013). The issue of undeclared ingredients in halal and kosher food production: A focus on processing aids. *Comprehensive Reviews in Food Science and Food Safety*, 12(2), 228–233.

Al-Qaradawi, Y. (1994). *The Lawful and the Prohibited in Islam: (Al-Halal Wal Haram Fil Islam)*. Plainfield, Indian: American Trust Publication.

Apple, J. K., Kegley, E. B., Maxwell, C. V., Rakes, L. K., Galloway, D., and Wistuba, T. J. (2005). Effects of dietary magnesium and short-duration transportation on stress response, postmortem muscle metabolism, and meat quality of finishing swine. *Journal of Animal Science*, 83(7), 1633–1645.

Hambrecht, E., Eissen, J. J., Nooijen, R. I. J., Ducro, B. J., Smits, C. H. M., Den Hartog, L. A., and Verstegen, M. W. A. (2004). Preslaughter stress and muscle energy largely determine pork quality at two commercial processing plants. *Journal of Animal Science*, 82(5), 1401–1409.

Hasnah, S. H. (2011). Consumption of functional food model for Malay Muslims in Malaysia. *Journal of Islamic Marketing*, 2(2), 104–124.

Hassan, S. H., Dann, S., Annuar, K., Kamal, M., and De Run, E. C. (2016). Influence of the Halal Certification Mark in Food Product Advertisements in Malaysia. *The New Cultures of Food: Marketing Opportunities from Ethnic, Religious and Cultural Diversity*, Eds. Lindgreen, A. and Hingley, M. K. London, UK: Routledge Taylor & Francis Group, 243–262.

Isa, N.M. Baharuddin, A., Man, S., and Wei, C.L. (2014). Bioethics In The Malay-Muslim Community In Malaysia: A Study On The Formulation Of Fatwa On Genetically Modified Food By The National Fatwa Council. Developing World Bioetetics. John Wiley and Sons Ltd. Available from: https://pdfs.semanticscholar.org/64c0/790348019a7c76896a832b52bf4b1cb446fe.pdf.

Kannan, G., Kouakou, B., Terrill, T. H., and Gelaye, S. (2003). Endocrine, blood metabolite, and meat quality changes in goats as influenced by short-term, preslaughter stress. *Journal of Animal Science*, 81(6), 1499–1507.

Krishnan, S., Omar, C. M. C., Zahran, I., Syazwan, N., and Alyaa, S. (2017). The awareness of Gen Z's toward Halal Food Industry. *Management*, 7(1), 44–47.

Ljungberg, D., Gebresenbet, G., and Aradom, S. (2007). Logistics chain of animal transport and abattoir operations. *Biosystems Engineering*, 96(2), 267–277.

Mathew, V. N. (2014). Acceptance on halal food among non-Muslim consumers. *Procedia-Social and Behavioral Sciences*, 121, 262–271.

MS1500, M.S. (2009). *Halal Food—Production, Preparation, Handling and Storage—General Guideline*. Cyberjaya, Malyasia: Department of Standards Malaysia, 1–13.

Nakyinsige, K., Che Man, Y. B., Sazili, A. Q., Zulkifli, I., and Fatimah, A. B. (2012a). Halal meat: A niche product in the food market. In *2nd International Conference on Economics, Trade and Development IPEDR*, Singapore: IACSIT Press, Vol. 36, 167–173.

Nakyinsige, K., Man, Y. B. C., and Sazili, A. Q. (2012b). Halal authenticity issues in meat and meat products. *Meat Science*, 91(3), 207–214.

Norhabibah, C. H. (2011). *The confidence level of purchasing product with Halal logo among consumers,* Doctoral dissertation, Universiti Utara Malaysia, Kedah, Malaysia.

Omar, E. N. and Jaafar, H. S. (2011). Halal supply chain in the food industry-A conceptual model. In *2011 IEEE Symposium on Business, Engineering and Industrial Applications (ISBEIA),* Bayview Hotel Langkawi, Malaysia, 384–389.

Papazyan, T. T. and Surai, P. F. (2007). EU clearance of Sel-Plex®: Expanding the possibilities for new nutraceutical foods. In *Nutritional Biotechnology in the Feed and Food Industries: Proceedings of Alltech's 23rd Annual Symposium. The New Energy Crisis: Food, Feed or Fuel?* Eds. Lyons, T. P., Jacques, K. A. and Hower, J. M. Stamford, UK: Alltech UK, 193–201.

Regenstein, J. M. and Chaudry, M. M. (2002). Kosher and halal issues pertaining to edible films and coating. In *Protein Based Film and Coating*, Gennadios, A., Ed. Boca Raton, FL: CRC Press, 601–620.

Schaefer, A. L., Dubeski, P. L., Aalhus, J. L., and Tong, A. K. W. (2001). Role of nutrition in reducing antemortem stress and meat quality aberrations. *Journal of Animal Science*, 79(E-Suppl), E91–E101.

Shafie, S. and Othman, M. N. (2006). Halal Certification: An international marketing issues and challenges. In *Proceeding at the International IFSAM VIIIth World Congress*, Berlin, Germany, 28–30.

Siti Hasnah, H., Dann, S., Annuar, M. K., and De Run, E. C. (2009). Influence of the halal certification mark in food product advertisement in Malaysia. In: *The New Culture of food. Marketing Opportunity from Ethnic, Religious and Cultural Diversity*, Eds. Lindgreen A. and Hingley M. K. Surrey, England: Gower Publishing Limited, chapter 14.

Soong, S. F. V. (2007). Managing halal quality in food service industry. *Professional Paper, Hotel Administration*. Las Vegas: University of Nevada. Available online: https://digitalscholarship.unlv.edu/thesesdissertations/701/.

Wibowo, M. W. and Ahmad, F. S. (2016). Non-Muslim consumers' halal food product acceptance model. *Procedia Economics and Finance*, 37, 276–283.

Yahaya, C. K. H. C. K., Kassim, M., Bin Mazlan, M. H., and Bakar, Z. A. (2011). A framework on halal product recognition system through smartphone authentication. In *Advances in Automation and Robotics*, Ed. Lee, G., Vol. 1. Berlin, Heidelberg: Springer, 49–56.

31 HACCP and Halal

Mian N. Riaz, Sibte Abbas, and Munir M. Chaudry

CONTENTS

Introduction	319
HACCP	320
Haram	321
Najis	322
Mashbooh	322
Makrooh	322
Halal HACCP Implementation with Halal Products	322
References	324

INTRODUCTION

The word halal comes from the Arabic language meaning "lawful" or "permitted." It generally refers to things or actions that are permissible or lawful under Islamic law (Al-Hanafi, 2006). Islamic law provides a legal framework for the regulation of public as well as some private facets of life of people living in accordance with Islam (Bonne and Verbeke, 2008). The opposite of halal is haram, which means unlawful or prohibited. Halal and haram are the terms applied to every aspect of life but mostly focused on food products and ingredients, food contact materials, meat products, pharmaceuticals, cosmetics, and personal care products. Muslims must ensure that all of the previously mentioned commodities must be halal. Often these products contain by-products or ingredients from animal sources or ethanol that are not permissible for Muslim consumption (Regenstein et al., 2003).

There are a few things which are not clearly categorized as halal or haram. Such things are often referred to as mashbooh, meaning doubtful or questionable. Further information is needed to categorize them as halal or haram. In the light of the Quran (Holy book of Islam), Hadiths and Sunnah (teachings and actions of the Holy Prophet Muhammad, PBUH), the foods are categorized as halal or haram mainly depending on their sources and the methods of production. All foods from the following sources are considered haram:

1. Swine/pork and its by-products
2. Animals NOT properly slaughtered according to Islamic method
3. Animals dead before slaughtering
4. Carnivorous animals and birds of prey
5. Blood and blood by-products
6. Alcoholic drinks and intoxicants
7. Foods contaminated with any materials mentioned above

Some foods having ingredients like enzymes, gelatin, flavors, and emulsifiers are questionable (mashbooh) as these ingredients may come from a haram source. Utmost care is required in slaughtering and processing of meat and poultry according to Islamic requirements. For the slaughter of animals, this is commonly known as Zabiha or Dhabiha, which refers to the slaughtering of an animal or bird by a Muslim according to Islamic requirements.

Halal slaughter includes the cutting of an animal in a way so it bleeds, bleeding results in the loss of life, and to invoke the name of God while slaughtering (Malaysian Standard 1500, 2009). The specific conditions for zabiha require that the animal must be slaughtered using a sharpened blade so that it does not suffer for a long time (Grandin, 2010). The animal must not see the blade or smell the blood from a previous slaughter. This actually explains the importance of life given by God. Therefore, by uttering the name of God at the time of slaughtering, one implies that the slaughter of the animal is being done at the command of God (Casoli et al., 2005; Micera et al., 2010; Nakyinsige et al., 2012). Conventionally, slaughtering was a manual process, usually known as "hand slaughter," in which an individual administers the cutting especially for large animals such as cattle and sheep. But this process has been subjected to automation owing to the large volume of animals and chickens that need to be slaughtered in the commercial meat and poultry industry. The automation, mainly for poultry, is called "machine slaughter" and uses an automated knife to do the cutting. The halal industry has grown over the past few years, with increasing demand for halal products from both Muslim and non-Muslim consumers. With increased awareness about food safety, food safety systems like GMP (good manufacturing practices) and GHP (good hygienic practices) are being used along with more attention to the personal hygiene of food industry workers and food handlers (Malaysian Standard 1500, 2009).

The Hazard Analysis and Critical Control Points (HACCP) system is a systematic approach to assure the production of safe and wholesome food by identifying and controlling various hazards (i.e., microbiological, chemical, or physical) that could pose a serious threat to the manufacturing process. The main objective of HACCP is to assure the safety of food materials from farm to fork, that is, from raw material to the end product through identification, evaluation, and control of potential hazards. Basic food hygiene conditions and practices (prerequisites) must be in place prior to implementing a HACCP system (Bas et al., 2006).

HACCP

The basic elements and characteristics of HACCP include the following:

1. Identify the potential food safety hazards
2. Determine preventive measures for the hazards identified
3. Implement those preventive measures to enhance food safety
4. Train personnel to identify food safety hazards and implement preventive actions
5. Keep a record of the measures taken

HACCP and Halal

To implement HACCP in a food plant, a comprehensive review of the plant and process is needed that permits the formal HACCP program to be developed and initiated by the following:

1. Conduct a hazard analysis
2. Determine the CCP (critical control points)
3. Establish critical limits for each CCP
4. Establish a monitoring system
5. Establish corrective actions
6. Establish validation and verification procedures
7. Establish documentation to support all of the above items (Figure 31.1)

Halal authenticity is needed to provide consumer confidence and assurance that a food indicated as halal is halal and can be bought without any hesitation. The standards for halal foods and the HACCP system have a common objective, that is, to provide safe and wholesome food. However, the halal food production requires some additional requirements to meet the specific criteria expected including some form of Muslim supervision. After manufacture, the food may not come in direct contact with any mashbooh or haram materials including packaging while it is being handled and transported (Chan, 2008; Soesilowati, 2011). Halal standards and HACCP together can provide a complete system for analyzing any food operation to identify potential food safety hazards and halal violations. The religious violations fall into four categories.

HARAM

Some foods, such as pork and alcohol are always forbidden. Some objects, foods, or actions that are normally halal can, under some conditions, become haram. Examples include halal foods and drinks consumed during the fasting time of the holy month

FIGURE 31.1 A simple quality management structure that describes the basic system for the production of all foods.

of Ramadan, and cattle or other halal animals that are not slaughtered in the Islamic way or without the name of Allah (God) being invoked.

Najis

Food should not contain any ingredients that are najis (unclean) according to Islamic law.

Mashbooh

Foods that are labeled as "mashbooh" should normally be avoided.

Makrooh

In Islam, the word "makrooh" is defined as anything that is undesirable or inappropriate. Muslims believe that makrooh foods are determined by their own innate sense of right and wrong. For example, most Muslims consider it makrooh to eat the meat of a horse, donkey, or mule.

HALAL HACCP IMPLEMENTATION WITH HALAL PRODUCTS

In addition to critical control points (CCP) in HACCP, it is often beneficial to establish a set of halal control points (HCP) that operate in parallel to the HACCP system, which is often required by secular law. These HCP should be designed to monitor, control, and remove any halal violations. Potential religious issues can be categorized into two sections: Tolerable and non-tolerable. Potential hazards that cannot be tolerated include stages or points in slaughtering or processing that clearly contradict the requirements of Islamic law. Examples of hazards that cannot be tolerated include: the presence of meat, fat, gelatin, or any other part of a pig; the presence of gushed blood in a manufactured product; the absence of acceptable religious cleansing (taharah) of production lines; and the use of stunning that led to the death of an otherwise halal animal or the method of slaughter of the animal was not in accordance with the prescribed conditions for halal slaughter (Karaman et al., 2012).

Potential religious issues that can be tolerated include forgetting (non-deliberately) to utter the name of Allah at the time of slaughter of an animal (in some schools) and forgetting to direct the bird or animal at the time of slaughter toward the Qibla (it is the direction a Muslim face when prays). These HCP can as a practical matter be incorporated into the company's HACCP management system. The identification of the HCP follows the same procedures as were used to determine the CCP. Thus, the HCP is defined as a point, step, or procedure in halal food production manufacturing at which a control can be applied and, as a result, halal food cross-contamination can be prevented or eliminated. Some practical guidance for a successful halal production system include:

- Halal and non-halal ingredients should be separated and stored in dedicated areas
- Cases/packages of a halal food products should not be placed on the same pallet as non-halal products

HACCP and Halal

- Processing of halal product without appropriate cleaning after production of non-halal product results in cross-contamination
- Addition of a non-halal ingredient (e.g., lard, non-halal gelatin) into halal products must be avoided
- Rework of non-halal products is not allowed. The common steps in creating both a HACCP program and halal control program include:
  - Assembling a HACCP/food quality team
  - Discussing the description of all products including which are and which are not halal
  - Identifying intended use
  - Constructing a process flow diagram
  - Conducting an on-site verification of the flow diagram

One then uses this information as previously for HACCP:

1. Identify and analyze all possible hazards or halal issues
2. Determine CCP and HCP
3. Establish control measures
4. Establish a monitoring system for all CCP and HCP
5. Establish corrective action for deviations that may occur
6. Establish validation and verification procedures
7. Establish record keeping and documentation including the disposal of product after an HCP failure

Like HACCP, the halal program needs appropriate documentation such as:

- Product description and halal ingredient specifications
- An HCP decision form
- Monitoring and corrective action form
- Record keeping and verification form
- Validation form and verification documentation
- A halal program summary form is recommended
- Additional forms—as necessary for various type of food processing operations

A Basic Structure for Halal Process Control

It is particularly important that the halal processing control system be in place for plants producing both halal and non-halal products in the same area. The basic job of a halal certification body is to ensure and certify that all the ingredients and production methods being used are halal, which permits such companies to serve both their domestic market and the export market with confidence. Halal foods are also becoming popular among non-Muslims who are health-conscious and want to eat halal products owing to the fact that these products are manufactured in a clean environment and in a sympathetic manner with respect to the treatment of animals at slaughter. Adding a halal control program to a HACCP program gives an additional layer of protection for those wishing to consume halal food products. Therefore,

HACCP coupled with halal standards can become a very effective approach for the halal industry, providing the consumers with halal food having higher standards of safety and quality (Bas et al., 2007; Karaman et al., 2012).

REFERENCES

Aamababmat Al-Hanafi. (2006). KashafIstilahatulFanun. Beirut: Daru AlKutubul Alilmiya.
Baş, M., Ersun, A. Ş., and Kıvanç, G. (2006). Implementation of HACCP and prerequisite programs in food businesses in Turkey. *Food Control, 17*(2), 118–126.
Baş, M., Yüksel, M., and Çavuşoğlu, T. (2007). Difficulties and barriers for the implementing of HACCP and food safety systems in food businesses in Turkey. *Food Control, 18*(2), 124–130.
Bonne, K. and Verbeke, W. (2008). Religious values informing halal meat production and the control and delivery of halal credence quality. *Agriculture and Human Values*, 25(1), 35–47.
Casoli, C., Duranti, E., Cambiotti, F., and Avellini, P. (2005). Wild ungulate slaughtering and meat inspection. *Veterinary Research Communications*, 29(2), 89–95.
Chan, E. S. (2008). Barriers to EMS in the hotel industry. *International Journal of Hospitality Management*, 27(2), 187–196.
Grandin, T. (2010). Auditing animal welfare at slaughter plants. *Meat Science*, 86(1), 56–65.
Karaman, A. D., Cobanoglu, F., Tunalioglu, R., and Ova, G. (2012). Barriers and benefits of the implementation of food safety management systems among the Turkish dairy industry: A case study. *Food Control*, 25(2), 732–739.
Micera, E., Albrizio, M., Surdo, N. C., Moramarco, A. M., and Zarrilli, A. (2010). Stress-related hormones in horses before and after stunning by captive bolt gun. *Meat science*, 84(4), 634–637.
MS1500, M.S. (2009). Halal food—Production, preparation, handling and storage—General guideline. Cyberjaya: Department of Standards Malaysia, 1–13.
Nakyinsige, K., Che Man, Y. B., Sazili, A. Q., Zulkifli, I., and Fatimah, A. B. (2012). Halal meat: A niche product in the food market. In *2nd International Conference on Economics, Trade and Development IPEDR*, Singapore: ACSIT Press, Vol. 36, 167–173.
Regenstein, J. M., Chaudry, M. M., and Regenstein, C. E. (2003). The kosher and halal food laws. Comprehensive reviews in food science and food safety, 2(3), 111–127.
Soesilowati, E. S. (2011). Business opportunities for halal products in the global market: Muslim consumer behaviour and halal food consumption. *Journal of Indonesian Social Sciences and Humanities*, 3, 151–160.

Epilogue

There are several emerging issues that have not been covered in this book nor discussed in detail with regard to their halal status. Process and product development technology is progressing at a fast pace and the need for religious clarification is greater than ever. Some of the emerging issues are discussed below.

DEVELOPMENT OF FERMENTATION PRODUCTS LIKE "CLEAN MEAT"

The term clean meat refers to a product manufactured in a laboratory or factory that is similar to genuine animal meat but grown by farming cells rather than slaughtering an animal. Clean meat, is also called cultured meat, *in vitro* meat, synthetic meat, vat-grown or lab-grown meat, and several other terms, has been used for meat grown *in vitro* using animal cell culture systems. Clean meat production is a form of cellular agriculture. The process of producing clean meat uses many of the same tissue engineering methods traditionally used in regenerative medicine. Each company has its own proprietary techniques for producing clean meat, although the overall process is similar. Some of the advantages of clean meat are that this technology is supposed to be much friendlier to the environment than conventional livestock production. Researchers estimate that the production of 1,000 kg of cultured meat would require 7%–45% less energy, 99% less land, and 82%–96% less water depending on the species of meat when compared with the same mass of regular meat. Perhaps the most prominent benefit of clean meat would be in the area of animal welfare. Scientists believe that clean meat would be safer for human consumption than traditional meat. Because clean meat is grown in a closed, controlled environment, it can be produced without antibiotics and risk of contamination with pathogenic bacteria like *Salmonella* and *E. coli*, as long as subsequent processing is done well. Several clean meat prototypes have gained attention recently; however, clean meat has not yet been commercialized at the time this is being written. One of the major challenges will be bringing clean meat from the laboratory to the industrial scale, and obtaining consumer acceptance. Another complicating factor is that, so far, clean meat prototypes have focused on ground meat, sausages, and nugget-like products. Structured meats, such as steaks and chicken breasts, will require further innovations. One more challenge that clean meat faces is that clean meat prototypes have lacked fat, which imparts flavor, aroma, and texture to meat. For that reason, many companies are trying to determine the best way to incorporate fat into clean meat. Lastly, there is no guarantee that consumers will accept the products. Scientists still need to provide information about the different clean meat manufacturing processes and their final product to religious scholars so they can determine its halal status.

PRODUCTS FROM INSECTS AND OTHER NON-CONVENTIONAL SOURCES

The second issue is the use of insects as a source of food as well as its protein, which can be used in human food and animal feed applications. Edible insects contain high-quality protein, vitamins and amino acids for humans and animals. The issue of insect consumption has been controversial in the Muslim community. While some denominations and scholars prohibit all insects, others allow most of them, and others only allow some very specific ones, based on the fact that the Prophet Muhammad (PBUH) did eat locusts. They also divide them into two categories, insects with blood and insects without blood. Those insects without blood have been classified as either pure or impure; the ones found in fruits, vegetables, wood, etc. are considered pure, while those that are parasites are considered impure. With new technology and market demand, insect extracts are finding their way into a number of human food products. Islamic jurists have discussed the eating of insects as a source of food, however, insect extracts and powders have not been discussed. For some Muslim consumers the derivative or byproducts of insects added to food will not be a major issue, as long as it is not harmful to a human's health. Most Muslim scholars agree that eating locusts is acceptable. (Note that crickets, which have been the major insect to date used for human food, are not automatically covered by these rulings). Insects are increasingly considered a good source of animal protein as well. Insect protein could replace some of the expensive fishmeal and soybean meal in animal feed and fish farming.

These and other evolving issues and subjects will be discussed in future publications. It is the hope of the authors that this book will help both Muslim consumers to be better consumers and the food industry to better understand how to serve this growing market in the West and provide new opportunities for export to the global Muslim community.

Appendix A
Key Terminology from Other Languages

Ahadith[a]	Plural of Hadith
Ahlul-Kitab[a]	People of the book, the ones who have received the scripture
Allah[a]	The One God
Allahu Akbar[a]	The God is Greater
As-Salat[a]	Daily and other prayers by a Muslim
Bismillah[a]	In the name of God
Chalef[b]	A special knife for Kosher kill
Chometz[b]	Prohibited grains in Jewish laws
Dhabh[a]	Islamic method of slaughtering an animal by slicing the neck
Dhabiha[a]	Slaughtered according to the Islamic method
Dhakaat[a]	Purification of meat by proper slaughtering of an animal
Fiqh	Deep understanding or "full comprehension"
Hadith[a]	Saying of the Prophet Muhammad
Haj[a]	Pilgrimage to Mecca by a Muslim
Halaal/Halal[a]	Lawful or permitted in Islamic laws
Halacha[b]	System of Jewish laws
Haraam/Haram[a]	Unlawful or prohibited in Islamic laws
Istihala[a]	Distinct chemical change in a product
Jalalah/Jalallah[a]	Condition of an animal, eating filth, and living in filth
Janaba[a]	A condition of being unclean
Khamr[a]	Fermented, frothing, alcoholic drinks
Kosher[b]	Lawful or permitted in Jewish laws
Makrooh[a]	Disliked or detested
Mashbooh[a]	Doubtful or questionable
Nahr[a]	Islamic method of slaughtering an animal by spearing or stabbing on the neck
Najis[a]	Unclean or dirty
Pareve[b]	A neutral non-dairy, non-meat Kosher product
Qibla[a]	It is the direction a Muslim face when prays
Quran[a]	The Muslim Holy Scripture
Shari'ah[a]	The Islamic jurisprudence or laws
Shochet[a]	A Jewish slaughter man

(Continued)

Sunnah[a]	Traditions (sayings and actions) of Prophet Muhammad
Taharah[a]	Cleanliness or Purification
Tasmiyyah[a]	Invoking the name of God
Treif[b]	Not kosher or prohibited in the Jewish law
Umra[a]	Short version of Haj
Zabh[c]	A variant pronunciation of Dhabh
Zabiha/Zabeeha[c]	A variant pronunciation of Dhabiha

[a] Arabic.
[b] Hebrew.
[c] Urdu.

Appendix B
Permitted Food Additives in the European Union[1,2]

Coloring materials	Used to make food more colorful and attractive, or to replace color lost in processing	
Yellow and orange colors		Halal status
E-100	Curcumin	Halal
E-101	Riboflavin, riboflavin phosphate (vitamin B_2)	Halal
E-102	Tartrazine (=FD&C Yellow no 6)	Halal
E-104	Quinoline yellow	Halal
107	Yellow 2G	Halal
E-110	Sunset yellow FCF or orange yellow S (=FD&C Yellow no 6)	Halal
Red colors		
E-120	Cochineal or carminic acid	Doubtful
E-122	Carmoisine or azorubine	Halal
E-123	Amaranth	Halal
E-124	Ponceau 4R or cochineal red A	Doubtful
E-127	Erythrosine BS (=FD&C Red no 3)	Halal
E-128	Red 2G	Halal
E-129	Allura red (=FD&C Red no 40)	Halal
Blue colors		
E-131	Patent blue V	Halal
E-132	Indigo carmine or indigotine (=FD&C Blue no 2)	Halal
E-133	Brilliant blue FCF (=FD&C Blue no 1)	Halal
Green colors		
E-140	(i) Chlorophylls, the natural green color of leaves, (ii) chlorophyllins	Halal
E-141	Copper complexes of (i) chlorophylls, (ii) chlorophyllins	Halal
E-142	Green S or acid brilliant green BS	Halal
Brown and black colors		
E-150a	Plain caramel (made from sugar in the kitchen)	Halal
E-150b	Caustic sulfite caramel	Halal
E-150c	Ammonia caramel	Halal
E-150d	Sulfite ammonia caramel	Halal

(*Continued*)

Coloring materials	Used to make food more colorful and attractive, or to replace color lost in processing	
E-151	Black PN or brilliant black BN	Halal
E-153	Carbon black or vegetable carbon (charcoal)	Halal
E-154	Brown FK	Halal
E-155	Brown HT (chocolate brown HT)	Halal
Derivatives of carotene		Halal status
E-160(a)	(i) Mixed carotenes, (ii) ß-carotene	Halal
E-160(b)	Annatto, bixin, norbixin	Halal
E-160(c)	Paprika extract, capsanthin or capsorubin	Halal
E-160(d)	Lycopene	Halal
E-160(e)	ß-Apo-8'-carotenal (vitamin A active)	Halal
E-160(f)	Ethyl ester of ß-apo-8'-carotenoic acid	Halal
Other plant colors		
E-161(a)	Flavoxanthin	Halal
E-161(b)	Lutein	Halal
E-161(c)	Cryptoxanthin	Halal
E-161(d)	Rubixanthin	Halal
E-161(e)	Violaxanthin	Halal
E-161(f)	Rhodoxanthin	Halal
E-161(g)	Canthaxanthin	Halal
E-162	Beetroot red or betanin	Halal
E-163	Anthocyanins (the pigments of many plants)	Halal
Inorganic compounds used as colors		
E-170	(i) Calcium carbonate (chalk), (ii) calcium hydrogen carbonate	Halal
E-171	Titanium dioxide	Halal
E-172	Iron oxides and hydroxides	Halal
E-173	Aluminum	Halal
E-174	Silver	Halal
E-175	Gold	Halal
E-180	Pigment rubine or lithol rubine BK	Halal
Preservatives	*Compounds that protect foods against microbes that cause spoilage and food poisoning. They increase the safe storage life of foods*	
Sorbic acid and its salts		
E-200	Sorbic acid	Halal
E-201	Sodium sorbate	Halal
E-202	Potassium sorbate	Halal
E-203	Calcium sorbate	Halal
Benzoic acid and its salts		
E-210	Benzoic acid (occurs naturally in many fruits)	Halal
E-211	Sodium benzoate	Halal
E-212	Potassium benzoate	Halal
E-213	Calcium benzoate	Halal

(*Continued*)

Appendix B

Coloring materials	Used to make food more colorful and attractive, or to replace color lost in processing	
E-214	Ethyl *p*-hydroxybenzoate	Halal
E-215	Ethyl *p*-hydroxybenzoate sodium salt	Halal
E-216	Propyl *p*-hydroxybenzoate	Halal
E-217	Propyl *p*-hydroxybenzoate sodium salt	Halal
E-218	Methyl *p*-hydroxybenzoate	Halal
E-219	Methyl *p*-hydroxybenzoate sodium salt	Halal
Sulfur dioxide and its salts		Halal status
E-220	Sulfur dioxide (also used to prevent browning of raw peeled potatoes)	Halal
E-221	Sodium sulfite	Halal
E-222	Sodium hydrogen sulfite	Halal
E-223	Sodium metabisulfite	Halal
E-224	Potassium metabisulfite	Halal
E-226	Calcium sulfite	Halal
E-227	Calcium hydrogen sulfite	Halal
E-228	Potassium hydrogen sulfite	Halal
Biphenyl and its derivatives		
E-230	Biphenyl or diphenyl (for surface treatment of citrus fruits)	Halal
E-231	Orthophenylphenol (2-hydroxybiphenyl) (for surface treatment of citrus fruits)	Halal
E-232	Sodium orthophenylphenol (sodium biphenyl-2-yl oxide)	Halal
Other preservatives		
E-233	2-(Thiazol-4-yl) benzimidazole (thiobendazole) (for surface treatment of citrus fruits and bananas)	Halal
E-234	Nisin	Halal
E-235	Natamycin (NATA) (for surface treatment of cheeses and dried cured sausages)	Halal
E-239	Hexamethylene tetramine (hexamine)	Halal
E-242	Dimethyl dicarbonate	Halal
E-912	Montan acid esters (for surface treatment of citrus fruits)	Doubtful
E-914	Oxidized polyethylene wax (for surface treatment of citrus fruits)	Halal
Pickling salts		
E-249	Potassium nitrite	Halal[a]
E-250	Sodium nitrite	Halal[a]
E-251	Sodium nitrate	Halal[a]
E-252	Potassium nitrate (saltpetre)	Halal[a]
Acids and their salts	*Used as flavorings and as buffers to control the acidity of foods, in addition to their anti-microbial properties*	
E-260	Acetic acid	Halal
E-261	Potassium acetate	Halal

(*Continued*)

Coloring materials	Used to make food more colorful and attractive, or to replace color lost in processing	
E-262	(i) Sodium acetate, (ii) sodium hydrogen acetate (sodium diacetate)	Halal
E-263	Calcium acetate	Halal
E-270	Lactic acid	Halal
E-280	Propionic acid	Halal
E-281	Sodium propionate	Halal
E-282	Calcium propionate	Halal
		Halal status
E-283	Potassium propionate	Halal
E-284	Boric acid (as preservative in caviar)	Halal
E-285	Sodium tetraborate (borax) (as preservative in caviar)	Halal
E-290	Carbon dioxide	Halal
E-296	Malic acid	Halal
E-297	Fumaric acid	Halal

Antioxidants Compounds used to prevent fatty foods going rancid, and to protect fat-soluble vitamins (A, D, E, and K) against the damaging effects of oxidation

Vitamin C and derivates

E-300	l-Ascorbic acid (vitamin C)	Halal
E-301	Sodium l-ascorbate	Halal
E-302	Calcium l-ascorbate	Halal
e-304	(i) Ascorbyl palmitate, (ii) ascorbyl stearate	Doubtful
E-315	Erythorbic acid (*iso*-ascorbic acid)	Halal
E-316	Sodium erythorbate (sodium *iso*-ascorbate)	Halal

Vitamin E

E-306	Natural extracts rich in tocopherols	Halal
E-307	Synthetic α-tocopherol	Halal
E-308	Synthetic γ-tocopherol	Halal
E-309	Synthetic δ-tocopherol	Halal

Other antioxidants

E-310	Propyl gallate	Halal
E-311	Octyl gallate	Halal
E-312	Dodecyl gallate	Doubtful
E-320	Butylated hydroxyanisole (BHA)	Halal
E-321	Butylated hydroxytoluene (BHT)	Halal
E-322	Lecithins	Doubtful

More acids and their salts	Used as flavorings and as buffers to control the acidity of foods, in addition to other special uses	

Salts of lactic acid (E270)

E-325	Sodium lactate	Halal
E-326	Potassium lactate	Halal
E-327	Calcium lactate	Halal
E-585	Ferrous lactate	Halal

(*Continued*)

Appendix B

Coloring materials	Used to make food more colorful and attractive, or to replace color lost in processing	
Citric acid, its salts and esters		
E-330	Citric acid (formed in the body, and present in many fruits)	Halal
E-331	(i) Monosodium citrate, (ii) disodium citrate, (iii) trisodium citrate	Halal
E-332	(i) Monopotassium citrate, (ii) dipotassium citrate, (iii) tripotassium citrate	Halal
E-333	(i) Monocalcium citrate, (ii) dicalcium citrate, (iii) tricalcium citrate	Halal
E-1505	Triethyl citrate	Halal
Tartaric acid and its salts		Halal status
E-334	l(+)Tartaric acid (tartaric acid occurs naturally; as well as their properties as acids, tartrates are often used as sequestrants and emulsifying agents)	Halal
E-335	(i) Monosodium tartrate, (ii) disodium tartrate	Halal
E-336	(i) Monopotassium tartrate (cream of tartar), (ii) dipotassium tartrate	Halal
E-337	Sodium potassium tartrate	Halal
Phosphoric acid and its salts		
E-338	Phosphoric acid	Halal
E-339	(i) Monosodium phosphate, (ii) disodium phosphate, (iii) trisodium phosphate	Halal
E-340	(i) Monopotassium phosphate, (ii) dipotassium phosphate, (iii) tripotassium phosphate	Halal
E-341	(i) Monocalcium phosphate, (ii) dicalcium phosphate, (iii) tricalcium phosphate	Halal
E-450	Diphosphates: (i) disodium diphosphate, (ii) trisodium diphosphate, (iii) tetrasodium diphosphate, (iv) dipotassium diphosphate, (v) tetrapotassium diphosphate, (vi) dicalcium diphosphate, (vii) calcium dihydrogen diphosphate	Halal
E-451	Triphosphates: (i) pentasodium triphosphate, (ii) pentapotassium triphosphate	Halal
E-452	Polyphosphates: (i) sodium polyphosphate, (ii) potassium polyphosphate, (iii) sodium calcium polyphosphate, (iv) calcium polyphosphate	Halal
E-540	Dicalcium diphosphate	Halal
E-541	Sodium aluminum phosphate, acidic	Halal
E-542	Edible bone phosphates (bone meal, used as anticaking agent)	Doubtful
E-544	Calcium polyphosphates (used as anticaking agent)	Halal
E-545	Ammonium polyphosphates (used as anticaking agent)	Halal

(*Continued*)

Coloring materials	Used to make food more colorful and attractive, or to replace color lost in processing	
Salts of malic acid (E-296)		
E-350	Sodium malate	Halal
E-351	Potassium malate	Halal
E-352	Calcium malate	Halal
Other acids and their salts		
E-353	Metatartaric acid	Halal
E-354	Calcium tartrate	Halal
E-355	Adipic acid	Halal
E-356	Sodium adipate	Halal
E-357	Potassium adipate	Halal
E-363	Succinic acid	Halal
E-370	1,4-Heptonolactone	Halal
E-375	Nicotinic acid	Halal
E-380	Triammonium citrate	Halal
E-381	Ammonium ferric citrate	Halal
E-385	Calcium disodium EDTA	Halal
Emulsifiers and stabilizers	Used to enable oils and fats to mix with water, to give a smooth and creamy texture to food, and slow the staling of baked goods. Many of these compounds are also used to make jellies.	
Alginates		Halal status
E-400	Alginic acid (derived from seaweed)	Halal
E-401	Sodium alginate	Halal
E-402	Potassium alginate	Halal
E-403	Ammonium alginate	Halal
E-404	Calcium alginate	Halal
E-405	Propane-1,2-diol alginate	Halal
Other plant gums		
E-406	Agar (derived from seaweed)	Halal
E-407	Carrageenan (derived from the seaweed Irish moss)	Halal
E-410	Locust bean gum (carob gum)	Halal
E-412	Guar gum	Halal
E-413	Tragacanth	Halal
E-414	Gum acacia (gum Arabic)	Halal
E-415	Xanthan gum	Halal
E-416	Karaya gum	Halal
E-417	Tara gums	Halal
E-418	Gellan gums	Halal
Fatty acid derivatives		
E-430	Polyoxyethylene (8) stearate	Doubtful
E-431	Polyoxyethylene (40) stearate	Doubtful
E-432	Polyoxyethylene (20) sorbitan monolaurate (Polysorbate 20)	Doubtful

(Continued)

Appendix B

Coloring materials	Used to make food more colorful and attractive, or to replace color lost in processing	
E-433	Polyoxyethylene (20) sorbitan mono-oleate (Polysorbate 80)	Doubtful
E-434	Polyoxyethylene (20) sorbitan monopalmitate (Polysorbate 40)	Doubtful
E-435	Polyoxyethylene (20) sorbitan monostearate (Polysorbate 60)	Doubtful
E-436	Polyoxyethylene (20) sorbitan tristearate (Polysorbate 65)	Doubtful
Pectin and derivatives		
E-440	(i) Pectin, (ii) amidated pectin (pectin occurs in many fruits, and is often added to jam to help it set)	Halal
Other compounds		
E-322	Lecithins	Doubtful
E-442	Ammonium phosphatides	Doubtful
E-444	Sucrose acetate isobutyrate	Doubtful
E-445	Glycerol esters of wood rosins	Doubtful
Cellulose and derivatives		Halal status
E-460	(i) Microcrystalline cellulose, (ii) powdered cellulose	Halal
E-461	Methyl cellulose	Halal
E-463	Hydroxypropyl cellulose	Halal
E-464	Hydroxyproplymethyl cellulose	Halal
E-465	Ethylmethyl cellulose	Halal
E-466	Carboxymethylcellulose, sodium carboxymethylcelluslose	Halal
Salts or esters of fatty acids		
E-470a	Sodium, potassium, and calcium salts of fatty acids	Doubtful
E-470b	Magnesium salts of fatty acids	Doubtful
E-471	Mono- and diglycerides of fatty acids	Doubtful
E-472a	Acetic acid esters of mono- and diglycerides of fatty acids	Doubtful
E-472b	Lactic acid esters of mono- and diglycerides of fatty acids	Doubtful
E-472c	Citric acid esters of mono- and diglycerides of fatty acids	Doubtful
E-472d	Tartaric acid esters of mono- and diglycerides of fatty acids	Doubtful
E-472e	Mono- and diacetyl tartaric esters of mono- and diglycerides of fatty acids	Doubtful
E-472f	Mixed acetic and tartaric acid esters of mono- and diglycerides of fatty acids	Doubtful
E-473	Sucrose esters of fatty acids	Doubtful
E-474	Sucroglycerides	Doubtful
E-475	Polyglycerol esters of fatty acids	Doubtful
E-476	Polyglycerol esters of polycondensed esters of castor oil) polyglycerol polyricinoleate)	Doubtful
E-477	Propane-1,2-diol esters of fatty acids	Doubtful

(*Continued*)

Coloring materials	Used to make food more colorful and attractive, or to replace color lost in processing	
E-478	Lactylated fatty acid esters of glycerol and propane-1,2-diol	Doubtful
E-479b	Thermally oxidized soya bean oil interacted with mono- and diglycerides of fatty acids	Doubtful
E-481	Sodium stearoyl-2-lactylate	Doubtful
E-482	Calcium stearoyl-2-lactylate	Doubtful
E-483	Stearyl tartrate	Doubtful
E-491	Sorbitan monostearate	Doubtful
E-492	Sorbitan tristearate	Doubtful
E-493	Sorbitan monolaurate	Doubtful
E-494	Sorbitan mono-oleate	Doubtful
E-495	Sorbitan monopalmitate	Doubtful
E-1518	Glyceryl triacetate (triacetin)	Doubtful

Acids and salts use buffers, emulsifying salts, sequestrants, stabilizers, and raising agents for special purposes anti-caking agents

Carbonates		Halal status
E-500	(i) Sodium carbonate, (ii) sodium bicarbonate (sodium hydrogen carbonate), (iii) sodium sesquicarbonate)	Halal
E-501	(i) Potassium carbonate, (ii) potassium bicarbonate (potassium hydrogen carbonate)	Halal
E-503	(i) Ammonium carbonate, (ii) ammonium hydrogen carbonate	Halal
E-504	(i) Magnesium carbonate, (ii) magnesium hydrogen carbonate (magnesium hydroxide carbonate)	Halal
Hydrochloric acid and its salts		
E-507	Hydrochloric acid (ordinary salt is sodium chloride)	Halal
E-508	Potassium chloride (sometimes used as a replacement for ordinary salt)	Halal
E-509	Calcium chloride	Halal
E-510	Ammonium chloride	Halal
E-511	Magnesium chloride	Halal
E-512	Stannous chloride	Halal
Sulfuric acid and its salts		
E-513	Sulfuric acid	Halal
E-514	(i) Sodium sulfate, (ii) sodium hydrogen sulfate	Halal
E-515	(i) Potassium sulfate, (ii) potassium hydrogen sulfate	Halal
E-516	Calcium sulfate	Halal
E-517	Ammonium sulfate	Halal
E-518	Magnesium sulfate	Halal
E-520	Aluminum sulfate	Halal
E-521	Aluminum sodium sulfate	Halal
E-522	Aluminum potassium sulfate	Halal
E-523	Aluminum ammonium sulfate	Halal

(Continued)

Appendix B

Coloring materials	Used to make food more colorful and attractive, or to replace color lost in processing	
Alkalis	Used as bases to neutralize acids in foods	
E-524	Sodium hydroxide	Halal
E-525	Potassium hydroxide	Halal
E-526	Calcium hydroxide	Halal
E-527	Ammonium hydroxide	Halal
E-528	Magnesium hydroxide	Halal
E-529	Calcium oxide	Halal
E-530	Magnesium oxide	Halal
Other salts		
E-535	Sodium ferrocyanide	Halal
E-536	Potassium ferrocyanide	Halal
E-538	Calcium ferrocyanide	Halal
E-540	Dicalcium diphosphate	Halal
E-541	Sodium aluminum phosphate, acidic	Halal
Compounds used as anticaking agents, and other uses		Halal status
E-542	Edible bone phosphate (bone meal)	Doubtful
E-544	Calcium polyphosphates	Halal
E-545	Ammonium plyphosphates	Halal
Silicon salts		
E-551	Silicon dioxide (silica, sand)	Halal
E-552	Calcium silicate	Halal
E-553a	(i) Magnesium silicate, (ii) magnesium trisilicate	Halal
E-553b	talc	Halal
E-554	Sodium aluminum silicate	Halal
E-555	Potassium aluminum silicate	Halal
E-556	Calcium aluminum silicate	Halal
Other compounds		
E-558	Bentonite	Halal
E-559	Kaolin (aluminum silicate)	Halal
E-570	Fatty acids	Doubtful
E-572	Magnesium stearate	Doubtful
E-574	Gluconic acid	Halal
E-575	Glucono-δ-lactone	Halal
E-576	Sodium gluconate	Halal
E-577	Potassium gluconate	Halal
E-578	Calcium gluconate	Halal
E-579	Ferrous gluconate	Halal
E-585	Ferrous lactate	Halal
Compounds used as flavor enhancers		
E-620	l-Glutamic acid (a natural amino acid)	Doubtful
E-621	Monosodium glutamate (MSG)	Doubtful

(*Continued*)

Coloring materials	Used to make food more colorful and attractive, or to replace color lost in processing	
E-622	Monopotassium glutamate	Doubtful
E-623	Calcium diglutamate	Doubtful
E-624	Monoammonium glutamate	Doubtful
E-625	Magnesium diglutamate	Doubtful
E-626	Guanylic acid	Halal
E-627	Disodium guanylate	Halal
E-628	Dipotassium guanylate	Halal
E-629	Calcium guanylate	Halal
E-630	Inosinic acid	Halal
E-631	Disodium inosinate	Halal
E-632	Dipotassium inosinate	Halal
E-633	Calcium inosinate	Halal
E-634	Calcium 5'-ribonucleotides	Halal
E-635	Disodium 5-ribonucleotides	Halal
E-636	Maltol	Halal
E-637	Ethyl maltol	Halal
E-640	Glycine and its sodium salt (a natural amino acid)	Doubtful
E-900	Dimethylpolysiloxane	Halal
Compounds used as glazing agents		Halal status
E-901	Beeswax	Halal
E-902	Candelilla wax	Halal
E-903	Carnauba wax	Halal
E-904	Shellac	Doubtful
E-912	Montan acid esters	Doubtful
E-914	Oxidized polyethylene wax	Halal
Compounds used to treat flour		
E-920	l-Cysteine hydrochloride (a natural amino acid)	Doubtful
924	Potassium bromate	Halal
925	Chlorine	Halal
926	Chlorine dioxide	Halal
927	Azodicarbamide	Halal
Propellant gases		
E-938	Argon	Halal
E-939	Helium	Halal
E-941	Nitrogen	Halal
E-942	Nitrous oxide	Halal
E-948	Oxygen	Halal
Sweeteners and sugar alcohols		
E-420	(i) Sorbitol, (ii) sorbitol syrup	Halal
E-421	Mannitol	Halal
E-422	Glycerol	Doubtful
E-927a	Azodicarbonamide	Halal

(*Continued*)

Appendix B

Coloring materials	Used to make food more colorful and attractive, or to replace color lost in processing	
E-927b	Carbamide	Halal
E-950	Acesulfame K	Halal
E-951	Aspartame	Doubtful
E-952	Cyclamic acid and its sodium and calcium salts	Halal
E-953	Isomalt	Halal
E-954	Saccharine and its sodium, potassium, and calcium salts	Halal
E-957	Thaumatin	Halal
E-959	Neohesperidin didihydrochalcone	Halal
E-965	(i) Maltitol, (ii) maltitol syrup	Halal
E-966	Lactitol	Halal
E-967	Xylitol	Halal
Miscellaneous compounds		
E-999	Quillaia extract	Halal
E-1105	Lysozyme	Doubtful
E-1200	Polydextrose	Halal
E-1201	Polyvinylpyrrolidone	Halal
E-1202	Polyvinylpolypyrrolidone	Halal
E-1505	Triethyl citrate	Halal
E-1518	Glyceryl triacetate (triacetin)	Doubtful
Modified starches		Halal status
E-1404	Oxidized starch	Halal
E-1410	Monostarch phosphate	Halal
E-1412	Distarch phosphate	Halal
E-1413	Phosphated distarch phosphate	Halal
E-1414	Acetylated distarch phosphate	Halal
E-1420	Acetylated starch	Halal
E-1422	Acetylated starch adipate	Halal
E-1440	Hydroxypropyl starch	Halal
E-1442	Hydroxypropyl distarch phosphate	Halal
E-1450	Starch sodium octanoyl succinate	Halal

[a] Due to their effect on health, some Islamic scholars consider these preservatives as doubtful.

©Reference: *Bender's Dictionary of Nutrition and Food Technology* by David A. Bender and Arnold E. Bender, seventh edition 1999, Woodhead Publishing Ltd and CRC Press LLC © 1999, Woodhead Publishing Ltd. (Information shown in this document was modified from Table 6 to reflect halal status of different food ingredients.

NOTES

1. Those without the prefix E- are permitted in UK but not throughout the EU.
2. The status of ingredients given here is for the pure form. Many of the ingredients are standardized with other food ingredients, status of which may be doubtful, making the listed ingredient doubtful. Ask the supplier for a full disclosure of the composition.

Appendix C
Ingredient List

Ingredient Name	Description	Halal Status
Acetone peroxide	A dough conditioner, maturing and bleaching agent.	Halal
Acesulfame potassium	A synthetic sweetener.	Halal
Acetylated monoglyceride	An emulsifier manufactured by the inter-esterification of edible fats with triacetin.	Doubtful
Acidophilus	A bacterial starter culture used to develop flavor	Halal
Adipic acid	An acidulant and flavoring agent	Halal
Agar or Agar-Agar	A gum obtained from red seaweeds	Halal
Albumin	Any of several water-soluble proteins from egg white, blood serum, and milk.	Doubtful
Algin and Alginates	A gum, a term used for derivatives of alginic acid, which are obtained from brown seaweed.	Halal
All-purpose flour	A flour that is intermediate between long-patent flours (bread flour and cake flour).	Halal
Allspice	A spice made from the dried, nearly ripe berries of *Pimenta officinalis,* a tropical evergreen tree.	Halal
Almond oil	The oil of the bitter almond after the removal of hydrocyanic acid.	Halal
Almond paste	A paste made by cooking sweet and bitter almonds that have been ground and blanched in combination with sugar	Halal
Alum	A preservative, the inclusive term for several aluminum-type compounds such as aluminum sulfate and aluminum potassium sulfate.	Halal
Aluminum acetate	A general purpose food additive.	Halal
Aluminum ammonium sulfate	A general purpose food additive that functions as a buffer and neutralizing agent.	Halal
Aluminum calcium silicate	An anticaking agent used in vanilla powder.	Halal
Aluminum caprate	The aluminum salt of capric acid.	Doubtful
Aluminum caprylate	The aluminum salt of caprylic acid.	Doubtful
Aluminum laurate	The aluminum salt of lauric acid.	Doubtful
Aluminum myristate	The aluminum salt of myristic acid.	Doubtful
Aluminum nicotinate	The aluminum salt of nicotinic acid.	Halal
Aluminum oleate	The aluminum salt of oleic acid.	Doubtful
Aluminum oxide	A dispersing agent.	Halal

(*Continued*)

Ingredient Name	Description	Halal Status
Aluminum palmitate	The aluminum salt of palmitic acid.	Doubtful
Aluminum sodium sulfate	A general purpose food additive that functions as a buffer, neutralizing agent, and firming agent.	Halal
Aluminum stearate	The aluminum salt of stearic acid.	Doubtful
Aluminum sulfate	A starch modifier and firming agent.	Halal
Amidated pectin	The low-methoxyl pectin resulting from deesterification with ammonia.	Halal
Ammonium alginate	A gum that is the ammonium salt of alginic acid.	Halal
Ammonium bicarbonate	A double strengthener and leavening agent.	Halal
Ammonium carbonate	A leavening agent and pH control agent that consists of ammonium bicarbonate and ammonium carbamate.	Halal
Ammonium caseinate	The ammonium salt of casein.	Halal
Ammonium chloride	A dough conditioner and yeast food.	Halal
Ammonium hydroxide	An alkaline, clear, colorless solution of ammonia.	Halal
Ammonium phosphate	A general purpose food additive.	Halal
Ammonium sulfate	A dough conditioner, firming agent, and processing aid.	Halal
Ammonium sulfite	An additive used in the production of caramel.	Halal
Anise	A spice that is the dried, ripe fruit of *Pimpinella anisum,* a small herb.	Halal
Annatto	A color source of yellowish to reddish-orange color obtained from the seed coating of the tree *Bixa orellanna.*	Halal
Arabic gum	A gum obtained from breaks or wounds in the bark of *Acacia* trees.	Halal
Arabinogalactan	A gum, the plant extract obtained from larch trees.	Halal
Arginine	A nonessential amino acid that exists as white crystals or powder.	Doubtful
Arrowroot	A starch obtained from *Maranta arundinacea,* a perennial that produces starchy rhizomes.	Halal
Artificial coloring	Any of the FD&C synthetic coloring materials.	Doubtful
Artificial flavor	Any substance whose function it is to impart flavor and that is not derived from a spice, fruit, vegetable, edible yeast, bark, bud, herb, root, leaf, or other plant material, meat, seafood, eggs, poultry, dairy, or fermentation products.	Doubtful
Ascorbic acid	It is termed vitamin C, a water-soluble vitamin that prevents scurvy, helps maintain the body's resistance to infection, and is essential for healthy bones and teeth.	Halal
Ascorbyl palmitate	An antioxidant formed by combining ascorbic acid with palmitic acid.	Doubtful
Ascorbyl stearate	An antioxidant used in peanut oil.	Doubtful
Aspartame	A synthetic sweetener that is a dipeptide, synthesized by combining the methyl ester of phenylalanine with aspartic acid.	Doubtful

(Continued)

Appendix C

Ingredient Name	Description	Halal Status
Aspartic acid	A nonessential amino acid that exists as colorless or white crystals of acid taste.	Doubtful
Azodicarbonamide	A dough conditioner that exists as a yellow to orange-red crystalline powder.	Halal
Babassu oil	The oil obtained from the nut of the babassu palm.	Halal
Baking powder	A leavening agent that consists of a mixture of sodium bicarbonate, one or more leavening agents such as sodium aluminum phosphate or monocalcium phosphate, and an inert material such as starch.	Halal
Baking soda	Bicarbonate of soda, chemically known as sodium bicarbonate.	Halal
Baker's yeast	Dried microorganism *Saccharomyces cerevisiae*.	Halal
Baker's yeast glycan	The dried cell walls of yeast, *Saccharomyces cerevisiae*, used as an emulsifier and thickener in salad dressing.	Halal
Barley	A cereal grain of which there are winter and spring types. It is used in malting of barley.	Halal
Basil	A spice obtained from the dried leaves and tender stems of *Ocimum basilicum L.*	Halal
Bay leaves	A spice flavoring that consists of the dried leaves obtained from the evergreen tree *Laurus nobilis,* also called sweet bay or laurel.	Halal
Beeswax	The purified wax obtained from the honeycomb of the bee, used to glaze candy and in chewing gum, and confections.	Halal
Beet extract	A natural red colorant obtained from beets, used in yogurt, beverages, candies, and desserts.	Halal
Bentonite	A general purpose additive that is used as a pigment and colorant.	Halal
Benzoic acid	A preservative that occurs naturally in some foods such as cranberries, prunes, and cinnamon.	Halal
Benzoyl peroxide	A colorless, crystalline solid with a faint odor of benzaldehyde resulting from the interaction of benzoyl chloride and a cooled sodium peroxide solution, used in specified cheeses.	Halal
Beta-apo-8'-catotenal	A colorant that is a carotenoid producing a light to dark orange hue.	Halal
Beta-carotene	A colorant that is a carotenoid producing a yellow to orange hue.	Halal
Biotin	A water-soluble vitamin that is a nutrient and dietary supplement.	Doubtful
Birch	An artificial flavoring used in soft drinks.	Halal
Bixin	A carotenoid that is the main coloring component of annatto. It is obtained from the *Bixa orellana* tree.	Halal

(Continued)

Ingredient Name	Description	Halal Status
Bleached flour	Flour that has been whitened by the removal of the yellow pigment. The bleaching can be obtained during the natural aging of the flour or can be accelerated by chemicals that are usually oxidizing agents.	Halal
Bran	The seed husk or outer coatings of cereals, such as wheat, rye, and oats that is separated from the flour.	Halal
Bread flour	A hard-wheat flour, which generally is obtained from straight or long patent flours.	Halal
Bromated flour	A white flour to which potassium bromate is added. It is used in baked goods.	Halal
Bromelin	A proteolytic enzyme obtained from pineapple, used in tenderizing beef.	Halal
Brominated vegetable oil (BVO)	A vegetable oil whose destiny has been increased to that of water by being combined with bromine. Flavoring oils are dissolved in the brominated oil which can then be added to fruit drinks.	Halal
Brown sugar	A sweetener that consists of sucrose crystals covered with a film of cane molasses which gives it the characteristic color and flavor.	Halal
Buckwheat	A member of the grass family that is usually marketed as a cereal grain, buckwheat grain.	Halal
Bulgur	A precooked cracked wheat that retains the bran and germ fraction of the grain.	Halal
Butter	A source of milk fat obtained from cream by a churning process which results in the production of butter fat and buttermilk.	Halal
Butter, clarified	Butter that has undergone purification by the removal of solid particles or impurities that may affect the color, odor, or taste.	Halal
Buttermilk	The product that remains when fat is removed from milk or cream in the process of churning into butter. Cultured buttermilk is prepared by souring buttermilk.	Halal
Buttermilk, dried	The powder form of buttermilk.	Halal
Butter oil	The clarified fat portion of milk, cream, or butter obtained by the removal of the nonfat constituent.	Halal
Butylated hydroxyanisole (BHA)	An antioxidant that imparts stability to fats and oils.	Halal
Butylated hydroxytoluene (BHT)	An antioxidant that functions similarly to butylated hydrxyanisole (BHA) but is less stable at high temperatures.	Halal
Butyric acid	A fatty acid that is commonly obtained from butter fat.	Halal
Caffeine	A white powder or needles that are odorless and have a bitter taste, occurring naturally in tea leaves, coffee, cocoa, and cola nuts.	Halal
Cake flour	A soft wheat flour.	Halal

(Continued)

Appendix C

Ingredient Name	Description	Halal Status
Calciferol	A fat-soluble vitamin, termed vitamin D2.	Halal
Calcium acetate	The calcium salt of acetic acid which functions as a sequestrant and mold control agent.	Halal
Calcium alginate	The calcium salt of alginic acid which functions as a stabilizer and thickener.	Halal
Calcium aluminum silicate	An anticaking agent that permits the free flow of dry ingredients.	Halal
Calcium ascorbate	The salt of ascorbic acid which is a white to slightly yellow crystalline powder.	Halal
Calcium bromate	A dough conditioner, maturing and bleaching agent which exists as a white crystalline powder	Halal
Calcium caprate	The calcium salt of capric acid, used as a binder, emulsifier, anticaking agent, and as a general additive.	Doubtful
Calcium caprylate	The calcium salt of caprylic acid, used as a binder, emulsifier, anticaking agent, and as a general additive.	Doubtful
Calcium carbonate	The calcium salt of carbonic acid which is used as an anticaking agent and dough strengthener.	Halal
Calcium caseinate	The calcium salt of casein.	Halal
Calcium chloride	A general purpose food additive.	Halal
Calcium citrate	The calcium salt of citric acid used as a sequestrant, buffer, and firming agent.	Halal
Calcium diacetate	The salt of acetic acid that is used as a preserevative and sequestrant.	Halal
Calcium gluconate	A white crystalline powder used as a firming agent, formulation aid, sequestrant, and stabilizer.	Halal
Calcium glycerophosphate	A nutrient and dietary supplement used in baking powder.	Halal
Calcium hydroxide	A general food additive made by adding water to calcium oxide (lime).	Halal
Calcium hydroxyphosphate	A food additive used in table salt.	Halal
Calcium iodate	Calcium salt of iodine used as a dough conditioner in bread.	Halal
Calcium lactate	The calcium salt of lactic acid.	Halal
Calcium lactobionate	The calcium salt of lactobionic acid used as a firming agent in dry pudding mixes.	Halal
Calcium laurate	The calcium salt of lauric acid used as a binder, emulsifier, and anticaking agent.	Doubtful
Calcium metaphosphate	A sequestrant used as a general additive.	Halal
Calcium myristate	The calcium salt of myristic acid used as a binder, emulsifier, and anticaking agent.	Doubtful
Calcium oleate	The calcium salt of oleic acid used as a binder, emulsifier, and anticaking agent.	Doubtful
Calcium oxide	A general food additive.	Halal

(*Continued*)

Ingredient Name	Description	Halal Status
Calcium palmitate	The calcium salt of palmitic acid used as a binder, emulsifier, and anticaking agent.	Doubtful
Calcium pantothenate	A nutrient and dietary supplement used in special dietary foods.	Halal
Calcium pectinate	The salt of pectin which is obtained from citrus or apple fruit.	Halal
Calcium peroxide	A dough conditioner used in bakery products.	Halal
Calcium phosphate	A phosphate existing in several forms in nature, used as an anticaking agent and mineral supplement.	Halal
Calcium phytate	A chelating agent used to bind metallic ions in foods to prevent discoloration and off-flavor.	Halal
Calcium propionate	The salt of propionic acid used as a preservative.	Halal
Calcium pyrophosphate	A nutrient and dietary supplement.	Halal
Calcium saccharate	A derivative of calcium hydroxide and sugar used to improve the whipping ability of whipping cream.	Halal
Calcium saccharin	A sweetener that is the calcium form of saccharin. It is about 500 times as sweet as sucrose.	Halal
Calcium silicate	An anticaking agent used in baking powder.	Halal
Calcium sorbate	The calcium salt of sorbic acid used in cheese.	Halal
Calcium stearate	The calcium salt of stearic acid used as an anticaking agent, binder, and emulsifier.	Doubtful
Calcium stearoyl lactylate	The calcium salt of lactic acid and stearic acid used as a dough conditioner, whipping agent, and emulsifier.	Doubtful
Calcium sulfate	A general additive made by the high-temperature calcining of gypsum.	Halal
Canthaxanthin	A synthetic red colorant similar to carotenoids.	Halal
Caramel color	A dark-brown color obtained by controlled oxidation of starch and other carbohydrates.	Halal
Carbon dioxide	A gas obtained during fermentation of glucose used in the carbonation of beverages.	Halal
Carbonated water	A beverage made by absorbing carbon dioxide in water.	Halal
Carboxymethylcellulose (CMC)	A water-soluble gum cellulose used as a thickener, stabilizer, binder, film former, and suspending agent.	Halal
Cardamom	A spice, dried, ripe seed of *Elettaria cardamomum*.	Halal
Carmine	The red colorant aluminum lake of carminic acid which is the coloring pigment obtained from dried bodies of the female insect *Coccus cacti*.	Doubtful
Carnauba wax	A general purpose additive obtained from leaf buds and leaves of the Brazilian wax palm *Copernicia cerifera*.	Halal
Carob	A cocoa substitute obtained from the pods of the carob tree *Ceratonia Siliqua*.	Halal
Carotene	A colorant and provitamin used in ice cream, cheese, and other dairy products.	Halal
Carrageenan	A seaweed extract obtained from red seaweed.	Halal

(*Continued*)

Appendix C

Ingredient Name	Description	Halal Status
Casein	A milk protein which is prepared commercially from skim milk by the precipitation with lactic, hydrochloric, or sulfuric acid.	Halal
Caseinates	Sodium or calcium salts of casein that are produced by neutralizing acid casein with calcium or sodium hydroxide.	Halal
Celery seed	A spice made from the dried, ripe fruit of the herb *Apium graveolens*.	Halal
Cellulose	A carbohydrate polymer used as a fiber source and bulking agent in low-calorie foods.	Doubtful
Cheese	The product obtained by the coagulation of the milk protein by suitable enzymes and/or bacteria	Doubtful
Cheese culture	Several bacteria used in the coagulation of the milk protein.	Halal
Chervil	A spice derived from the plant *Anthriscus cerefolium*.	Halal
Chewing gum base	A formulation containing masticatory substances such as chicle and several other GRAS substances used in the manufacture of chewing gum.	Doubtful
Chia oil	The oil from the seed of plants of the genus *Salvia* used in the preparation of soft drinks.	Halal
Chicle	A natural masticatory substance of vegetable origin that is used in chewing gum base.	Halal
Chilte	A substance of vegetable origin used in chewing gum base.	Halal
Chives	A spice from the *Allium* plant similar to green onions.	Halal
Chloropentafluoroethane	A propellant and aerating agent for foamed or sprayed foods.	Halal
Chlorophyll	A green pigment from plants used in sausage casings and shortening.	Halal
Cholic acid	An emulsifier used as an emulsifying agent in egg white.	Halal
Choline	A substance of the vitamin B complex family.	Doubtful
Cider vinegar	The product made by the alcoholic and subsequent acetic fermentation of apple juice used in salad dressings, mayonnaise, and sauces.	Halal
Cinnamon	A spice made from the dried bark of the evergreen tree *Cinnamomum cassia*.	Halal
Citric acid	An acidulant and antioxidant produced by extraction from lemon and lime juice.	Halal
Citrus oil	A flavorant obtained by pressing the oil from the rind of citrus fruits.	Halal
Clove	A spice that is the unripened bud from the clove tree *Eugenia caryophyllata*.	Halal
Cochineal	A red colorant extracted from the dried bodies of the female insect *Coccus cacti*. The coloring is carminic acid in which the water-soluble extract is cochineal.	Doubtful

(Continued)

Ingredient Name	Description	Halal Status
Cocoa butter	The fat obtained by pressing chocolate liquor.	Halal
Cocoa liquor	Viscous liquid obtained by the grinding of cocoa nibs from the cocoa bean. It is also termed chocolate liquor.	Halal
Cocoa powder	The powder produced by grinding cocoa presscake obtained from the cocoa nibs.	Halal
Coconut	The fruit of coconut palm which produces coconut meat and coconut oil.	Halal
Collagen	A protein that is the principal constituent of connective tissue used in casings and personal care products.	Doubtful
Copper gluconate	Salt of gluconic acid used as a dietary supplement.	Halal
Copra	A dried coconut meat used as a flavorant.	Halal
Coriander	A spice that is the dried, ripe fruit of *Coriandrum sativum* L.	Halal
Corn bran	A dry-milled product of high fiber content obtained from corn.	Halal
Corn flour	A finely ground flour made from milling and sifting maize or corn.	Halal
Cornmeal	A ground corn of granular form.	Halal
Corn oil	The oil obtained from the germ of the corn plant.	Halal
Cornstarch	The starch made from the endosperm of corn.	Halal
Cornstarch, modified	A starch produced by treating cornstarch with dilute mineral acid or other chemicals.	Halal
Corn syrup	A corn sweetener liquid composed of maltose, dextrin, dextrose, and other polysaccharides.	Halal
Corn syrup solids	The dry form of corn syrup.	Halal
Cottonseed oil	The oil obtained from the seeds of the cotton plant.	Halal
Cranberry extract	A natural red colorant from cranberries.	Halal
Cream	Portion of milk that is high in milkfat, with 18% to 40% fat.	Halal
Cream of Tartar	The potassium salt of tartaric acid used in baked goods, candy, and puddings.	Halal
Cumin	A spice that is the dried, ripe fruit of *Cuminum cyminum* L.	Halal
Cupric/cuprous aspartate	A salt of aspartic acid with copper.	Doubtful
Cupric/cuprous carbonate	A copper salt of carbonic acid.	Halal
Cupric/cuprous chloride	A copper salt.	Halal
Cupric/cuprous citrate	A copper salt of citric acid.	Halal
Cupric/cuprous glycerophosphate	A copper salt of glycerine and phosphate.	Halal
Cupric/cuprous nitrate	A copper salt of nitric acid.	Halal
Cupric/cuprous sulfate	A copper salt of sulfuric acid.	Halal
Cuprous iodide	A salt of copper and iodine used in table salt.	Halal
Curry powder	A blend of spices used as seasoning in curries, sauces, and meats. Typical spices include coriander, ginger, clove, cinnamon, red pepper, and cumin.	Halal

(Continued)

Appendix C

Ingredient Name	Description	Halal Status
Cyanocobalamin	Vitamin B12, a water-soluble vitamin found in meat, fish, and milk.	Doubtful
Cysteine	A nonessential amino acid used to increase elasticity. Can be made from human hair, duck feather or synthetically.	Doubtful
Cystine	A nonessential amino acid used as a nutrient and dietary supplement.	Doubtful
Deoxycholic acid	An emulsifier used in dried egg whites.	Halal
Dextran	A gum obtained by fermentation of sugar with bacteria.	Halal
Dextrin	A compound formed from acid or enzymes.	Halal
Dextrose	A corn sweetener commercially made from starch.	Halal
Diacetyl tartaric acid esters of mono- and diglycerides	A hydrophylic emulsifier used in oil-in-water emulsions.	Doubtful
Diammonium phosphate	A leavening agent used in the production of cookies and crackers.	Halal
Dicalcium phosphate	A mineral supplement and dough conditioner	Halal
Diglyceride	An emulsifier prepared by esterification of two fatty acids with glycerol, or by interesterification between glycerol and triglycerides.	Doubtful
Dill seed	A spice from the dried, ripe fruit of the plant *Anethum graveolens L.*	Halal
Dill weed	A spice from the leaf of the dill plant.	Halal
Dimethylpolysiloxane	An antifoaming agent used in fats and oils.	Halal
Dioctyl sodium sulfosuccinate	An emulifiying agent used as a flavor potentiator in canned milk.	Doubtful
Dipotassium phosphate	The potassium salt of phosphoric acid used as a stabilizing salt, buffer, and sequestrant.	Halal
Disodium calcium EDTA	A sequestrant and chelating agent.	Halal
Disodium dihydrogen EDTA	A sequestrant and chelating agent.	Halal
Disodium 5'-guanylate (DSG)	A flavor enhancer that belongs to the same family as monosodium glutamate.	Halal
Disodium 5'-inosinate	A flavor enhancer which performs as does disodium guanylate.	Halal
Disodium malate	An acidulant and a sodium salt of malic acid.	Halal
Disodium phosphate	The sodium salt of phosphoric acid that functions as a protein stabilizer and mineral supplement.	Halal
Distilled monoglyceride	An emulsifier containing a minimum of 90 percent monoglyceride derived from edible fat and glycerine.	Doubtful
Dodecyl gallate	An antioxidant used in cream cheese, instant mashed potatoes, margarine, fats, and oils.	Halal
Dough conditioner	A blend of minerals used in baked goods.	Halal
Durum flour	The fine powder obtained from durum wheat.	Halal

(Continued)

Ingredient Name	Description	Halal Status
Edta (ethylenediaminetetraacetate)	A sequestrant and chelating agent.	Halal
Egg albumen	The protein fraction of egg, also termed egg white.	Halal
Egg yolk	The yellow portion of the egg used as an emulsifier in mayonnaise and salad dressing.	Halal
Enriched bleach flour	Flour that has been whitened by removal of the yellow pigments and fortified with vitamins and minerals.	Halal
Erythorbic acid	A food preservative that is a strong reducing agent.	Halal
Ester gum	A density adjuster prepared from glycerol of non-animal sources and refined wood rosin of pine trees.	Halal
Ethoxyquin	An antioxidant used in the preservation of color in chili powder, ground chili, and paprika.	Halal
Ethoxylated mono- and diglycerides	An emulsifier prepared by the glycerolysis fo edible vegetable fats and reacting with ethylene oxide.	Doubtful
Ethyl maltol	A synthetic flavor enhancer related to maltol.	Halal
Ethyl vanillin	A flavoring agent that is a synthetic vanilla flavor.	Halal
Eugenol	A flavoring obtained from clove oil and also found in carnation and cinnamon leaves.	Halal
Farina	Wheat granules from which the bran and germ has been removed.	Halal
Fat	Water-insoluble material of plant or animal origin, consisting of triglycerides that are semisolid at room temperature.	Doubtful
Fatty acids	A mixture of aliphatic acids of plant or animal origin. A fatty acid may be used as a lubricant, a binder, a food processing defoamer, and an emulsifier.	Doubtful
FD&C blue # 1	A colorant also called brilliant blue, used in candies, baked goods, soft drinks, and desserts.	Halal
FD&C blue # 2	A colorant, also called indigotine, used in candies, confections, and baked goods.	Halal
FD&C green # 3	A colorant, also called fast green FCF, used in cereals, soft drinks, beverages, and desserts.	Halal
FD&C red # 3	A colorant also called erythrocine, used in candies, confections, and cherry dyeing.	Halal
FD&C red # 4	A colorant also called ponceau SX, used only in maraschino cherries.	Halal
FD&C red # 40	A colorant also called Allura red AC, used in beverages, desserts, candy, confections, cereals, and ice cream.	Halal
FD&C yellow # 5	A colorant also called tartrazine, used in beverages, baked goods, pet foods, desserts, candy, confections, cereal, and ice cream.	Halal
FD&C yellow # 6	A colorant also called sunset yellow FCF, used in beverages, bakery goods, dessert confections, and ice cream.	Halal

(Continued)

Appendix C

Ingredient Name	Description	Halal Status
Fennel	A spice that is the dried, ripe fruit of the plant *Foeniculum vulgare* Mil.	Halal
Fenugreek	The seeds of the herb *Trigonella foenumgraecum*, used as a spice and flavoring.	Halal
Ferric ammonium citrate	A nutrient and dietary supplement that is a source of iron, containing 17% iron.	Halal
Ferric/ferrous ammonium sulfate	A nutrient and dietary supplement.	Halal
Ferric chloride	A nutrient and dietary supplement.	Halal
Ferric fructose	A nutrient and dietary supplement.	Halal
Ferric glycerophosphate	A nutrient and dietary supplement.	Halal
Ferric nitrate	A nutrient and dietary supplement.	Halal
Ferric orthophosphate	A nutrient and dietary supplement.	Halal
Ferric oxide	A nutrient and dietary supplement.	Halal
Ferric pyrophophate	A nutrient and dietary supplement.	Halal
Ferric/ferrous sulfate	A nutrient and dietary supplement.	Halal
Ferrous carbonate	A nutrient and dietary supplement.	Halal
Ferrous citrate	A nutrient and dietary supplement.	Halal
Ferrous fumarate	A nutrient and dietary supplement.	Halal
Ferrous gluconate	A nutrient and dietary supplement and a coloring adjunct.	Halal
Ferrous lactate	A nutrient and dietary supplement.	Halal
Ferrous sulfate	A nutrient and dietary supplement.	Halal
Ferrous tartrate	A nutrient and dietary supplement.	Halal
Flavoring	Any ingredients, natural or artificial, single or in a mixture, that impart flavor to a food. Components of a flavoring are generally not revealed to the consumer.	Doubtful
Folic acid	It is a B complex vitamin found in liver, nuts, and green vegetables.	Doubtful
Fructose (fruit sugar)	A sweetener found naturally in fresh fruit and honey. It is obtained by the inversion of sucrose with the enzyme invertase and by the isomerization of dextrose.	Halal
Fructose corn syrup	A corn sweetener derived from the isomerization of glucose in the syrup to fructose by the enzyme isomerase.	Halal
Fumaric acid	An acidulant used in desserts, pie fillings, and candy.	Halal
Furcellaran	A gum extract of the red alga *Furcellaria fastigiata*. It is used in milk puddings, flans, jelly, jam, and other products.	Halal
Garlic	A spice that is cloves of the herb *Allium sativium*.	Halal
Garlic salt	A seasoning that is a mixture of garlic powder and salt.	Halal
Gelatin	A protein of animal origin that functions as a gelling agent. It is obtained from collagen derived from beef bones and calf skin, pork skin, fish skin, or poultry skin.	Doubtful

(Continued)

Ingredient Name	Description	Halal Status
Ghatti	A gum that is a plant exudate form the *Anogeissus latifolia* tree. It is used in buttered syrup and as a stabilizer.	Halal
Ginger	A spice that is the dried rhizome of the ginger plant, *Zingiber officinale*.	Halal
Glacial acetic acid	A strong acidulant preservative flavoring.	Halal
Gluconic acid	A mild organic acid which is the hydrolyzed form of glucono-delta-lactone.	Halal
Glucono-delta-lactone (GDL)	A mild acidulent used in sausages, frankfurter, and dessert mixes.	Halal
Glutamic acid	An amino acid used as a flavor enhancer, nutrient, dietary supplement, and salt substitute.	Doubtful
Glutamic acid hydrochloride	An amino acid used as a flavoring agent.	Doubtful
Glycerin or Glycerol	A polyol used as a humectant, crystallization modifier, and plasticizer in candy, baked goods, and other products.	Doubtful
Glycerol Ester of Wood Rosin	Same as Ester Gum	Halal
Glycerol-lacto-stearate	An emulsifier that is a monoglyceride esterified with lactic acid used in whipped toppings, shortenings, cake mixes, and coatings.	Doubtful
Glyceryl-Lacto Esters of Fatty Acids	Emulsifiers that are the lactic acid esters of mono- and diglycerides, used as emulsifiers and plasticizers in toppings, cakes, and icings.	Doubtful
Glyceryl monolaurate	A monoglyceride emulsifier produced by the esterification of glycerine and lauric acid used in baked goods, whipped toppings, frosting, and glazes	Doubtful
Glyceryl triacetate	A triglyceride of acetic acids used as a humectant and solvent	Doubtful
Glycine	A nonessential amino acid used as a nutrient and dietary supplement in artificially sweetened soft drinks.	Doubtful
Glycocholic acid	An emulsifier used in dried egg whites	Halal
Glycyrrhizin	A flavorant and foaming agent derived from licorice root.	Halal
Graham flour	See Whole Wheat Flour	
Grain vinegar	An acidulant, also called distilled or spirit vinegar, made by the fermentation of dilute distilled alcohol.	Halal
Grape color extract	A solution of grape pigment made from Concord Grapes used for coloring nonbeverage foods	Halal
Grape seed oil	The oil obtained from grape seeds	Halal
Grape skin extract	A neutral red colorant used in soft drinks and candies	Halal
Guaiac gum	An antioxidant and preservative made from wood resin	Halal
Guar	A gum that is obtained from the seed kernel of the guar plant *Cyamopsis tetragonoloba*. It is a thickener and stabilizer used in ice cream, baked goods, and sauces	Halal

(Continued)

Appendix C

Ingredient Name	Description	Halal Status
Gum base	The component of chewing gum that is insoluble in water and remains after chewing. It is prepared from several ingredients such as chicle, crown gum, petroleum wax, lanolin, polythylene, polyvinyl acetate, rubber, paraffin, and antioxidants.	Doubtful
Heptyl paraben	A preservative and antimicrobial agent.	Halal
High-fructose corn syrup (HFCS)	A sweetener made from corn syrup through isomerization of the glucose in the syrup to fructose by the enzyme isomerase.	Halal
Honey	A sweetener that is a natural syrup, made by honey bees, through the action of the enzyme honey invertase on nectar gathered by bees.	Halal
Horseradish	A spice obtained from the horseradish plant used in sauces.	Halal
Hydrochloric acid	An inorganic acid used as an acidulant and neutralizing agent.	Halal
Hydrogenated vegetable oil	Oil that has been hydrogenated to modify the texture from a liquid to a semisolid or solid through the chemical addition of hydrogen.	Halal
Hydrolyzed vegetable protein (HVP)	A flavor enhancer obtained from vegetable proteins such as wheat gluten, corn gluten, defatted soy flour through acid hydrolysis. It is used to improve flavors in soups, dressings, meats, snack foods, and crackers.	Halal
Hydroxylated soybean lecithin	An emulsifier and clouding agent obtained by treating soybean lecithin with peroxide. It is used in dry-mixed beverages, margarine, and baked goods.	Halal
Hydroxypropyl cellulose	A gum obtained from the reaction of alkali cellulose with propylene oxide, used in whipped toppings as a stabilizing and foaming aid.	Halal
Hydroxypropyl methylcellulose	A gum formed by the reaction of propylene oxide and methyl chloride with alkali cellulose. It is used in bakery goods, dressings, breaded foods, and salad dressings.	Halal
Invert sugar	A sweetener that is a mixture of dextrose (glucose) and fructose. It is used in candy and icings.	Halal
Iron, reduced	It is a nutrient used in cereals.	Halal
Isoamyl butyrate	A synthetic flavoring agent used in dessert gels, puddings, and baked goods.	Halal
Isoamyl formate	A synthetic flavoring agent used in dessert gels, puddings, candy, and ice cream.	Halal
Isoamyl hexanoate	A synthetic flavoring agent used in desserts, candy, and ice cream.	Halal
Isobutyl cinnamate	A synthetic flavoring agent used in beverages, ice cream, candy, and baked goods.	Halal
Isobutyl formate	A synthetic flavoring agent used in beverages, ice cream, candy, and baked goods.	Halal

(Continued)

Ingredient Name	Description	Halal Status
Isopropyl citrate	An antioxidant made by reacting citric acid with isopropyl alcohol. It is used in vegetable oils.	Halal
Karaya	A gum, the dried exudate from the *Sterculia urens* tree. It is used in baked goods, denture adhesives, toppings, and frozen desserts.	Halal
Kelp	A brown seaweed product used as a source of iodine and as a flavor enhancer.	Halal
Kola nut	The seed of Cola species. It is used in beverages as a flavoring.	Halal
L-cysteine	See Cysteine	Doubtful
L-glutamine	An amino acid. Isolated from sugar beet juice. Can be detected in most plants and animals (including bacteria)	Doubtful
L-taurine	An amino acid, usually synthetic.	Doubtful
Lactic acid	An acidulant that is a natural organic acid present in milk. It is used as a flavor agent, preservative, and acidity adjuster in foods.	Halal
Lactose	A milk sugar that occurs in most mammalian milk. It is usually obtained from cows' milk and is used in baked goods for flavor, browning, and tenderizing and in dry mixes as an anticaking agent.	Halal
Lactylated fatty acid esters of glycerol and propylene glycol	An emulsifier made by the reaction of propylene glycol ester with lactic acid. It is used in whipped toppings and coffee whiteners.	Doubtful
Lard	A fat rendered from hogs, consisting principally of oleic and palmitic fatty acids and is used in many food and nonfood products.	Haram
Lauric acid	A fatty acid usually from coconut oil and other vegetable fats. It is used as a lubricant, binder, and defoaming agent.	Doubtful
Leavening agents	Mild acids that chemically react with alkaline sodium bicarbonate to produce carbon dioxide gas. Leavening agents include tartaric acid, monocalcium phosphate, sodium acid pyrophosphate, sodium aluminum phosphate, and acidic acid.	Halal
Lecithin	An emulsifier that is a mixture of phosphatides commercially obtained from soybeans and egg yolk. It is extensively used in foods.	Halal
Lemon oil	An oil obtained from lemon fruit used to impart lemon flavor.	Halal
Licorice extract	A flavoring agent made from dried root portions of *Glycyrrhiza glabra*. It is used in candy, baked goods, beverages, and tobacco.	Halal
Limonene	An antioxidant and flavoring agent obtained from citrus oils.	Halal

(*Continued*)

Appendix C

Ingredient Name	Description	Halal Status
Locust bean gum	A gum obtained from the plant seed of a locust bean tree, *Ceratonia siligua*. It is used in processed cheese, ice cream, bakery products, soups, and pies.	Halal
Mace	A spice comprised of skin covering of the nutmeg *Myristica fragrans* Houtt.	Halal
Magnesium carbonate	An anticaking agent and general purpose food additive.	Halal
Magnesium caseinate	The magnesium salt of caseinate. It is used in bakery goods, drinks, and dietary supplements.	Halal
Magnesium chloride	A dietary supplement and foods additive.	Halal
Magnesium hydroxide	A general purpose foods additive.	Halal
Magnesium laurate	The magnesium salt of lauric acid used as an emulsifier, and anticaking agent.	Doubtful
Magnesium myristate	The magnesium salt of myristic acid used as emulsifier and anticaking agent.	Doubtful
Magnesium oleate	The magnesium salt of oleic acid used as an emulsifier and anticaking agent.	Doubtful
Magnesium oxide	A nutrient and dietary supplement.	Halal
Magnesium palmitate	The magnesium salt of palmitic acid used as an emulsifier, and anticaking agent	Doubtful
Magnesium silicate	An insoluble salt used as an anticaking agent.	Halal
Magnesium stearate	The magnesium salt of stearic acid used as a lubricant, binder, emulsifier, and anticaking agent.	Doubtful
Magnesium sulfate	A nutrient and dietary supplement.	Halal
Maleic acid	An organic acid used as a preservative, for fats and oils.	Halal
Malic acid	An acid mainly from apples used in soft drinks, dry-mix beverages, puddings, jellies, and fruit filling.	Halal
Malt	A source of the enzyme alpha-amylase which hydrolyzes starch to fermentable sugars such as dextrins and maltose, produced by the controlled sprouting of grains, usually barley. It is used as a supplement to flour.	Halal
Malt barley flour	The barley produced under the controlled sprouting of the barley grain and milling.	Halal
Malt extract	An extract of water-soluble enzymes from barley evaporated to form a concentrate that contains D-alpha-amylase enzyme.	Halal
Malt flour	The flour prepared by the drying and grinding of barley or wheat sprouted under controlled conditions.	Halal
Maltodextrin	A product obtained from the partial acid or enzymatic hydrolysis of starch used in crackers, puddings, and candies.	Halal
Maltol	A flavor enhancer or modifier. Occurs naturally in chicory, cocoa, coffee, and cereals. It is used in vanilla, and chocolate-flavored foods and beverages.	Halal
Maltose	A sugar formed by the enzymatic action of enzymes on starch. It is used in bread, instant foods, and pancake syrups.	Halal

(*Continued*)

Ingredient Name	Description	Halal Status
Malt syrup	The syrup obtained from barley by extraction and evaporation used as a malt flavor, source of malt and protein, and is used in bakery goods.	Halal
Malt vinegar	A vinegar made by the alcoholic and subsequent acetous fermentation of malted barley and/or cereals. Used as an acidulant and preservative in foods.	Halal
Manganese carbonate	A source of manganese that functions as a nutrient and dietary supplement.	Halal
Manganese chloride	A nutrient and dietary supplement.	Halal
Manganese oxide	A nutrient and dietary supplement.	Halal
Manganese sulfate	A nutrient and dietary supplement.	Halal
Mannitol	A polyhydric alcohol used as a sweetener, humectant, and bulking agent in sugarless candy, chewing gum, cereal, and pressed mints.	Halal
Maple sugar	A sweetener obtained by concentrating the sap of the maple sugar tree used in syrups and candy.	Halal
Maple syrup	A sweetener made by concentrating the sap of the sugar maple tree used in syrups and candies.	Halal
Margarine	A butter like substance made by emulsifying oils and milk. Vegetable oils or mixtures of vegetable oils and animal fat may be used.	Doubtful
Marjoram	A spice obtained from leaves of the herb *Majorana bortensis* Moench.	Halal
Methylcellulose	A gum composed of cellulose used in coatings and pie fillings.	Halal
Methylethylcellulose	A gum that is the methyl ether of ethyl cellulose used in whipped toppings, meringues, and aerated confectionary products.	Halal
Methylparaben	An antimicrobial agent used to prevent yeast and mold growth.	Halal
Methyl salicylate (wintergreen oil)	A synthetic flavoring agent used in chewing gum, candy, beverages, and baked goods.	Halal
Microcrystalline cellulose	A water insoluble gum used in tablets, capsules, and shredded cheese as a non-nutritive filler, and anticaking agent.	Halal
Milkfat or butter fat	The fat of milk concentrated as cream to make butter; used in bakery products, confections, and frozen desserts.	Halal
Milk powder	The dry, whole milk that is produced by a spray or roller-drying process used in soup mixes, dessert mixes, and for reconstituting.	Halal
Milk solids-not-fat	The dry form of skim milk used in ice cream mix, baked goods, and desserts.	Halal
Mint	A spice derived from the mint family.	Halal

(*Continued*)

Appendix C

Ingredient Name	Description	Halal Status
Modified starch	Starch treated with certain chemicals to modify the physical characteristics of the native starch for improved solubility and texture. It is used as a thickener, binder, and stabilizer.	Halal
Molasses	The by-product of the manufacture of sugar from sugar cane containing sucrose and invert sugar; used as a flavoring and sweetener.	Halal
Monoammonium glutamate	A flavor enhancer obtained from glutamic acid used in low-salt diets.	Doubtful
Monoammonium phosphate	An acidulant used as a leavening agent and yeast nutrient in bread.	Halal
Mono- and diglycerides	A mixed emulsifier containing monoglycerides and diglycerides made by reacting glycerol with fats or oils. It is used in numerous food applications.	Doubtful
Monocalcium phosphate	An acid leavening agent and nutritional supplement; widely used in foods.	Halal
Monoglyceride	An emulsifier prepared by the direct esterification of fatty acids with glycerol or by the interesterification between glycerol and other triglycerides.	Doubtful
Monoglyceride citrate	A sequestrant that is a mixture of glyceryl mono-oleate and citric acid; used as an antioxidant synergist.	Doubtful
Monopotassium phosphate	A mild acid neutralizing agent and sequestrant. It is used in low-sodium products, milk, and meat products.	Halal
Monosodium glutamate (msg)	A flavor enhancer sodium salt of glutamic acid in meats, soups, and sauces as a flavor intensifier.	Doubtful
Monosodium phosphate	An acidulant, buffer, and sequestrant used in cheese and carbonated beverages.	Halal
Mustard	A spice made from the dried, ripe seed of several varieties of the family Cruciferae used as a flavorant in baked goods, sauces, and salad dressings.	Halal
Mustard flour	The ground dehusked mustard seed used in salad dressings and sauces and as a condiment.	Halal
Myristic acid	A fatty acid obtained from coconut oil and other fats used as a lubricant and defoaming agent.	Doubtful
Natamycin	A preservative used as a coating on the surface of cheeses to prevent the growth of mold or yeast.	Halal
Niacin	A water-soluble B vitamin used as a nutrient and dietary supplement.	Halal
Niacinamide	A water-soluble B vitamin and nutrient and dietary supplement.	Halal
Nitrous oxide	A noncombustible gas used as a propellant in certain dairy and vegetable fat whipped toppings contained in pressurized containers, also called laughing gas.	Halal
Nutmeg	A spice obtained from the nutmeg tree *Myristica fragrans*.	Halal
Oat flour	Fine-mesh ground dehulled oats.	Halal

(Continued)

Ingredient Name	Description	Halal Status
Oatmeal	Coarse ground dehulled oats.	Halal
Oleic acid	An unsaturated fatty acid used as a lubricant and defoamer.	Doubtful
Oleoresins	Spice extractions containing the volatile and nonvolatile flavor components used in seasonings for foods.	Halal
Oleoresin paprika	A colorant extraction containing the volatile and nonvolatile flavor components of paprika.	Halal
Oxystearin	A modified fatty acid composed of the glycerides of partially oxidized stearic and other fatty acids.	Doubtful
Palmitic acid	A fatty acid composed principally of palmitic acid with varying amounts of stearic acid; used as a lubricant, binder, and defoaming agent.	Doubtful
Palm kernel oil	An oil obtained from palm kernels. It is used interchangeably with coconut oil used in margarine and confectionary.	Halal
Palm oil	The oil obtained from the fruit of the palm tree; used in margarine and shortenings.	Halal
Pantothenic acid	A water-soluble B vitamin found in liver, eggs, and meat.	Doubtful
Papain	A meat tenderizer that is a protein-digesting enzyme obtained from the papya fruit.	Halal
Paprika	A spice and colorant made from the ground, dried, ripe fruit of sweet red peppers.	Halal
Parabens	Antimicrobial agents that are methyl or propyl esters of para-hyroxybenzoic acid used in baked goods, beverages, and food color.	Halal
Parboiled rice	The rice that is cooked to gelatinized, then dried. It is used in soups and rice dinners.	Halal
Parsley	A herb made from the dried leaves of *Petroselinum hortense* used for garnishing and seasoning.	Halal
Pastry flour	A flour obtained from soft wheat.	Halal
Peanut oil	The oil obtained from peanuts.	Halal
Pectin	A water-soluble plant gum obtained from citrus peel and apple pomace.	Halal
Pepper	A spice from the vine *Piper nigrum* L., which produces green, red, black and white berries.	Halal
Pepper, cayenne	A hot spice related to paprika, bell peppers of the Capsicum family.	Halal
Pepper, red	The ripe pod spice of the genus *Capsicum*.	Halal
Petrolatum	A mixture of semisolid hydrocarbons obtained from petroleum used in bakery products, dehydrated fruits and vegetables and egg white solids as a release agent and defoamer.	Halal
Petroleum wax	A wax from petroleum in chewing gum base as a protective coating on raw fruits and vegetables.	Halal

(*Continued*)

Appendix C

Ingredient Name	Description	Halal Status
Phosphoric acid	A strong inorganic acid produced by burning phosphorus used as a flavoring acid in cola and root beer beverages to provide desirable acidity.	Halal
Polydextrose	A reduced calorie bulking agent made by condensation of dextrose.	Halal
Polyethylene glycol	The polymers of ethylene oxide and water used as flavoring adjuvants in carbonated beverages.	Halal
Polyglycerol esters of fatty acids	Emulsifiers that are mixed partial esters formed by reacting polymerized glycerols with edible fats, oils, or fatty acids. It is used in cake mixes, whipped toppings, and in flavors and colors as a solubilizer.	Doubtful
Polyoxyethylene sorbitan fatty acid esters	Emulsifiers made by reacting ethylene oxide with sorbitan esters; used in oil and water emulsions.	Doubtful
Polyoxyethylene (20) sorbitan monooleate (polysorbate 80)	An emulsifier produced by reacting oleic acid with sorbitol to yield a product which is reacted with ethylene oxide.	Doubtful
Polyoxyethylene (20) sorbitan monostearate (polysorbate 60)	An emulsifier manufactured by reacting stearic acid with sorbitol to yield a product which is reacted with ethylene oxide.	Doubtful
Polyoxyethylene (20) sorbitan tristearate (polysorbate 65)	An emulsifier manufactured by reacting stearic acid with sorbitol to yield a product which is then reacted with ethylene oxide.	Doubtful
Polyoxyl (40) stearate	An emulsifier and antifoaming agent.	Doubtful
Poppy seed	A seed spice of *Papaver somniferum* L. plant.	Halal
Potassium alginate	A gum that is the potassium salt of alginic acid.	Halal
Potassium bicarbonate	An alkali and leavening agent.	Halal
Potassium bisulfite	A preservative and an antioxidant.	Halal
Potassium bromate	A dough conditioner	Halal
Potassium carbonate	A general purpose food additive and alkali.	Halal
Potassium chloride	A nutrient and dietary supplement used as a salt substitute.	Halal
Potassium citrate, monohydrate	A sequestrant food additive.	Halal
Potassium hydroxide	A water-soluble alkali and a food additive.	Halal
Potassium iodate	A nutrient made by reacting iodine with potassium hydroxide.	Halal
Potassium iodine	A nutrient and dietary supplement.	Halal
Potassium metabisulfite	A chemical preservative and antioxidant.	Halal
Potassium nitrate	A preservative and color fixative in meats.	Halal
Potassium nitrite	A color fixative in meats.	Halal
Potassium oleate	The potassium salt of oleic acid used as an emulsifier, and anticaking agent.	Doubtful
Potassium palmitate	The potassium salt of palmitic acid used as a binder, emulsifier, and anticaking agent.	Doubtful

(Continued)

Ingredient Name	Description	Halal Status
Potassium sorbate	A preservative, the potassium salt of sorbic acid used in cheese, bread, beverages, margarine, and dry sausage.	Halal
Potassium stearate	The potassium salt of stearic acid used as a plasticizer in chewing gum base.	Doubtful
Potassium sulfite	A preservative and antioxidant sulfite salt.	Halal
Potassium tripolyphosphate	A phosphate used as a moisture binder in meat as an emulsifier and a sequestrant.	Halal
Potato starch	A starch obtained from potatoes.	Halal
Powdered sugar	A sweetener obtained by grinding sugar and adding cornstarch.	Halal
Pregelatinized starch	Starch that has been heat processed to permit swelling in cold water used in instant puddings.	Halal
Propane	An aerating agent used as a propellant and aerating agent for foamed or sprayed foods.	Halal
Propionic acid	The acid source of the propionates used as a mold inhibitor.	Halal
Propylene glycol	A humectant and flavor solvent.	Halal
Propylene glycol alginate	A gum that is the propylene glycol ester of alginic acid obtained from kelp.	Halal
Propylene glycol mono- and di-esters	An emulsifier that consists of propylene glycol esters of fatty acids such as palmitic and stearic.	Doubtful
Propylene glycol monostearate	An emulsifier that is a propylene glycol ester of stearic acid.	Doubtful
Propyl gallate	A synthetic antioxidant used to retard fat and oil rancidity.	Halal
Pyridoxine	A water-soluble B vitamin found in liver, eggs, and meat.	Doubtful
Pyridoxine hydrochloride	A water-soluble B vitamin.	Doubtful
Quince seed	A gum produced from the fruit of the quince tree *Cydonia oblonga*.	Halal
Quinine	A flavorant naturally obtained from the cinchona tree used as a bitter flavoring in beverages such as quinine water.	Halal
Raisin	A dried grape used as a fruit and as an ingredient in cereals.	Halal
Rape seed oil	The oil derived from seeds of *Brassica campestris* and related plants.	Halal
Rennet	A milk coagulant that is the concentrated extract of rennin enzyme obtained from calves' stomachs (calf rennet) or adult bovine stomachs (bovine rennet).	Doubtful
Riboflavin	The water-soluble vitamin B2.	Halal
Rice bran oil	An oil extracted from rice bran.	Halal
Rice bran wax	A refined wax obtained from rice bran.	Halal
Rice flour	The flour made from different varieties of long-, medium-, and short-grain rice.	Halal
Rice starch	The starch obtained from rice used in puddings.	Halal
Roselle	A natural red colorant obtained from the roselle flower extract from hibiscus.	Halal

(*Continued*)

Ingredient Name	Description	Halal Status
Rosemary	A herb made from the dried leaves of *Rosmarinus officinalis* L. an evergreen shrub.	Halal
Rum ether (ethyl oxyhydrate)	A synthetic flavoring agent used in beverages, candy, and ice cream.	Halal
Rye flour	The flour obtained by milling rye.	Halal
Saccharin	A non-nutritive synthetic sweetener which is 300 to 400 times sweeter than sucrose.	Halal
Safflower oil	A vegetable oil obtained from the safflower seeds.	Halal
Saffron	A spice and colorant obtained from the dried stigmas of the *Crocus sativus* L.	Halal
Sage	A herb made from the dried leaves of the shrub *Salvia officinalis* L.	Halal
Sago starch	The starch obtained from the sago palm used in puddings.	Halal
Savory	A herb that is the dried leaves and flowering tops of the plants *Satureia hortensis* L.	Halal
Self-rising flour	White flour with added sodium bicarbonate and other salts.	Halal
Semen cydonia	A product from quince seeds.	Halal
Semolina	The purified coarse ground durum wheat.	Halal
Sesame oil	The oil obtained from sesame seeds.	Halal
Sesame seed	The seed of the plant *Sesamum indicum* L.	Halal
Shallot	A spice *Allium ascalonicum*, a member of the onion family	Halal
Shortening	Any animal or vegetable fat or oil used in baked goods.	Doubtful
Silicon dioxide	An anticaking agent of mineral source.	Halal
Smoke flavoring	A flavorant obtained from burning hardwoods. It is used and mixed with other ingredients.	Doubtful
Sodium acetate	A source of acetic acid.	Halal
Sodium acid pyrophosphate (SAPP)	A leavening agent, preservative, sequestrant, and buffer.	Halal
Sodium alginate	A gum obtained as a sodium salt of alginic acid obtained from seaweed.	Halal
Sodium aluminum phosphate	An emulsifier and leavening agent.	Halal
Sodium aluminum sulfate	A leavening agent.	Halal
Sodium ascorbate	An antioxidant and a vitamin C.	Halal
Sodium benzoate	A preservative that is the sodium salt of benzoic acid.	Halal
Sodium bicarbonate	A leavening agent.	Halal
Sodium bisulfate	A strong acid and food additive	Halal
Sodium bisulfite	A preservative that prevents discoloration and inhibits bacterial growth.	Halal
Sodium calcium alginate	A gum that is the sodium and calcium salt of alginic acid used as a thickener.	Halal
Sodium calcium aluminosilicate	An anticaking agent.	Halal

(*Continued*)

Ingredient Name	Description	Halal Status
Sodium caprate	The sodium salt of capric acid used as an emulsifier and anticaking agent.	Doubtful
Sodium caprylate	The sodium salt of caprylic acid used as an emulsifier and anticaking agent.	Doubtful
Sodium carbonate	An alkali and food additive.	Halal
Sodium caseinate	The sodium salt of casein, a milk protein.	Halal
Sodium citrate	A buffer and sequestrant made from citric acid.	Halal
Sodium diacetate	A preservative, sequestrant, acidulant, and flavoring agent.	Halal
Sodium erythorbate	An antioxidant that is the sodium salt of erythorbic acid.	Halal
Sodium hexametaphosphate	A sequestrant and moisture binder.	Halal
Sodium hydroxide	An alkali and a neutralizing agent.	Halal
Sodium iron pyrophosphate	A nutrient and dietary supplement.	Halal
Sodium lactate	A humectant; the sodium salt of lactic acid.	Halal
Sodium laurate	The sodium salt of lauric acid used as an emulsifier.	Doubtful
Sodium lauryl sulfate	An emulsifier and whipping aid.	Doubtful
Sodium metabisulfite	A preservative and antioxidant.	Halal
Sodium myristate	The sodium salt of myristic acid. It is used as a binder, emulsifier and anticaking agent.	Doubtful
Sodium nitrate	The salt of nitric acid; used as an antimicrobial agent and preservative.	Halal
Sodium nitrite	The salt of nitrous acid; used as an antimicrobial agent and preservative	Halal
Sodium oleate	The sodium salt of oleic acid; used as an emulsifier and anticaking agent.	Doubtful
Sodium palmitate	The sodium salt of palmitic acid used as an emulsifier and anticaking agent.	Doubtful
Sodium polyphosphate	A sequestrant and emulsifier.	Halal
Sodium potassium tartrate	A buffer and sequestrant.	Halal
Sodium propionate	A preservative that is the salt of propionic acid.	Halal
Sodium silicate	A preservative for eggs.	Halal
Sodium silicoaluminate	An anticaking and conditioning agent used to prevent caking.	Halal
Sodium sorbate	A preservative that is the salt of sorbic acid.	Halal
Sodium stearate	The sodium salt of stearic acid used as a plasticizer in chewing gum base.	Doubtful
Sodium stearoyl fumarate	A dough conditioner for yeast-raised baked goods.	Doubtful
Sodium stearyl fumarate	A dough conditioner for yeast-raised baked goods.	Doubtful
Sodium tartrate	A sequestrant and stabilizer.	Halal
Sodium tetrametaphosphate	A sequestrant and emulsifier.	Halal
Sodium thiosulfate	A sequestrant, antioxidant.	Halal
Sodium tripolyphosphate	A binder stabilizer and sequestrant.	Halal

(Continued)

Appendix C

Ingredient Name	Description	Halal Status
Sorbic acid	A preservative used against yeasts and molds.	Halal
Sorbitan monostearate	An emulsifier that is a sorbitan fatty acid ester, being a sorbitol-derived analog of glycerol monostearate.	Doubtful
Sorbitol	A humectant and sugarless sweetener.	Halal
Soy flour	Flour obtained from defatted soybean.	Halal
Soybean oil	The oil obtained from the seed of the soybean legume.	Halal
Soybean protein concentrate	The concentrate obtained by processing soybean flour to remove the soluble carbohydrates.	Halal
Soybean protein isolate	The isolate prepared from soybean flour by extracting the protein.	Halal
Spice	A variety of dried plant products that exhibit an aroma and flavor to foods.	Halal
Stannous chloride	An antioxidant and preservative.	Halal
Stearic acid	A fatty acid composed of a mixture of solid organic principally stearic acid and palmitic acid.	Doubtful
Stearoyl lactylate	A dough conditioner, emulsifier, and whipping agent.	Doubtful
Stearyl citrate	An antioxidant made by reacting citric acid with stearyl alcohol.	Doubtful
Succinylated monoglycerides	Emulsifiers and dough conditioners.	Doubtful
Succinic acid	An acidulant prepared by the hydrogenation of maleic or fumaric acid.	Halal
Sulfur dioxide	A preservative.	Halal
Sulfuric acid	An acidulant.	Halal
Sunflower oil	A vegetable oil obtained from sunflower seeds.	Halal
Tallow	Animal fat from mutton or beef.	Doubtful
Tarragon	The dried leaves and flowering tops of the herb *Artemisia dracunculus* L.	Halal
Tartaric acid	An acidulant and flavoring.	Halal
Tertiary butylhydroquinone (TBHQ)	A synthetic antioxidant.	Halal
Tetrasodium pyrophosphate	A coagulant, emulsifier, and a sequestrant	Halal
Textures soy flour	Soy flour that is heat processed and extruded.	Halal
Textured vegetable protein	A vegetable protein that is heat processed and extruded to form meat analogs.	Halal
Thiamine	Water Soluble vitamin B1.	Halal
Thyme	The herb composed of leaves of the plant *Thymus Vulgaris* L.	Halal
Titanium dioxide	A white pigment used as a color additive.	Halal
Tocopherol	A fat-soluble vitamin E obtained from vegetable oils	Halal
Tofu	A soybean curd product.	Halal
Tragacanth	A gum produced from the plant of the genus *Astragalus* used in salad dressings, and sauces	Halal

(Continued)

Ingredient Name	Description	Halal Status
Tricalcium phosphate	An anticaking agent.	Halal
Tricalcium silicate	An anticaking agent.	Halal
Triethyl citrate	A sequestrant used in lemon drinks	Halal
Trihydroxybutyrophenone (THBP)	An antioxidant	Halal
Tripotassium citrate	A buffer and a sequestrant	Halal
Tripotassium phosphate	An emulsifier and alkaline buffer used in low-sodium products	Halal
Trisodium citrate	A buffer and sequestrant	Halal
Trisodium phosphate	An emulsifier and buffer	Halal
Turbinado sugar	Washed raw sugar of large crystals	Halal
Turmeric	A spice and colorant obtained from rhizome or root of *Curcuma Longa*.	Halal
Tyrosine	An amino acid usually isolated from silk waste.	Doubtful
Vanilla extract	A flavorant made from vanilla bean extract.	Doubtful
Vanilla flavor, artificial	A flavorant composed of vanillin, and ethyl vanillin	Doubtful
Vanillin	A flavorant made from synthetic or artificial vanilla.	Doubtful
Vegetable gum	Gums obtained from plant source	Halal
Vegetable oils	Oils obtained from a vegetable source, including soy beans, peanuts, cottonseeds, and plants.	Halal
Vinegar	An acidulant and flavorant produced by successive alcoholic and acetous fermentations.	Halal
Vitamin K	A fat-soluble vitamin.	Halal
Waxy maize starch	The starch protein of waxy corn.	Halal
Waxy rice flour	A flour obtained from waxy rice.	Halal
Wheat flour	A fine powder obtained by milling wheat	Halal
Wheat germ	The oil-containing portion of the wheat kernel.	Halal
Wheat gluten	The water-insoluble complex protein fraction separated from wheat flours.	Halal
Wheat starch	A starch obtained from wheat.	Halal
Whey	The portion of milk remaining after coagulation and removal of curd.	Doubtful
Whey powder	The solid fraction or dry form of whey.	Doubtful
Whey protein isolate	Proteins isolated from whey.	Doubtful
Whey protein concentrate	Whey powder where some of the non-protein has been removed.	Doubtful
Whole milk solids	The product resulting from the drying or desiccation of milk.	Halal
Whole wheat flour	The flour obtained by grinding cleaned wheat	Halal
Wine vinegar	The vinegar made by the alcoholic and acetous fermentation of the juices.	Halal
Worcestershire sauce	A sauce consisting of many ingredients.	Doubtful
Xanthan gum	A gum obtained by microbial fermentation from the *Xanthomonas campestris* organism.	Halal
Xylitol	A sweetener that is as sweet as sucrose.	Halal

(*Continued*)

Appendix C

Ingredient Name	Description	Halal Status
Yeast extract	A flavor enhancer obtained from the yeast cells of *Saccharomyces cerevisiae*.	Halal
Yeast food	A complete food used in doughs.	Doubtful
Yellow prussiate of soda	An anticaking agent and crystallizing agent.	Halal
Yogurt	A custard-like or soft gel product	Doubtful
Zein	A corn protein produced from corn gluten meal	Halal
Zinc acetate	A nutrient and dietary supplement	Halal
Zinc carbonate	A nutrient and dietary supplement	Halal
Zinc chloride	A nutrient and dietary supplement	Halal
Zinc gluconate	A source of zinc that functions as a nutrient and dietary supplement.	Halal
Zinc oxide	A nutrient and dietary supplement	Halal
Zinc stearate	A nutrient and dietary supplement	Doubtful
Zinc sulfate	A nutrient and dietary supplement used in frozen substances.	Halal

Appendix D
Codex Alimentarius[1]

GENERAL GUIDELINES FOR USE OF THE TERM "HALAL"

The Codex Alimentarius Commission accepts that there may be minor differences in opinion in the interpretation of lawful and unlawful animals and in the slaughter act, according to the different Islamic Schools of Thought. As such, these general guidelines are subjected to the interpretation of the appropriate authorities of the importing countries. However, the certificates granted by the religious authorities of the exporting country should be accepted in principle by the importing country, except when the latter provides justification for other specific requirements.

SCOPE

1. These guidelines recommend measures to be taken on the use of Halal Claims in food labelling.
2. These guidelines apply to the use of the term *Halal* and equivalent terms in claims as defined in the General Standard for the Labelling of Prepackaged Foods and include its use in trademarks, brand names and business names.
3. These guidelines are intended to supplement the Draft Revision of the Codex General Guidelines on Claims and do not supersede any prohibition contained therein.

DEFINITION

HALAL FOOD MEANS FOOD PERMITTED UNDER THE ISLAMIC LAW AND SHOULD FULFIL THE FOLLOWING CONDITIONS

1. does not consist of or contain anything which is considered to be unlawful according to Islamic Law;
2. has not been prepared, processed, transported or stored using any appliance or facility that was not free from anything unlawful according to Islamic Law; and
3. has not in the course of preparation, processing, transportation or storage been in direct contact with any food that fails to satisfy 2.1.1 and 2.1.2 above.

NOTWITHSTANDING SECTION A.2.1 ABOVE

1. *Halal* food can be prepared, processed or stored in different sections or lines within the same premises where non-halal foods are produced, provided

that necessary measures are taken to prevent any contact between Halal and non-halal foods;

[1] The Draft Guidelines were advanced to Step 8 subject to the advice of the Executive Committee on whether they fall out-side of the mandate of the Commission and are contrary to the statements of principles adopted by the Commission at its 21st Session concerning the Role of Science in the Codex Decision-Making Process and the Extent to which Other Factors are Taken into Account.

> 2. *Halal food can be prepared, processed, transported or stored using facilities which have been previously used for non-halal foods provided that proper cleaning procedures, according to Islamic requirements, have been observed.*

CRITERIA FOR USE OF THE TERM "HALAL"

Lawful Food

The term Halal may be used for foods which are considered lawful. Under the Islamic Law, all sources of food are lawful *except* the following sources, including their products and derivatives which are considered unlawful:

Food of Animal Origin

a. Pigs and boars.
b. Dogs, snakes and monkeys.
c. Carnivorous animals with claws and fangs such as lions, tigers, bears and other similar animals.
d. Birds of prey with claws such as eagles, vultures and other similar birds.
e. Pests such as rats, centipedes, scorpions and other similar animals.
f. Animals forbidden to be killed in Islam i.e., ants, bees and woodpecker birds.
g. Animals which are considered repulsive generally like lice, flies, maggots and other similar animals.
h. Animals that live both on land and in water such as frogs, crocodiles and other similar animals.
i. Mules and domestic donkeys.
j. All poisonous and hazardous aquatic animals.
k. Any other animals not slaughtered according, to Islamic Law.
l. Blood

Food of Plant origin

Intoxicating and hazardous plants except where the toxin or hazard can be eliminated during processing.

Drink

a. Alcoholic drinks.
b. All forms of intoxicating and hazardous drinks.

Appendix D

A.3.1.4 Food Additives
All food additives derived from Items A.3.1.1 through A.3.1.3

SLAUGHTERING

All lawful land animals should be slaughtered in compliance with the rules laid down in the Codex Recommended Code of Hygienic Practice for Fresh Meat [2] and the following requirements:

1. The person should be a Muslim who is mentally sound and knowledgeable of the Islamic slaughtering procedures.
2. The animal to be slaughtered should be lawful according to Islamic law.
3. The animal to be slaughtered should be alive or deemed to be alive at the time of slaughtering.
4. The phrase "Bismillah" (In the Name of Allah) should be invoked immediately before the slaughter of each animal.
5. The slaughtering device should be sharp and should not be lifted off the animal during the slaughter act.
6. The slaughter act should sever the trachea, oesophagus and main arteries and veins of the neck region.

PREPARATION, PROCESSING, PACKAGING, TRANSPORTATION AND STORAGE

All food should be prepared, processed, packaged, transported and stored in such a manner that it complies with Sections A.2.1 and A.2.2 above and the Codex General Principles on Food Hygiene and other relevant Codex Standards.

ADDITIONAL LABELLING REQUIREMENTS

1. When a claim is made that a food is *Halal*, the word *Halal* or equivalent terms should appear on the label.
2. In accordance with the Draft Revision of the Codex General Guidelines on Claims, claims on *Halal* should not be used in ways which could give rise to doubt about the safety of similar food or claims that *Halal* foods are nutritionally superior to, or healthier than, other foods.

The Codex General Guidelines for the Use of the Term Halal were adopted by the Codex Alimentarius Commission as its 22nd session, 1997. They have been sent to all member Nations and Associate Members of FAO and WHO as an advisory text, and it is for individual governments to decide what use they wish to make of the Guidelines.

NOTE
1. www.fao.org/DOCREP/005/Y2770E/y277e08.htm.

Index

Abu Darda, 67
Abu Hanifah, 19
Accreditation, 139
Acetic acid, 203
Agar, 179
Ahlul Kitab, 9, 262–263
AIFDC, *see* Assessment Institute for Foods, Drugs, and Cosmetics (AIFDC)
Air chilling, 117
Alcohol, 13–14, 23–24, 196–197, 201–206
    beef stroganoff, 202, 205
    beverages, 202
    consumptions, 204–205
    cooking, 202–203
    cosmetics, 203
    cough syrups, 203
    flavorant in cooking, 202–203
    flavorings, 216–217
    government regulators, 205
    halal certification of items, 205–206
    industrial chemical, 202–203
    intoxicants, 201
    liquor-flavored chocolates, 202, 205
    mouthwash, 203
    perfumes, 203
    pharmaceuticals, 203
    production of vinegar, 215
    residual alcohol, 205
    retention, 203
    rum cakes, 202, 205
    SD alcohol, 203
    topical products, 203
    uses of, 203–204
    vanilla bean, 203
    vinegar, 203
    wine, 202, 205
Alcoholic beverages, 202–206
    beer, 205
    compound, 202
    distilled, 202
    fermented, 202
    fortified, 202
    liquor, 206
    rum, 205
    spirit, 202
    tequila, 205
    whisky, 205
    wine, 205
Allantoin, 235
Ambergris, 235
Amino acids, 214

Amyl alcohol, 215
Animal breeding, 79–80
Animal fat and protein, 24
Animal hair, 235
Animal housing, 77–78
Animals and birds
    halal
        buffaloes, 271
        camels, 271
        chickens, 114, 271
        doves, 271
        ducks, 114, 271
        emus, 271
        geese, 271
        goat, 271
        lamb, 271
        ostriches, 271
        partridges, 271
        pigeons, 114, 271
        quail, 114, 267
        sheep, 271
        sparrows, 271
        turkeys, 114, 271
    non-halal
        boars, 360
        crocodiles, 360
        dogs, 360
        donkeys, 360
        eagles, 114
        falcons, 271
        flies, 360
        frogs, 360
        kites, 271
        lice, 360
        lions, 360
        maggots, 360
        monkeys, 360
        mules, 360
        ospreys, 271
        pigs, 360
        poisonous aquatic animals, 360
        snakes, 360
        swine, 271
        tigers, 360
        vultures, 114
Animal-based enzymes, 167, 215
Animal-based ingredients, 129
Animal-based materials, 195–196
Animal Products Regulations (2000), 141
Arachidonic acid, 235
Arachidyl propionate, 236

Animal feeds, 229–231
Animal welfare
 biosecurity, 80–81
 breeding, 79–80
 contemporary issues, 76–77
 coping, 76
 feeding, 78–79
 history of, 73–75
 housing, 77–78
 issues, 77
 science of, 75–76
 slaughter handling, 82–83
 standards, 77
 transport, 81
Anticaking agents, 157, 207, 241
Anti-dusting agents, 216
Antioxidants, 208
Assessment Institute for Foods, Drugs, and Cosmetics (AIFDC), 222, 247
Auditing, 139
Autolyzed yeast extracts, 13

Bacon bits, 214, 216
 artificial, 214
 natural, 214
Bacterial cultures, 157
Baking, 164, 168, 194
Batters, 216
Beans, 22
Beef tallow, 218
Beer, 13, 23, 202, 214, 273
Beer batter, 214
Benzoic acid, 209
Beta-carotene, 222, 236
Bible, 260
Bioengineered enzymes, 169–170
Biosecurity, 80–81
Biotechnology, 194–195, 225–228, 274–275
Birds, 114–117
Bismillah Allahu Akbar, 19, 109, 113, 118, 121
Bleeding, 13, 20, 96, 113, 118, 320
Blessing, 109, 116, 122, 265–266, 278
Blood, 12–13, 18
 albumin, 271
 sausage, 271
Blowfish, 147
Boar bristles, 236
Bodek, 266
Bones, 179
Boning establishment, 119
Botanicals, 222, 235
Bovine spongiform encephalopathy (BSE), 230
Brandy, 273
Bread, 164
Breadcrumbs, 216
Breading, 135, 216
Breakfast cereals, 163–164

British Retail Consortium (BRC) standards, 300
Broth, 170
BRC, see British Retail Consortium (BRC) standards
BSE, see Bovine spongiform encephalopathy (BSE)
Buffalo, 155
Butter, 21, 156

Cadaver, 116
Cakes, 165
Campylobacter spp., 136
Canadian Food Inspection Agency (CFIA), 141
Captive bolt stunning, 112
Carbon dioxide stunning, 113
Carotene, 236
Carrageenan, 179
Carrion, 13
Cattle hides, 22, 179
Casings, 121
Casting, 82
CCP, see Critical control points
Cellulose casings, 121
Cellulose gum, 179
Certification, 139
Cheese, 21, 23, 156–157
 cheddar, 157
 colby, 157
 cottage, 157
 flavors, 217
 making, 157
 mozzarella, 157
 shredded, 157
Chewing gum, 165
Chymosin, 157, 225, 227
Citric acid, 217
Civet oil, 214, 216, 222
Clean meat
 benefits animal welfare, 325
 safer for human consumption, 325
 without antibiotic usage, 325
 halal status of, 325
Coating, 135
Codex Alimentarious (Codex) international food standards (FAO/WHO), 294, Appendix D
Collagen casings, 121
Colors, 217
 dyes, 217
 lakes, 217
 synthetic, 217
 water-soluble, 217
Contamination protocols, 198
Cookies, 165
Cooking, 202, 203
Cornstarch, 170
Corn-syrup, 168

Index

Cosmetics, 70, 203
Cow, 155
Crescent M symbol, 250
Cream, 21, 156
Critical control points (CCP), 322–324
Crumbing, 135
Cultured meat, *see* Clean meat
Cultured milk, 158
Cured meats, 135

Dairy equipment (DE), 267
Dairy ingredients, 218
Dairy products, 157–158
Dead animals, 13
DE, *see* Dairy equipment (DE)
Debaryomyces hansenii, 128
Dhabh slaughtering, 19–20
Dhabiha, 9, 320
Dhakaat, 18
Diacetyl tartaric esters of mono- and
    di-glycerides (DATEM), 23
Dolphins, 147
Double rail restraint, 82
Doughnuts, 165
Dragging, *see* Hoisting conscious cattle
Dry bones, 178
Dry milk powder, 156
Dyes, 217

E. coli, 270, 325
E-numbers, 242, 291
Edible printing dyes, 243
Eels, 146
Eggs, 14, 21
Electric prod, 81, 89, 94, 100, 111
Electrical stunning, 112, 115–116
Electrified water, 21
Electrocution, 115–116, 118
Emulsifiers, 23, 164
Encapsulation materials, 215
    cellulose, 215
    gelatin, 215
    shellac, 215, 217
Encrusting, 135
Energy drinks, 302
Enterohaemorrhagic Escherichia coli
    (EHEC), 136
Enzyme-modified products, 159
Enzymes, 23, 157, 167, 215
    acid hydrolysis method, 168
    animal-derived material, 215
    bioengineered, 169–170
    breads, 168
    broth, 170
    cheese, 159
    classifications, 169
        hydrolases, 169

isomerases, 169
ligases, 169
lyases, 169
oxidoreductases, 169
transferases, 169
halal and, 170–174
juice, 169
label or not to label, 174
meat extracts, 168
protein hydrolysis, 168
sugar, 168
uses in food, 168–169
vegetable oil, 169
Ethanol, 196, 201, 203, 221, 235
Exsanguination, 108

FAME, *see* Fatty acid methyl esters (FAME)
FAO/WHO Codex Alimentarius, 292
Fatty acid methyl esters (FAME), 294
Fatty chemicals, 156, 223
FDA, *see* US Food and Drug Administration (FDA)
Federation of Islamic Associations of New
    Zealand (FIANZ), 140
Feed, 78–79
FEMA, *see* Flavor and Extract Manufacturers
    Association (FEMA)
Fermented alcoholic drinks, 201
FIANZ, *see* Federation of Islamic Associations
    of New Zealand (FIANZ)
Fish, 14, 21, 145–146
Five freedoms, 75–76
Flavoring, 185–186
    civet oil, 214, 216, 222
    extraction, 189–190
    materials, 185
    reaction flavors, 190
Flavors, 24–25, 186, 202
    ambergris, 222
    amino acids, 193
    castoreum, 222
    cheese, 159
    delivery systems, 192
    emulsions, 191–192
    engineered flavor systems, 191
    essential oils, 192
    fruit juices and concentrates, 192
    grill, 217
    halal and, 194
    lipids and fatty acids, 193
    liquor, 214
    oleoresins, 192
    oranges, 203
    production processes, 197–198
        cleaning, 198
        cross-contamination, 198
        equipment segregation, 197–198
        HACCP controls, 198

Index

raw material receipt, 197
  storage, 198
simple mixtures, 188–189
smoke, 217
spray-dried, 191
sugars and starches, 193
synthetic chemicals, 192
systems, 186
vanilla beans, 203, 206
wine, 202, 214
Flavor and Extract Manufacturers Association (FEMA), 208, 216
Flour, 163
Food, 7
Food additives, 207–211, Appendix B
  acidity regulators, 207
  acids, 207
  anticaking agents, 207
  antioxidants, 208
  bulking agents, 208
  color retention agents, 208
  control points for halal, 209–211
  definition of, 207
  emulsifiers, 208–209
  flavors, 208
  flour treatment agents, 208
  food colorings, 208
  glazing agents, 208
  halal certifier, 207
  halal status, 207
  humectants, 208
  hydrocolloids, 209
  leavening agents, 208
  market, 209
  microorganism, 208
  minerals, 208–209
  natural, 207
  nutritional quality, 208
  packaging materials, 211
  plant stanol esters, 209
  prebiotics, 209
  preservatives, 208
  probiotics, 209
  processing aids, 209–211
  sequestrants, 208
  soya ingredients, 209
  stabilizers, 208
  surface active agents, 208
  sweeteners, 208–209
  synthetic, 207
  thickeners, 208
  tracer gasses, 208
  types of, 207
  vitamins, 208–209
Food coatings, 164
Food and Drug Administration (FDA), 202, 207, 210, 216, 222, 230, 235

Food ingredients, 213–218, Appendix B, Appendix C
  amino acids, 214
  animal oils, 217
  bacon bits, 214
  batters, 216
  breadcrumbs, 216
  breading, 216
  civet oil, 214
  colorants, 217
  condiments, 216
  curing agents, 217
  dairy ingredients, 217–218
  dressings, 216
  encapsulation materials, 215
  fried batter products, 214
  fruit coatings, 217–218
  fusel oil derivatives, 215
  gelling, 215
  grill flavors, 217
  liquor, 214
  manufacturing, 216
  meat, 217
  minor ingredients, 215
  modified starches, 215
  poultry, 217
  production of vinegar, 215
  sauces, 216
  seasonings, 216
  smoke flavors, 217
  spices, 216
  stabilizing agents, 215
  texturizing, 215
  thickening, 215
  vegetable coatings, 217–218
  wine extracts, 214
Food industry, 167
Food model, 309–315
Food Safety and Inspection Service (FSIS), 26, 141
Food Standards Code (2002), 141
  groups, 14–15
  halal foods, 229–231
  ingredients, 22
  nutraceutical ingredients, 223
  permissibility, 9
  principles, 8–9
  prohibitions, 12–13
Food technology magazine, 208
Fresh bones, 178–179
Fried goods, 165
Frozen desserts, 158
FSIS, see Food Safety and Inspection Service

Gas chromatograph, 294
Gas stunning, 113
GCC, see Gulf Cooperation Council (GCC)

Index

Gelatin, 22–23, 177, 215, 222, 273–274
    calf skins, 178
    cattle bones, 22, 177
    cattle hides, 177
    cereals, 163
    cosmetics, 181
    dietetic, 181
    fish, 14, 21–22
    fish skins, 177
    foods, 180
    halal and, 179
    in Islam, 177–178
    kosher, 22
    medicinal, 181
    pharmaceuticals, 181
    pigskin, 177
    poultry skins, 177
    preparation, 178–179
    production process, 179–180
    sources of, 178
    therapeutic uses, 181
    vegetable substitutes, 179
Generally recognized as safe (GRAS), 208
Genetically modified organisms (GMO), 225; *see also* Biotechnology
Genetics, 226
GFSI, *see* Global Food Safety Initiative (GFSI) standards
Global Food Safety Initiative (GFSI) standards, 300
Global halal economy, 61–70
    adoption spectrum, 66–67
    challenges, 68–70
    classification, 63
    cosmetic sector, 69
    define, 63
    food sector, 66–68
    future of, 70
    key drivers, 65–66
    importance, 61
    Islamic values, 66
    pharmaceutical sector, 69
    size of, 63
Glucoamylase, 168
Glycerin, 23, 223
GMO, *see* Genetically modified organisms (GMO)
Good Hygienic Practices (GHP), 141
Good Manufacturing Practices (GMP), 141, 300
Grandin, Dr. Temple, 88
GRAS, *see* Generally recognized as safe (GRAS)
Green bones, *see also* Fresh bones
Gulf Cooperation Council (GCC), 29, 41

HACCP, *see* Hazard analysis and critical control points (HACCP), 141
Hadith, 7, 73, 146, 204

Halacha, 259
Halal, 1, 7–11, 66, 125, 204–211, 295–304
    allowed animals, 110
    animal feeds, 229–231
    awareness, 303–307
        consumers, 306
        food industry, 304–305
        methodology, 304
        religious knowledge, 304
        scientist, 305–306
        supply chain, 306–307
    biotechnology, 225–228
    birds, 114–117
    brand, 26
    certification, 247–257
    certification of items containing alcohol, 205–206
    certifying bodies, 126, 129, 139, 142
    cheese, 159
    cold stores, 119
    cosmetics, 234–239
    dairy products, 156
    deboning and processing rooms, 119
    facilities, 299–301
        potential hazards, 299–301
    edible coating, 244–245
    edible films, 244–245
    education scheme, 297–301
    enzymes, 21
    equipment, 119
    flavors, 194–197
    food additives, 207–211
    food ingredients, 213–218
    food model, 309–315
    foods, 1–2
    gelatin, 178
    generally accepted, 129
    globalization, 283–288
        acceptability, 284–285
        accreditation of certification bodies, 287–288
        discussions, 286
        early days, 283–284
        lack of recognition, 285–286
        multiple certifications, 285
        multiple standards, 287
        standards, 286–287
    global trading, 142
    GMO ingredients, 225–228
        donor gene affect, 227
        religious prohibition, 227
    guidelines, 25–26
    HACCP, 319–324
    halal certifier, 207
    implications for, 204–205
    ingredients, 129
    ingredients to watch, 222–223

beta-carotene, 222
flavors and colorant, 222
gelatin, 222
glycerin, 223
stearates, 223
tweens, 223
inherently, 129
labeling products, 241–242
labels, 243
laws, 7, 263
meat, 106, 119–120
nutritional food products, 223
opinions on, 205
organoleptic tests, 206
packaging containers, 244
packaging food, 244
packaging materials, 244
poultry, 106–110
printing on food products, 243
processed meats, 125–126, 130–134
processed products, 148
production processes, 197–198
production requirements, 221–223
prohibition of alcohol, 205
segregation of, 211
slaughter houses, 108
slaughter procedures, 108
slaughtering, 20–21
special concerns, 194–197
style, 26
types of products, 223; *see also* Nutraceutical ingredients
wholesome and, 126
Halal certification, 247–257
authorized to issue, 249
definition of, 248
markings, 252
process, 250–251
products certified, 250
all kinds of snacks, 250
beverages, 250
butter, 250
cereal-based products, 250
chocolate, 250
confectionary, 250
cooking oil, 250
cosmetics, 250
dairy products, 250
dietary supplements, 250
enzymes, 250
fish, 250
flavor, 250
frozen, 250
fruits, 250
greases, 250
gums, 250
juices, 250

lubricants, 250
margarine, 250
meals, 250
meat, 250
nutritional, 250
packaging materials, 250
personal care products, 250
pharmaceuticals, 250
poultry fresh, 250
prepared foods, 250
prepared meal, 250
processed products, 250
seafood products, 250
vegetables, 250
steps involved, 252
types of, 248–249
site certificate, registration, 248
time period, 249
yearly certification, 249
Halal control points (HCP), 4, 110–114, 126
casings, 121
cheese, 159–161
conventional fermentation process, 172–174
conventional process for extracting enzymes, 171–172
enzyme production, 170–171
equipment, 119–120
flour- and starch-based products, 165
gelatin, 181–182
meat slaughtering, 119–121
meat source, 119
non-meat ingredient, 120–121
packaging and labeling, 121
poultry slaughtering, 117–119
processing aids, 209–211
anti-clumping, filtering, 209
antifoaming, 209
bleaching compounds, 209
clarifying agents, 209
enzymes, 210
extraction aids, 209
packaging materials, 211
release agents, 209
spot checked, 211
processor, 140
seaproducts, 149
smoked fish, 153
Halal cosmetics,
guidelines for, 234–235
list of ingredients, 235–239
allantoin, 235
ambergris, 235
animal hair, 235
arachidonic acid, 235
arachidyl propionate, 236
beta carotene, 236
boar bristles, 236

Index

carotene, 236
castor, 236
castoreum, 236
chitosan, 236
cholesterol, 236
civet oil, 236
collagen, 236
colors, 237
cystine, 237
dexpanthenol, 239
dyes, 237
emu oil, 237
fatty acid mixtures, 237
fatty acids, 237
feathers, 237
gelatin, 237
glycerin/glycerol, 237
hyaluronic acid, 237
hydrolyzed animal protein, 238
hydrolyzed vegetable protein, 238
keratin, 238
lanolin, 238
lanolin acids, 238
lard, 238
l-cysteine, 237
mink oil, 238
musk (oil), 238
myristic acid, 238
oleic acid, 238
palmitic acid, 238
panthenol, 239
placenta, 239
polypeptides, 239
pristane, 239
pro-vitamin A, 236
pro-vitamin B-5, 239
snail slime, 239
stearic acid, 239
stearyl alcohol, 239
turtle oil, 239
vitamin B-complex factor, 239
wool fat, 238
wool wax, 238
Halal critical control points (HCCPs), 135
Halal food model, 309–315
    education, 315
        business, 315
        consumers, 315
        governments, 315
    need for development, 310–311
    strengthen, 312
        animals, 313
        at farm, 312–313
        plants, 313
        post-production mechanism, 314–315
        processing facilities, 314
        processing stage, 313
        production facilities, 314
        slaughtering facilities, 313–314
Halal laws, 263
    dietary laws, 268–275
    food processing, 273
    halal cooking, 273
    prohibited and permitted animals, 71
    prohibition of alcohol and intoxicants, 272–273
    prohibition of blood, 271–272
    sanitation, 273
Halal pharmaceuticals, 221
    ethanol, 221
Hand slaughter, 96, 320
Hanging, *see* Casting
Haram, 7–11, 207, 226
    alcoholic drinks, 319
    animals dead, 319
    birds of prey, 319
    carnivorous animals, 319
    donkey, 226
    intoxicants, 319
    pigs, 207, 225–226
    pork, 319
    swine blood, 230, 319
Hazard Analysis Critical Control Points (HACCP), 294, 319–324
    haram, 321–322
    implementation with halal products, 322
    makrooh, 322
    mashbooh, 322
    najis, 322
Hazard Analysis Critical Control Point (HACCP) framework, 198
HCP, *see* halal control points
Hemorrhaging, 115
Hens, 114
Hindquarter, 266
Honey, 129, 210
Hoisting conscious cattle, 83
Holding, 110–111, 118
Hormones, 114, 134
Hot wanding, 109
Hot-melt glues, 243
Human hairs, 237
Humane methods, 86
Humane restraining, 111, 118
Humane slaughter regulation, 112
Hydrolyzed collagen, 177, 292
Hydrolyzed vegetable proteins, 13

Icecream, 158
ICU, *see* Indonesian Council of Ulama (ICU)
IFANCA, *see* Islamic Food and Nutrition Council of America (IFANCA)
IFT, *see* Institute of Food Technologist (IFT)
IKIM, *see* Institute Kefahaman Islam Malaysia (IKIM)

Imam Malik, 19
Imam Shaf'ii, 19
Imitation seafood, 149
Incidental ingredients, 214
Incision, 19–20, 97,
Indonesian Council of Ulama (ICU), 247
Industrial chemical, 202–204
Insects, 196, 326
  extracts, 326
  as food, 326
  powders, 326
  types
    insects with blood, 326
    insects without blood, 326
  consumption in Muslim community, 326
Inspectors, 86, 91, 93, 119, 122, 265, 314
Institute Kefahaman Islam Malaysia (IKIM), 226
Institute of Food Technologist (IFT), 209
International Food Information Council, 227
International Standards Organization (ISO), 142, 300
Intoxicants, 13–14, 201
  ethanol, 201
  beverages, 201
In vitro meat, *see* Clean meat
Invocation, 113
Islam, 1, 7, 73
Islamic dietary laws, 39, 226, 248, 268, 304
Islamic Food and Nutrition Council of America (IFANCA), 4, 206, 249
Islamic jurisprudence, 14, 21
Islamic laws, 263, 270
Islamic philosophy, 76
Islamic requirements, 168, 178, 320
Islamic standard, 263
ISO, *see* International Standards Organization (ISO)
Isoamyl alcohol, 215

Jabatan Kemajuan Islam Malaysia (JAKIM), 226, 249
Jalalah, 231
Jellies, 177
Jewish milk, 267
Jews, 261
Jugular veins, 14, 19, 96 , 108, 116, 118
Juices, 192

Khamr, 21
Killing, 113
Knife, 83, 113, 118
Kosher, 22, 261–279
  allowed animals for, 263–264
  comparison, 276–279
  definition of, 261
  dietary laws, 264
  laws, 263

meat of animals, 262
prohibition of blood, 265–266
prohibition of mixing of milk and meat, 266–267
  passover requirements, 268
  special foods, 267
Kosher animals, 264
Kosher birds, 264
Kosherization, 268

L-cysteine, 164, 214
  batters, 214, 216
  doughnuts, 214
  pizza crust, 214
L. monocytogenes, 136, 141–142
Labeling, 241
Labeling requirements, 242
Lab-grown meat, *see* Clean meat
Lac-resin, 217, 245
Lactose, 23, 158
Lakes, 217
Lard, 218, 238
Latter Day Saints, 275
Leatherhead Food Research, 209
Leavening agents, 208
Liquid supplements, 223
Liquor, 214
  batter-coated fries, 214
  flavors, 158
  fried appetizers, 214
  onion rings, 214
Lobsters, 21, 146–147, 271
Locusts, 271–272

Machine-killed birds, 116
Mad cow disease, 230
Majelis Ulama Indonesia (MUI), 222, 247, 249
Majlis Ugama Islam Singapura (MUIS), 249
Makrooh, 9
Marine mammal, 147
Marshmallows, 165
Mashbooh, 9
Matzos, 268
Malaysian Department of Health, 222
Malaysian Department of Islamic Affairs, 247
Meat equipment (ME), 267
Meat, products, 14, 17–18, 24, 141
  halal certified, 106–110
  safety for human consumption, 107
Mecca, 7
Medical products, 177, 222
MENA (Middle East and North Africa) region, 30, 42
Metal cans, 244
Methyl pentadecanoic acid, 294
Microbial enzymes, 167
Microwave stunning, 109

Index

Milk, 14, 21, 156
Mineral-based ingredients, 129
Minor ingredients, 216
    anti-dusting agent, 216
    encapsulating agent, 216
    free-flow agent, 216
Modified starches, 215
    corn, 179, 215
    potato, 215
    tapioca, 215
    wheat, 215
Mold inhibitor, 14, 21, 157
Monkfish, 147, 260
Mono- and di-glycerides, 23, 156, 163
Mouthwash, 203, 205
Muhammad, 7
MUI, see Majelis Ulama Indonesia (MUI)
MUIS, see Majlis Ugama Islam Singapura (MUIS)
Multivitamins, 221
Mushroom-shaped hammer stunner, 112
Muslims, 1, 21
    age, 33
    demography, 29–30, 61
    in Europe, 34
    geographic spread, 30–33
    global halal trade, 39–44
    life expectancy, 33
    market, 241–245
        edible coating, 244–245
        edible films, 244–245
        labeling products, 241–245
        labels, 241–244
        packaging containers, 244
        packaging food, 244
        packaging materials, 244
        printing on food products, 244
    MENA, 33
    in North America, 34
    urbanization, 33
    US halal market, 34–39
    US trade, 44–50
Muslim World League (MWL), 249
Mussels, 147
MWL, see Muslim World League (MWL)

NAMI, see North American Meat Institute (NAMI)
Nahr, 19
Najis, 192
Natamycin, 157, 159
Natural bacon flavor, 120
Natural casings, 121
Natural food additives, 208–209
    beet root juice, 209
    benzoic acid, 209
    cranberries, 209

Non-halal materials, 291–296
    DNA for species identification, 294–296
    fatty acid composition, 293–294
    forensic science laboratory, 291
    gelatin testing, 292
    presence of alcohol, 292–293
Non-kosher animal, 97
Non-kosher fish, 147
North American Meat Institute (NAMI), 88
Nutraceutical ingredients, 223
    hardgel capsule, 223
    liquid supplements and drinks, 223
    softgel capsules, 223
    tablets, 223
Nutritional Labeling and Education Act of 1990, 218
Nutritional food products, 223
    liquids, 223
    one-piece capsules, 223
    powders, 223
    tablets, 223
    two-piece capsules, 223

Octopus ink, 146–147, 217
OIC, see Organization of Islamic Conference (OIC) countries
Orange skin, 203
Organization of Islamic Conference (OIC) countries, 33, 41
Organoleptic, 207
Ossein, 179
Oysters, 21, 147

Packaging, 114, 119, 121
Pan greases, 164
Pareve (parve, parev), 263–264
Passover, 268
Pastries, 165
Pectin, 179
Penicillium nalgiovense, 128
Pepsin, 225
Petroleum, 129, 192
Pharmaceuticals, 69, 203
Pigskin, 22
Plants, 14
    gums, 222
    ingredients, 129
    materials, 170, 203, 267, 279
    sources, 21, 173, 186, 223, 237, 294
Polysorbates, 23
Porcine, 295–296
Pork, 22
Pork extract, 120
Potassium chloride, 135
Poultry, 14, 24
Prawns, 267
Pre-slaughter handling, 87, 92, 96, 100, 107

Index

Preservatives, 157, 172, 174, 208
Processed meats, 126
   critical considerations, 127–128
   incidents and regulations, 136–137
   ISO, 142
   halal, 130–134
   wholesome aspects, 134–136
Prophet Muhammad (PBUH), 19–20, 66, 73, 75, 114, 145
Propionate, 159, 236
Propylene glycol, 217
Protein supplements, 230
Pro-vitamin A, 236

Quality assurance, 119, 197
Quran, 1–2, 7, 66, 73–75, 106, 108–109, 145

Rabbinical supervisor, 264–266
Ramadan, 7
Ready-to-eat (RTE) meats, 135, 141
Red wine vinegar, 242
Regulatory agencies, 173–174, 210, 263, 300
Releasing agents, 164, 210, 300
Religious symbols, 242
Rennet, 227
Residual alcohol, 205, 206
Restraint equipment, 107, 115
Ritual slaughter guidelines, 88–89
Royal jelly, 271, 275–276
RTE, *see* Ready to eat (RTE) meats
Rum, flavor, 158, 205

Salmonella spp., 136, 325
Salt, 135
Sanitation, 25, 299–301
Sauces, 216
Scallops, 147
SD alcohol, 203–204
Seafood, 14, 21, 145–149, 271, 278, *see also* Fish
Sea animals, 146
Seasonings, 216
Shark, 147, 156, 239, 264
Shelf-life, 186
Shellac, 215, 217
Shellfish, 148, 264, 275
Shirk, 225
Shochet, 97, 261
Shrimp, 146–147
Slaughtering, 13
   abominable acts, 19–20
   allowed animals, 106, 110
   bleeding, 118
   capital investment, 107
   fish, 147
   gelatin, 178
   halal, 20–21
   holding, 110–111, 118
   instrument, 18–19, 107, 110, 113, 116
   invocation, 19, 113, 118
   methods, 18, 82–83, 116
   person, 18, 108, 110, 113, 116, 118
   post-slaughter treatment, 114, 118
   stunning, 108–109, 111–112
   traditional practice, 82–83, 110
Slaying, 113
SMIIC, *see* Standards and Metrology Institute for Islamic Countries (SMIIC)
Snails, 147, 239
Sodium ascorbate, 217
Sodium benzoates, 157
Sodium erythorbate, 217
Sodium nitrate, 135
Softgel capsules, 223
Solvent, 169
Sour cream, 158
Soy peptone, 172
Soybean grits, 170
Spices, 216
Spinal cord, 19, 108, 113, 116
Squid ink, 147, 217
Squeeze box restraint, 82
Stabilizers, 158, 192, 208, 266
Standard of identity, 158, 189, 203
Standards and Metrology Institute for Islamic Countries (SMIIC), 41
Staphylococcus, 136
Starches, modified, 193, 215
State food laws
   California, 26
   Michigan, 26
   Minnesota, 25
   New Jersey, 25
   Texas, 26
Stearates, 223
Steel drums, 244
Stunning techniques, 108, 111–112, 115, 118
   captive bolt, 112
   carbon dioxide, 113
   electrical, 112, 115
   gas, 112
   mushroom-shaped hammer, 112
Sugars, 170
Sunnah, 7
Supplements, 221
   dietary, 221
   nutritional, 221
Surimi, 149
Swine, 13
Swordfish, 147
Synthetic colors, 217
Synthetic meat, *see* Clean meat

Tablets, 181, 221–223
Taenia solium, 12, 269

Index

Tasmiyyah, 19
Tayyab, 66, 77
Trichinella spiralis, 12, 269
Tartaric acid, 215
Terminology, 319–320, Appendix A
Testing, non-halal materials, 291–296
Texturizing ingredients, 215
Topical products, 203
Torah, 263–264, 266
Trachea, 14, 19, 96, 98, 108, 113, 116, 118, 271, 304, 367
Trademark, 4, 114, 119, 310
Triacylglycerol (TAG), 293
Trichinae, 270
Triglycerides (TG), 293
Turkeys, 100, 114–117, 271
Turtles, 146–147, 239
Tweens, 223

US Food and Drug Administration (FDA), 22, 206
United States Department of Agriculture (USDA), 202, 210, 226
USDA, *see* United States Department of Agriculture

V restraint, 82
Vanilla flavor, 158, 189, 203, 206
Vat-grown meat, *see* Clean meat
Vegan, 243, 275–276
Vegetable, 14
 capsules, 275
 ingredients, 223
 oil, 23, 156, 165, 169
 products, 21–22, 275
Vegetarian, 275–279
 animal products, 276
 food standards, 276
 kitchen and hygiene standards, 276
 lacto-ovo, 275
 pesco, 275
 vegans, 275
Vinegar, 203, 215

Wax, 22, 193, 236, 238, 244
Wax coating, 22
Whale, 146–147, 235, 239
Whey, 158, 168
 concentrate, 23, 158
 isolates, 23, 158
 lactose, 23
 protein, 23
Whey powder, 23
Wholesome processed meat, 125–126
 assurances, 139
 halal and, 126
 ingredients, 126
 processor, 139
 quality control (QA), 126
Wine, extracts, 23–24, 196, 202, 205, 214

Xanthan gum, 179

Yashon flour, 267
Yearly certification, 171
Yeast, 135, 164, 173
Yogurt, 158

Zabh, 18
Zabiha, 9, 320
Zein, 215

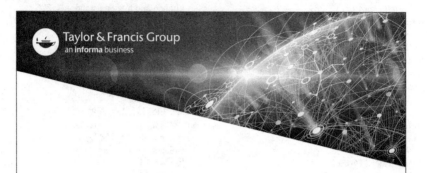